THE TRUTH OF

THE ORIGIN OF THE UNIVERSE

FULLY MATURED FUNCTIONAL CREATION AND THE MIRACULOUS FINELY TUNED YOUNG-EARTH

VOLUME-4

COLOUR EDITION

THE TRUTH OF THE ORIGIN OF THE UNIVERSE

FULLY MATURED FUNCTIONAL CREATION AND THE MIRACULOUS FINELY TUNED YOUNG-EARTH

Dr. Sabrie Soloman

Ph.D., Sc.D., MBA, PE

KHANNA BOOK PUBLISHING CO. (P) LTD.

Publisher of Engineering and Computer Books

4C/4344, Ansari Road, Darya Ganj, New Delhi-110002

Phone : 011-23244447-48 **Mobile:** +91-9910909320

E-mail : contact@khannabooks.com

Website : www.khannabooks.com

ISBN: 978-93-5538-772-1

**THE TRUTH OF THE ORIGIN
OF THE UNIVERSE (VOLUME-4)**
by **Dr. Sabrie Soloman**

First Edition: Dec 2024

Published by:
Khanna Book Publishing Co. (P) Ltd.
CIN: U22110DL1998PTC095547

Visit us at: www.khannabooks.com
Write us at: contact@khannabooks.com

To view complete list of books,
please scan the QR Code:

DEDICATIONS

William Thomson – The voice of his words of wisdom echoed through the core of the inner chambers of my heart, engraving channels of living knowledge, which generated floods of insight to illuminate my life`s journey.

William Thomson`s pure and innocent soul has reached the gates of heavens to bring me knowledge of Eternity. His living loving example compelled me to follow even his shadow to touch Eternity, while my feet are still on Earth!

His knowledge imparted on my mind inspiring me to write the "**The Origin of The Universe**" book – distinct from most other science books, which composed by worldwide prominent scientists highlighting unjustified and unproven hypotheses. Through **William Thomson** the factual truth of the Universe is recognized and logically revealed to my modest intellect to compose "**The Truth of The Origin of The Universe**," Thus, I humbly dedicate this book to him.

Joseph Rondinelli – While the size of the universe may not be comprehended, so also the goodness of the heart of my friend Joseph Rondinelli may not be grasped. Joseph permitted me to touch the limitless boundaries of the goodness of his loving nature. His friendship with me permitted me to observe a living loving example to follow.

Joseph Rondinelli is bountifully living loving those whom he cherished. He brought heavens highest values to my humble living earth. While I was temporarily incapacitated, he was my feet, hands, and even eyes sitting for hours by my side, debating our intellectual capacities to understand life under the domain of our **Creator, The Origin of Our Universe**.

When my mind acted as an interrupted volcano spouting novel thoughts of **The Origin of The Universe**, he came to my dwelling to rescue me out of the dense-smoke of my scattered knowledge --suffocating my mind. He instantly changed my environment to a more tranquil atmosphere enabling me once again recollect my thoughts, helping me to **Create Order out of Chaos**. For this and much more I am forever, indebted.

FOREWORD

Shortly before Frances R. Havergal, at age 36, wrote the hymn, "Lord, Speak to Me" she wrote in a letter, "… if I am to write any good, a great deal of living must go to a very little writing." Before putting pen to paper, a great deal of living has gone into what you are about to read.

When Dr. Soloman asked me to write a forward to his book on the origin of the universe, I was humbled and honored because this book is of unusual importance by an author with an exceptional life of service and greatly respected by all who know him and his work. I pray that those who read his 2,800 pages, 32 chapters will find its message clearly stated, giving clarity to a subject that has too long been delt with on a surface and non-scientific level.

Understanding the origin of the universe is key to understanding everything else and developing a truly Biblical worldview. A person's view on this subject is a window by which we view the world. Dr. Soloman gives us a verifiable frame to evaluate whether his findings are flawed or something that is understandably sensible to merit serious consideration.

When we tell others our view on evolution is based on scientific evidence rather than solely on the book of Genesis, we have a much stronger argument. Dr. Soloman builds the case that evolution, with his accompanying philosophy, is identified with a worldview at such a level that readers must sincerely struggle how the theory could possibly go against the evidence.

This book has the capacity to train a whole new generation to confront and survive the worldview challenges they encounter all their life. To those of you who read this book in its entirety will be able to survive in a post-modern culture without being brainwashed by prejudice, faulty thinking, and preconceived assumptions.

Dr. Soloman has made exciting reading for the ordinary person, but which is thoroughly researched by scholarship to present a deep understanding of the subject. The situation we often find ourselves in is flawed thinking and lazy scholarship. To that end, I invite you to read Dr. Soloman's book which will prove to not only be enjoyable reading, but one of the most prolific and insightful tomes you will ever read on the origins of the universe.

Dr. Peter W. Teague,
President Emeritus
Lancaster Bible College | Capital Seminary & Graduate School

PREFACE

The Universe Had a Beginning

The scientific consent 100 years ago was that the universe was everlasting. This idea started to unravel with the inferences of Albert Einstein's Theory of Relativity back in 1916, where his equations pointed to an expanding universe. Yet he didn't like that conclusion and so supplemented a constant to his equation that invalidated the expansion. Later, he admitted it had been the major mathematical blunder in his life.

Then in 1929 astronomer Edwin Hubble acknowledged he observed galaxies expanding outward, which meant they had been much closer together in the past. Einstein, fascinated, wanted to see the indication for himself and in 1931 he went to the Mt. Wilson Observatory in Los Angeles, California. Einstein examined through the telescope, observed the evidence and then concluded, "I now see *the necessity* of a beginning." This began a change in the scientific boldness toward the cosmos.

Years after, in 1965, two U.S. scientists distinguished the leftovers of the original burst of energy of the creation occurrence characteristically so-called the "Big Bang." They both won a Nobel Prize in physics. One of them, Arno Penzias, later professed, "The best data we have [about the beginning of the universe] are exactly what I would have predicted had I nothing to go on but the first five books of Moses, the Psalms and the Bible as a whole" ("Clues to Universe Origin Expected," *The New York Times*, March 12, 1978, p. 1).

With the indication at hand, what was written in Genesis 1:1 truly surprised many scientists by its precision: "In the beginning, God created the heavens and the earth." Here it says the universe of matter and energy *appeared at a certain point in time* and was all created by an Ultimate Creator who existed before all of this occurred. It was an enormous proof of God's existence, with no real substitute enlightenments for a universe that, according to modern physics, *appeared out of nothing*.

The Universe is Fine-Tuned for Life

Almost 50 years ago, in 1973, cosmologist Brandon Carter instituted that the independent constants or laws in physics have one exceedingly rare characteristic in common—*they are precisely the values needed to establish and sustain a universe capable of producing life*. This is another vast and virtually accepted proof for a universe that has been prudently designed.

Scientists have instituted some 30 constants or laws of physics that rule the universe. All are dissimilar to each other and yet are finely tuned to unbelievable proportions to make life possible. The indication points to "Someone" spending much too much time tuning all of these laws so they would work in unison.

Amazingly, the Bible exposed this truth long before any scientist revealed these facts. As Jeremiah 33:25 states, "But I, the Lord, have a covenant with day and night, and *I have made the laws* that *control* earth and sky" (*Good News Translation*).

The Origin of Life – The Genetic Code

Opposing to what many have been led to trust, scientists have no genuine explanation for how life arose.

Even the well-known atheist and evolutionist Richard Dawkins acknowledged concerning the appearance of life, "*Nobody knows how it happened*" (*Climbing Mount Improbable,* 1996, p. 282). Moreover, one of the pioneers of the DNA code, the atheist Francis Crick, determined, "An honest man, armed with all the knowledge available to us now, could only state that in some sense, *the origin of life* appears at the moment to be *almost a miracle,* so many are the environments which would have had to have been gratified to get it going" (*Life Itself: Its Origin and Nature,* 1981, p. 88).

In the past 60 years, biologists have found that life started with a massive amount of accurate information *already embedded* in the cell. The human genome unaided is a molecule with almost 3 billion genetic letters, all precisely well-organized to give instructions to the cell. Furthermore, scientists have never instituted inorganic matter to generate a coded system of information and the machinery to interpret it. From the most primitive cells to human beings, all have the same elementary operating system of mind-boggling intricacy, with codes, transmitters and receivers all working together.

Furthermore, the origin of life aenigma has a "chicken-and-egg question"—which came first, the chicken or the egg? In this case, to get life to transpire, you need *both* the complete genetic code and the proteins—the machine parts—that read the code and build new proteins. Deprived of the code, you can't shape proteins. And without proteins, you can't process the code. So how could both have risen at the same time?

Biological life Exquisitely Programmed - "Robotic Machines"

In order to comprehend what is fashioning inside a cell, a good illustration is picturing a large city swarming with life and movement.

Biochemist Michael Denton defines the cell this way: "To clutch the reality of life as it has been discovered by molecular biology, we must amplify a cell a billion times until it is twenty kilometers in diameter and looks like a giant airship large enough to cover a great city like London or New York. What we would then see would be an entity of unmatched intricacy and adaptive design . . .

"We would see around us, in every direction we looked, all sorts of robot-like machines. We would notice that the simplest of the functional components of the cell, the protein molecules, were astonishingly complex pieces of molecular machinery, each one consisting of about three thousand atoms arranged in highly organized 3-D spatial conformation.

"We would wonder even more as we watched the strangely purposeful activities of these weird molecular machines, particularly when we realized that, despite all our accumulated knowledge of physics and chemistry, the task of designing one such molecular machine—that is one single functional protein molecule—would be completely beyond our capacity at present" (*Evolution: A Theory in Crisis,* 1986, p. 329).

This is the reason biochemists have a hard time have faith in explaining that blind evolution can build such machinery—and obtain all the parts to work together from the start. Furthermore, to keep the human body operational, biologists calculate that "about *330 billion cells* are substituted *daily,* equivalent to about 1 percent of all our cells" (Mark Fischetti, "Our Bodies Replace Billions of Cells Every Day," *Scientific American,* April 1, 2021).

"We take life for granted," adds Douglas Ell, "because it is everywhere. Our planet is overrun by *biological machines.* There are at least 10 million different types (species) of machines; some estimate that tens of millions of other types (species) have not yet been discovered . . .

"Coordinated systems permit blue whales to dive thousands of feet below sea level without being crumpled and sing complex songs that travel across oceans. Other systems guide bees to do a dance that tells other bees where to find the best sources of pollen. There are systems for hiding, systems for fighting, systems for reproducing, systems for obtaining food, systems for communicating, and so on" (*Counting to God,* p. 110).

Such findings show that everything about life is programmed to the last detail and that virtually nothing has been left to chance. Does this superb design point to evolution or to God? The answer is clear.

The Earliest Evidence of Life

Though Darwin titled his book *On the Origin of Species by Means of Natural Selection,* he was never able to substantiate that assumption. Many people adopt that the theory of evolution, with its countless mutations and natural selection as the means of change, can account for the origin and development of all the living things on this planet.

Yet this is sleight of hand, since evolution can account for *micro*evolution, or changes *within* the species (such as dogs of varying sizes, shapes and colors), but not *macro*evolution, or changes from one kind of creature to another. Natural selection can tell you something about the *survival* of the species, but nothing about the *arrival* of the species. It undoubtedly cannot trace the origin of the approximately *10 million species* on earth. These are classified into some 33 main body types or phyla, such as sponges, worms, insects and mammals.

Darwin foretold that as more of the fossil record was revealed, it would display types of species gradually appearing, beginning with one or a few, and then multiplying from simple to more complex life forms. He wrote, "If numerous species . . . have really *started into life at once,* the fact would be *fatal* to the theory of evolution through natural selection" (*Origin of Species,* 1859, p. 305). Yet that is precisely what has been found—major body types appearing at what's considered the *beginning* of the fossil record rather than in deposits laid down later.

Scientists call this "the Cambrian Explosion," referring to major types of plants and animals suddenly appearing *fully formed* in that fossil layer. This is the *opposite* of what Darwin and evolutionists had claimed would be found—and they have no real explanation or answers. Of the 33 main body types, 23 of them (or 70%) appear at the recognized *beginning* stage of the fossil record.

The issue in question here, by analogy, would be as discovering together such diverse inventions as a washing machine, a refrigerator, a bicycle, a car and an airplane. While they do have some shared features, they have very discrete purposes and resolutions. Likewise, the major types of creatures found in the Cambrian layer, such as sponges, worms, trilobites and jawless fish, are very diverse, complex and appear suddenly, with no evidence of these main body types evolving from other creatures.

As paleontologist Niles Eldredge admitted: "If life had evolved into the wondrous profusion of creatures little by little, then there should be some fossiliferous record of those changes . . . But no one has found any evidence of such in-between creatures . . . All of the fossil evidence to date has failed to turn up any such missing links" (George Alexander, "Alternate Theory of Evolution Considered," *Los Angeles Times,* Nov. 19, 1978). Indeed, Darwin has been let down by the fossil record!

Earth Has "Just Right" Conditions to Sustain Life

In 1966, Carl Sagan presented the famous TV documentary series *Cosmos*. He believed in order to have life you just needed two conditions—a right kind of star and a planet at the right distance. This supposition showed to be totally baseless.

Now, more than 50 years later, scientists have come to the comprehension that *more than 200 conditions* have to be "just right" for life to exist and thrive. The probabilities are about 1 in $10^{2,685,000}$. The probability of that happening comes out at about **1 in $10^{2,685,000}$**, or 10 followed by **2,685,000** zeros. For comparison, the Universe only has 10^{80} atoms. The infographic finishes by letting you know that the probability of you existing as you is pretty much zero.

As author Eric Metaxas explains: "Today there are more than 200 known parameters necessary for a planet to support life—*every single one of which must be perfectly met, or the whole thing falls apart.* Without a massive planet like Jupiter nearby, whose gravity will draw away asteroids, a thousand times as many would hit Earth's surface. The odds against life in the universe are simply astonishing" ("Science Increasingly Makes the Case for God," *The Wall Street Journal,* Dec. 25, 2014).

The Bible tells us: "The Lord is God. He made the skies and the earth. He put the earth in its place. He did not want the earth to be empty when he made it. He created it *to be lived on.* I am the Lord. There is no other God" (*Isaiah 45:18, Easy-to-Read Version*).

The Universe Mathematically Designed

Amazingly, the universe has been discovered to be mathematically designed. It follows orderly laws that can be described in mathematical terms. Sir James Jeans, one of the great astronomers of the 20th century, remarked: "From the intrinsic evidence of his creation, the Great Architect of the Universe *now begins to appear as a pure mathematician* . . . The universe begins to look more like a great thought than like a great machine" (*The Mysterious Universe,* 1930, pp. 134, 137).

A substantial problem for evolutionists and atheists is this: *Evolution can't do math,* since it is founded on random variations and mutations, and math necessitates an intelligent agent who can first formulate a mathematical blueprint of laws before creating things so they will be logical. This is why the present cosmos can be traced back to mathematical rules.

As Einstein distinguished, "The most incomprehensible thing about the universe is that it is comprehensible." He meant that it could be understood in *mathematical* terms but that an explanation for that was outside math.

As far back as the early 1900s, scientists were discovering the laws that govern the subatomic realm, the tiny microcosm designated by quantum mechanics. It has very diverse rules than our macro world and appears to make room for such things as free will to arise.

Many scientists came to comprehend that *not all is determined by matter and energy.* Experiments illustrate that an observer can change a particle through means of observing it. The consequences are that we can determine the outcome of our lives by the choices we make.

It brings to mind what God said: "*Today I have given you a choice between life and death, success and disaster. I command you today to love the Lord your God. I command you to follow him and to obey his commands, laws, and rules. Then you will live . . .*" (*Deuteronomy 30:15-16, ERV*).

Science Points to The Existence of God

Cutting-edge science is continually showing more intricacy and deeper design, not only in the cosmos, but also in all living things.

The biblical patriarch Job once challenged skeptics to look at the design of the creatures around them and notice they witness to a Supreme Designer and Creator. He stated: "Even birds and animals have much they could *teach* you; ask the creatures of earth and sea for their wisdom. All of them know that *the Lord's hand made them*" (*Job 12:7-9, GNT*).

Therefore, by exploring all the evidence and grasping where it leads, we hope you will believe in God, the Creator and the "Origin of the Universe." *Keep* believing in Him, and earnestly seek His will for your life!

Sabrie Soloman

ACKNOWLEDGMENTS

The Author is extremely grateful for the opportunity to express his heartfelt acknowledgments to the individuals who have played a vital role in the successful publication of "The Truth of The Origin of The Universe" book by Khanna Book Publishing Company.

First and foremost, the author extends his deepest gratitude to the Director and Editor-in-Chief, Mr. Mukul Seth, for his unwavering dedication and expertise in shaping the book's narrative. His keen eye for detail and impeccable editing skills have significantly enriched the content and ensured its quality.

Also, the author expresses his sincere appreciation to the Editorial Team, comprising of Mr. Mukesh Sharma and Mr. Yogesh Kumar. Their valuable feedback, insightful suggestions, and meticulous attention to detail have greatly enhanced the overall coherence and readability of the manuscript.

A special mention goes out to the Marketing Team, led by Mr. Rakshit Khanna for his tireless efforts in promoting the book and reaching a wider audience. Their innovative marketing strategies and relentless pursuit of excellence have played a crucial role in generating interest and enthusiasm among readers.

Furthermore, I am truly grateful for the dedication and hard work of the Sales Team, consisting of Mr. Nand Lal, Mr. Surinder Kumar, and Mr. Roshan Lal. Their exceptional sales acumen and customer-centric approach have boosted sales and fostered positive relationships with clients and partners.

The author would also like to extend his heartfelt thanks to Creation Ministries International for its invaluable contributions to the research and development of the book. Their pioneering work in the field of creation science has laid the foundation for a deeper understanding of the origins of the universe.

Sabrie Soloman

ABOUT THE AUTHOR

Dr. Sabrie Soloman, Ph.D., Sc.D., MBA, PE – He is the Chairman & CEO of American SensoRx, Inc., USA; Founder of Advanced Manufacturing Technology Post Graduate Studies at Columbia University, NY, USA; Professor of Advanced Technology at Columbia University. Dr. Soloman authored numbers of technical books published and translated worldwide: Sensors Handbook (2 editions), Sensors and Control Systems in Manufacturing (2 editions); Affordable Automation; Introduction to Electromechanical Engineering; Modern Welding Technology; 3D Printing Technology; 3D Bioprinting Technology & Design to name a few. Dr. Soloman holds numerous Patents, Technical Awards, and several US Product Registrations. Dr. Soloman is considered an international authority on advanced manufacturing technology, robotics, biomedical engineering, pharmaceuticals, and automation in the microelectronic, automotive, beef, pork, poultry industries. He has been and continues to be instrumental in developing and implementing several industrial and modernization programs through the United Nations to Europe, Asia, and African Countries. He is the first to introduce and implement unmanned flexible synchronous/asynchronous manufacturing systems in the microelectronic and the meat industries, and the first to incorporate advanced vision technology in wide array of robot/micro-robot manipulators. Dr. Soloman was selected to deliver the US presidential closing address "Innovative Remote Sensors Technology," at the Universal Design Conference," New York, USA. Dr. Soloman was the President of the International Christian Union at New Castle-Upon-Tyne University. He debated the "Origin of the Universe" before numerous attendees presenting his case against intellectuals and proponents of Big Bang's Singularity in Great Britain.

CONTENTS

CHAPTER 24: LIFE IN A FINELY TUNED COSMOS 185-238

INTRODUCTION

YOUNG UNIVERSE

The concept of a young universe, filled with young planets and a finely tuned environment for life, has been the subject of much debate and discussion in the scientific community. While some theories suggest that the universe is billions of years old and the result of random chance and natural processes, there is a growing body of evidence that points to a different conclusion. This evidence suggests that the universe is indeed young and that it has been specifically designed to support life.

Young Planets

One of the key pieces of evidence for a young universe is the presence of young planets. The current models of planetary formation suggest that it should take hundreds of millions of years for planets to form from the dusty disks surrounding young stars. However, recent observations have shown that planets can form much more quickly than previously thought. In fact, some of the youngest planets discovered are only a few million years old, which is orders of magnitude younger than predicted billions by current theories.

Earth Magnetic Field

Furthermore, the magnetic field of the Earth provides additional evidence for a young universe. The Earth's magnetic field is believed to be generated by the motion of molten iron in the planet's core. However, this process is not sustainable over long periods of time, as the Earth's core would eventually cool and solidify, leading to the loss of the magnetic field. The fact that the Earth still has a strong magnetic field after billions of years is a strong indication that the planet is much younger than previously believed.

Finely Tuned Universe Supports Life

In addition to the evidence from young planets and Earth's magnetic field, there is also the concept of a finely tuned universe that supports life. The fundamental constants of nature, such as the strength of gravity and the speed of light, are carefully balanced to allow for the existence of life as we know it. If any of these constants were even slightly different, the universe would be inhospitable to life. This level of fine-tuning suggests that the universe was designed with a specific purpose in mind, rather than the result of random chance.

Isaac Newton Recognizes the Creator

When it comes to the question of the Creator behind this finely tuned universe, one cannot overlook the contributions of great minds such as Isaac Newton. Newton, the father of modern physics, was a devout Christian who believed that the laws of nature were evidence of a Creator. In his famous work Principia, Newton wrote, "This most beautiful system of the sun, planets, and comets, could only proceed from the counsel and dominion of an intelligent and powerful Being." Newton's belief in a Creator was central to his scientific work, and his discoveries laid the foundation for the modern understanding of the universe.

The Mysteries of the Created Universe

The evidence for a young universe, young planets, a finely tuned environment for life, and the existence of a Creator is compelling and worthy of further study. The idea that the universe is the result of random chance and natural processes is increasingly being called into question by new discoveries and observations. As we continue to explore the mysteries of the universe, it is important to consider all possible explanations and to follow the evidence where it leads. The truth of the origin of the universe may be more complex and awe-inspiring than we can imagine.

The Current Understanding of the Truth of The Origins of the Universe

In this book, we have delved into the various aspects of a young universe, young planets, Earth's magnetic field, the science behind a finely tuned universe, the role of the Creator, and the influence of figures like Isaac Newton. By examining these topics from a persuasive, analytical, and informative perspective, we have sought to provide a comprehensive overview of the current understanding of the truth of the origins of the universe. As we continue to push the boundaries of scientific knowledge, it is crucial to remain open-minded and receptive to new ideas and evidence that may challenge our existing beliefs. The truth of the origin of the universe may be more extraordinary than we can fathom, but by approaching the subject with curiosity and rigor, we can uncover the mysteries that lie at the heart of our existence.

Sabrie Soloman

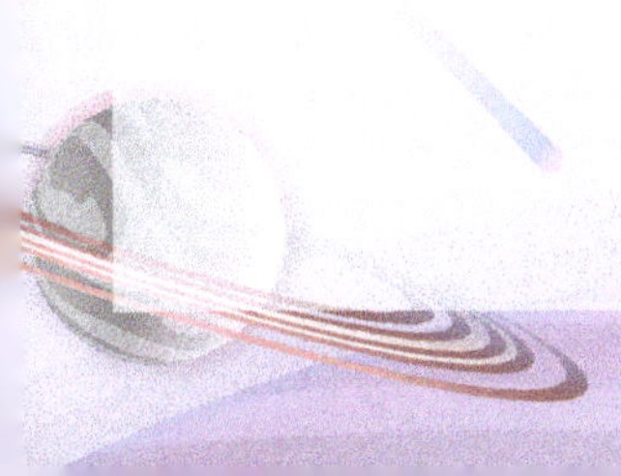

YOUNG UNIVERSE – YOUNG EARTH

AGE OF THE EARTH

101 Evidences for a Young Age of the Earth and the Universe

The Age of the Earth and Science

The widely accepted age of the universe is currently 13.77 billion years and for the solar system (including Earth) it is 4.543 billion years. However, no scientific method can *prove* the age of the earth and the universe, and that includes the ones we have listed here that strongly suggest that these accepted ages are in serious error. Although age indicators are called 'clocks' they aren't, because all ages result from calculations that necessarily involve making assumptions about the past. The starting time of the 'clock' has always to be assumed as well as the way in which the speed of the clock has varied over time. Further, it has to be assumed that the clock was never disturbed, Figure 20.1.

Fig.20.1: *The Age of the Earth - Curtsy – R.C. Sproul Jr*

There is no independent natural clock against which those assumptions can be tested, Figure 20.2. For example, the amount of cratering on the moon, based on currently observed cratering rates, would suggest that the moon is quite old. However, to draw this conclusion we have to assume that the rate of cratering has been the same in the past as it is now. And there are now good reasons for thinking that it might have been quite intense in the past, in which case the craters do not indicate an old age at all.

Fig.20.2: *Difficult to believe that Numerous Scientists are all wrong on the Age of the Earth*

Ages of millions of years are all calculated by assuming the rates of change of processes in the past were the same as we observe today—called the principle of uniformitarianism. If the age calculated from such assumptions disagrees with what they think the age should be, they conclude that their assumptions did not apply in this case, and adjust them accordingly. If the calculated result gives an acceptable age, the investigators publish it, Figure 20.3.

Fig.20.3: *There are many categories of evidence for the age of the earth and the cosmos that indicate they are much younger than is generally asserted today.*

Examples of *young* ages listed here are also obtained by applying the same principle of uniformitarianism. Long-age proponents will dismiss this sort of evidence for a young age of the earth by arguing that the assumptions about the past do not apply in these cases. In other words, age is not really a matter of scientific observation but an argument about our assumptions about the unobserved past.

The assumptions behind the evidences presented here cannot be proved, but the fact that such a wide range of different phenomena all *suggest* much younger ages than are currently generally accepted, provides a strong case for questioning the accepted ages.

Also, a number of the evidences, rather than giving any estimate of age, challenge the assumption of slow-and-gradual uniformitarianism, upon which all deep-time dating methods depend.

Many of these indicators for younger ages were discovered when creation scientists started researching things that were supposed to 'prove' long ages. The lesson here is clear: when the evolutionists throw up some new challenge to the Bible's timeline, don't fret over it. Sooner or later that supposed evidence will be turned on its head and will even be added to this list of evidences for a younger age of the earth. On the other hand, some of the evidences listed here might turn out to be ill-founded with further research and will need to be modified. Such is the nature of science, especially historical science, because we cannot do experiments on past events.

Science is based on observation, and the only reliable means of telling the age of anything is by the testimony of a reliable witness who observed the events. The Bible claims to be the communication of the only One who witnessed the events of Creation: The Creator himself. As such, the Bible is the only reliable means of knowing the age of the earth and the cosmos. In the end we believe that the Bible will stand vindicated and those who deny its testimony will be confounded.

BIOLOGICAL EVIDENCE FOR A YOUNG AGE OF THE EARTH

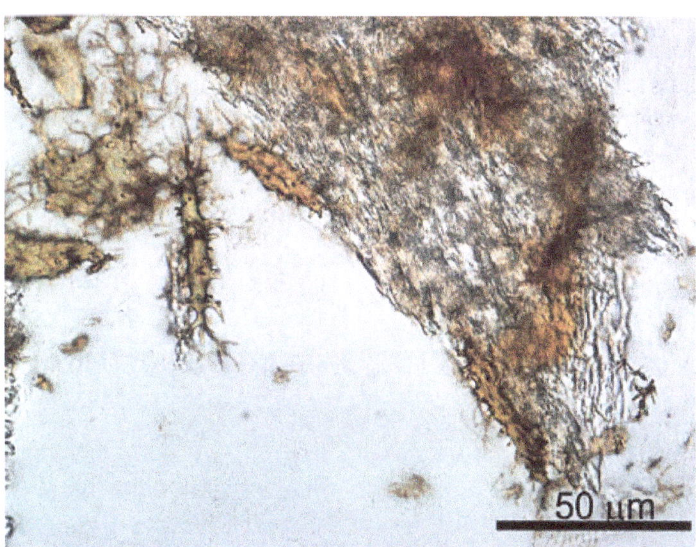

Fig.20.4: The finding of Pliable Blood Vessels, Blood Cells, Animal Proteins, and Even DNA in Dinosaur Bone is Consistent with an Age of Thousands of Years for the Fossils, not the 65+ million years Claimed by the Paleontologists - Curtsy - Image: Dr Mary Schweitzer

1. DNA in 'Ancient' Fossils. Figure 20.4 - DNA extracted from bacteria that are supposed to be 425 million years old brings into question that age, because DNA could not last more than thousands of years.
2. Lazarus Bacteria—bacteria revived from salt inclusions supposedly 250 million years old, suggest the salt is not millions of years old.
3. The decay in the human genome due to multiple slightly harmful mutations each generation is consistent with an origin several thousand years ago. Sanford, J., *Genetic entropy and the mystery of the genome*, Ivan Press, 2005. This has been confirmed by realistic modelling of population genetics, which shows that genomes are young, in the order of thousands of years. See Sanford, J., Baumgardner, J., Brewer,

W., Gibson, P. and Remine, W., Mendel's Accountant: A biologically realistic forward-time population genetics program, *SCPE* **8**(2):147–165, 2007.

4. The data for 'mitochondrial Eve' are consistent with a common origin of all humans several thousand years ago.
5. Very limited variation in the DNA sequence on the human Y-chromosome around the world is consistent with a recent origin of mankind, thousands not millions of years.
6. Many fossil bones 'dated' at many millions of years old are hardly mineralized, if at all. This contradicts the widely believed old age of the earth. See, for example, Dinosaur bones just how old are they really? Tubes of marine worms, 'dated' at 550 million years old, that are soft and flexible and apparently composed of the original organic compounds.
7. Dinosaur blood cells, blood vessels, proteins (hemoglobin, osteocalcin, collagen, histones) and DNA are not consistent with their supposed more than 65-million-year age, but make more sense if the remains are thousands of years old (at most).
8. Lack of 50:50 racemization of amino acids in fossils 'dated' at millions of years old, whereas complete racemization would occur in thousands of years.
9. Living fossils—jellyfish, graptolites, coelacanth, stromatolites, Wollemi pine and hundreds more. That many hundreds of species could remain so unchanged, for even up to billions of years in the case of stromatolites, speaks against the millions and billions of years being real.
10. Discontinuous fossil sequences. E.g., Coelacanth, Wollemi pine and various 'index' fossils, which are present in supposedly ancient strata, missing in strata representing many millions of years since, but still living today. Such discontinuities speak against the interpretation of the rock formations as vast geological ages—how could Coelacanths have avoided being fossilized for 65 million years, for example? See The 'Lazarus effect': rodent 'resurrection'!
11. The ages of the world's oldest living organisms, trees, are consistent with an age of the earth of thousands of years.

GEOLOGICAL EVIDENCE FOR A YOUNG AGE OF THE EARTH

Fig.20.5: Radical folding at Eastern Beach, near Auckland in New Zealand, indicates that the sediments were soft and pliable when folded, inconsistent with a long time for their formation. Such folding can be seen world-wide and is consistent with a young age of the earth – Curtsy - Photo by Don Batten

1. Scarcity of plant fossils in many formations containing abundant animal / herbivore fossils. E.g., the Morrison Formation (Jurassic) in Montana. See *Origins* **21**(1):51–56, 1994. Also, the Coconino sandstone in the Grand Canyon has many track-ways (animals), but is almost devoid of plants. Implication: these rocks are *not* ecosystems of an 'era' buried *in situ* over eons of time as evolutionists claim. The evidence is more consistent with catastrophic transport then burial during the massive global Flood of Noah's day. This eliminates supposed evidence for millions of years.
2. Thick, tightly bent strata without sign of melting or fracturing. E.g., the Kaibab Upwarp in Grand Canyon indicates rapid folding before the sediments had time to solidify (the sand grains were not elongated under stress as would be expected if the rock had hardened). This wipes out hundreds of millions of years of time and is consistent with extremely rapid formation during the biblical Flood. See *Warped earth* (written by a geophysicist).
3. Polystrate fossils—tree trunks in coal (*Araucaria* spp. king billy pines, celery top pines, in *southern hemisphere coal*). There are also polystrate tree trunks in the *Yellowstone fossilized forests* and *Joggins, Nova Scotia* and in many other places, Figure 20.6. Polystrate fossilized lycopod trunks occur in *northern hemisphere coal*, again indicating rapid burial / formation of the organic material that became coal.

Fig.20.6: *Polystrate fossils Young Age*

4. Experiments show that with conditions mimicking natural forces, coal forms quickly; in weeks for brown coal to months for black coal. It does not need millions of years. Furthermore, long time periods could be an impediment to coal formation because of the increased likelihood of the permineralization of the wood, which would hinder coalification.
5. Experiments show that with conditions mimicking natural forces, oil forms quickly; it does not need millions of years, consistent with an age of thousands of years.
6. Experiments show that with conditions mimicking natural forces, *opals form quickly*, in a matter of weeks, not millions of years, as had been claimed.
7. Evidence for rapid, catastrophic formation of coal beds speaks against the hundreds of millions of years normally claimed for this, including Z-shaped seams that point to a single depositional event producing these layers.
8. Evidence for rapid petrifaction of wood speaks against the need for long periods of time and is consistent with an age of thousands of years, Figure 20.7.

Fig.20.7: Colorful Crystal Patterns are Displayed in a Cross-section of Petrified Wood in the Petrified Forest National Park, Located in Northeast Arizona. Curtsy ARTERRA/GETTY IMAGES

9. Clastic dykes and pipes (intrusion of sediment through overlying sedimentary rock) show that the overlying rock strata were still soft when they formed. This drastically compresses the time scale for the deposition of the penetrated rock strata. See, Walker, T., Fluidization pipes: Evidence of large-scale watery catastrophe, *J. Creation (TJ)* 14(3):8–9, 2000.
10. Para(pseudo)conformities—where one rock stratum sits on top of another rock stratum but with supposedly millions of years of geological time missing, yet the contact plane lacks any significant erosion; that is, it is a 'flat gap'. E.g., Coconino sandstone / Hermit shale in the Grand Canyon (supposedly a 10-million-year gap in time). The thick Schnebly Hill Formation (sandstone) lies *between* the Coconino and Hermit in central Arizona. See Austin, S.A., *Grand Canyon, monument to catastrophe,* ICR, Santee, CA, USA, 1994 and Snelling, A., The case of the 'missing' geologic time, *Creation* 14(3):31–35, 1992.
11. The presence of ephemeral markings (raindrop marks, ripple marks, animal tracks) at the boundaries of Para conformities show that the upper rock layer has been deposited immediately after the lower one, eliminating many millions of years of 'gap' time. See references in Para (pseudo) conformities.
12. Inter-tonguing of adjacent strata that are supposedly separated by millions of years also eliminates many millions of years of supposed geologic time. The case of the 'missing' geologic time; Mississippian and Cambrian strata interbedding: 200 million years hiatus in question, *CRSQ* 23(4):160–167.
13. The lack of bioturbation (worm holes, root growth) at Para conformities (flat gaps) reinforces the lack of time involved where evolutionary geologists insert many millions of years to force the rocks to conform with the 'given' timescale of billions of years.
14. The almost complete lack of clearly recognizable soil layers anywhere in the geologic column. Geologists do claim to have found lots of 'fossil' soils (paleosols), but these are quite different to soils today, lacking the features that characterize soil horizons; features that are used in classifying different soils. Every one that has been investigated thoroughly proves to lack the characteristics of proper soil. If 'deep time' were correct, with hundreds of millions of years of abundant life on the earth, there should have been ample opportunities many times over for soil formation. See Klevberg, P. and Bandy, R., *CRSQ* **39**:252–68; *CRSQ* **40**:99–116, 2003; Walker, T., Paleosols: digging deeper buries 'challenge' to Flood geology, *J. Creation* **17**(3):28–34, 2003.
15. Limited extent of unconformities (unconformity: a surface of erosion that separates younger strata from older rocks). Surfaces erode quickly (e.g., Badlands, South Dakota), but there are very limited unconformities. There is the 'great unconformity' at the base of the Grand Canyon, but otherwise there are supposedly ~300 million years of strata deposited on top without any significant unconformity. This is again consistent with a much shorter time of deposition of these strata. See Para (pseudo) conformities.

16. The Arches National Park (USA) has over 2,000 rock arches. If 43 have collapsed since 1970 and the linked article was written in 2015, that's 45 years, giving a rate of collapse of ~1 per year, which means that all would be gone in ~2,000 years. This is thoroughly consistent with the biblical timeframe but not the evolutionary one of millions of years (5 million?). Historical records of the '12 Apostles' in southern Australia should allow a similar 'clock' to be calculated, albeit coarser than this USA park one, Figure 20.8. See A dangerous view.

Fig.20.8: The Arches National Park (USA)

17. The discovery that underwaters landslides ('turbidity currents') travelling at some 50 km/h can create huge areas of sediment in a matter of hours (Press, F., and Siever, R., *Earth*, 4th ed., Freeman & Co., NY, USA, 1986). Sediments thought to have formed slowly over eons of time are now becoming recognized as having formed extremely rapidly. See for example, A classic tillite reclassified as a submarine debris flow (Technical).
18. Flume tank research with sediment of different particle sizes show that layered rock strata that were thought to have formed over huge periods of time in lake beds actually formed very quickly. Even the precise layer thicknesses of rocks were duplicated after they were ground into their sedimentary particles and run through the flume. See Experiments in stratification of heterogeneous sand mixtures, Sedimentation Experiments: Nature finally catches up! and Sandy Stripes Do many layers mean many years? Also, very fine particles have been shown to settle far more quickly than previously thought, enabling the rapid formation of mudstone deposits.
19. Observed examples of rapid canyon formation; for example, *Providence Canyon* in southwest Georgia, Burlingame Canyon near Walla Walla, Washington, and Lower Loowit Canyon near Mount St Helens. The rapidity of the formation of these canyons, which look similar to other canyons that supposedly took many millions of years to form, brings into question the supposed age of the canyons that no one saw form.
20. Observed examples of rapid island formation and maturation, such as Surtsey, which confound the notion that such islands take long periods of time to form. See also, Tuluman—A Test of Time.
21. Rate of erosion of coastlines, horizontally. E.g., Beachy Head, UK, loses a meter of coast to the sea every six years.
22. Rate of erosion of continents vertically is not consistent with the assumed old age of the earth. See *Creation* **22**(2):18–21.

23. Existence of significant flat plateaux that are 'dated' at many millions of years *old* ('elevated paleoplains'). An example is Kangaroo Island (Australia). C.R. Twidale, a famous Australian physical geographer wrote: "the survival of these paleoforms is in some degree an embarrassment to all the commonly accepted models of landscape development." Twidale, C.R. On the survival of paleoforms, *American J. Science* **5**(276):77–95, 1976 (quote on p. 81). See Austin, S.A., *Did landscapes evolve? Impact* **118**, April 1983.
24. The recent and almost simultaneous origin of all the high mountain ranges around the world—including the Himalayas, the Alps, the Andes, and the Rockies—which have undergone most of the uplift to their present elevations beginning 'five million' years ago, whereas mountain building processes have supposedly been around for up to billions of years. See Baumgardner, J., Recent uplift of today's mountains. *Impact* **381**, March 2005.
25. Water gaps. These are gorges cut through mountain ranges where rivers run. They occur worldwide and are part of what evolutionary geologists call 'discordant drainage 'systems.' They are 'discordant' because they don't fit the deep time belief system. The evidence fits them forming rapidly in a much younger age framework where the gorges were cut in the recessive stage / dispersive phase of the global Flood of Noah's day. See Oard, M., Do rivers erode through mountains? Water gaps are strong evidence for the Genesis Flood, *Creation* **29**(3):18–23, 2007.

Fig.20.9: Erosion rates at places like Niagara Falls are consistent with a time frame of several thousand years since Noah's Flood.

26. *Erosion at Niagara Falls* and other such places is consistent with just a few thousand years since the biblical Flood. However, much of the Niagara Gorge likely formed very rapidly with the catastrophic drainage of glacial Lake Agassiz; see: Climate change, Niagara and catastrophe, Figure 20.9
27. River delta growth rate is consistent with thousands of years since the biblical Flood, not vast periods of time. E.g., 1. Mississippi—*Creation Research Quarterly (CRSQ)* **9**(2): *96–114*, 1972; *CRSQ* **14**(2): *87*, 1977; *CRSQ* **25**(3):*121–123*, 1988. E.g., 2 Tigris–Euphrates: *CRSQ* **14**(2): *87*, 1977.
28. Underfit streams. River valleys are too large for the streams they contain. Dury speaks of the "continent-wide distribution of underfit streams." Using channel meander characteristics, Dury concluded that past streams frequently had 20–60 times their current discharge. This means that the river valleys would have been carved very quickly, not slowly over eons of time. See Austin, S.A., *Did landscapes evolve? Impact* 118, 1983.
29. Amount of *salt in the sea*. Even ignoring the effect of the biblical Flood and assuming zero starting salinity and all rates of input and removal so as to maximize the time taken to accumulate all the salt,

the *maximum* age of the oceans, 62 million years, is less than 1/50 of the age evolutionists claim for the oceans. This suggests that the age of the earth is radically less also, Figure 20.10.

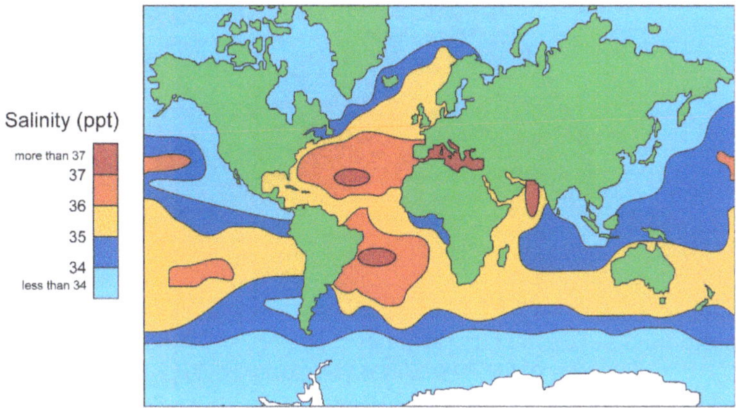

Fig.20.10: Ocean Salinity

30. The amount of sediment on the sea floors at current rates of land erosion would accumulate in just 12 million years; a blink of the eye compared to the supposed age of much of the ocean floor of up to 3 billion years. Furthermore, long-age geologists reckon that *higher* erosion rates applied in the past, which shortens the time frame. From a biblical point of view, at the end of Noah's Flood lots of sediment would have been added to the sea with the water coming off the unconsolidated land, making the amount of sediment perfectly consistent with a history of thousands of years.
31. Iron-manganese nodules (IMN) on the sea floors. The measured rates of growth of these nodules indicates an age of only thousands of years. Lalomov, A.V., 2006. Mineral deposits as an example of geological rates. *CRSQ* **44**(1):64–66. Related to this is the concentration of nickel in the oceans.
32. The age of placer deposits (concentrations of heavy metals such as tin in modern sediments and consolidated sedimentary rocks). The measured rates of deposition indicate an age of thousands of years, not the assumed millions. See Lalomov, A.V., and Tabolitch, S.E., 2000. Age determination of coastal submarine placer, Val'cumey, northern Siberia. *J. Creation (TJ)* **14**(3):83–90.
33. Pressure in oil / gas wells indicate the recent origin of the oil and gas. If they were many millions of years old, we would expect the pressures to equilibrate, even in low permeability rocks. "Experts in petroleum prospecting note the impossibility of creating an effective model given long and slow oil generation over millions of years (Petukhov, 2004). In their opinion, if models demand the standard multimillion-years geochronological scale, the best exploration strategy is to drill wells on a random grid." —Lalomov, A.V., 2007. Mineral deposits as an example of geological rates. *CRSQ* **44**(1):64–66.
34. Direct evidence that *oil is forming today* in the Guaymas Basin and in *Bass Strait* is consistent with a young earth (although not *necessary* for a young earth).
35. Rapid reversals in paleomagnetism undermine use of paleomagnetism in long ages dating of rocks and speak of rapid processes, compressing the long-age time scale enormously.
36. The pattern of magnetization in the magnetic stripes where magma is welling up at the mid-ocean trenches argues against the belief that reversals take many thousands of years and rather indicates rapid sea-floor spreading as well as rapid magnetic reversals, consistent with a young earth (Humphreys, D.R., Has the Earth's magnetic field ever flipped? *Creation Research Quarterly* **25**(3):130–137, 1988), Figure 20.11.

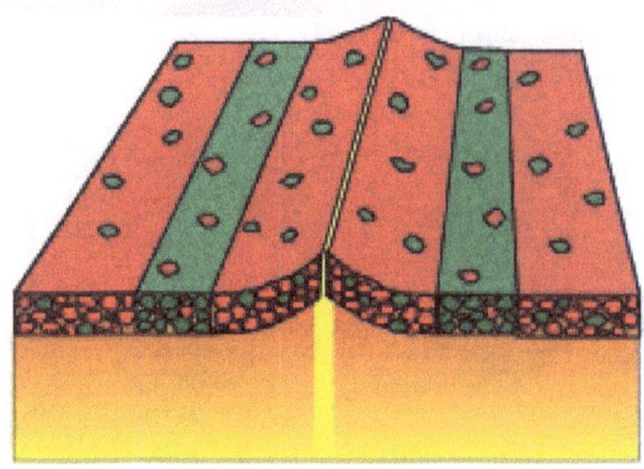

Fig.20.11: *Along the mid-ocean ridges, the detailed pattern of magnetic polarization, with islands of differing polarity, speaks of rapid changes in direction of Earth's magnetic field because of the rate of cooling of the lava. This is consistent with a young Earth.*

37. Measured rates of stalactite and stalagmite growth in limestone caves are consistent with a young age of several thousand years. See also *articles on limestone cave formation.*
38. The decay of the earth's magnetic field. Exponential decay, with fluctuations especially during and after the Flood, is evident from historical measurements and is consistent with the hypothesis of free decay since creation, suggesting an age of the earth of only thousands of years. For further evidence that it follows exponential decay with a time constant of 1611 years (±10) see: Humphreys, R., *Earth's magnetic field is decaying steadily—with a little rhythm*, *CRSQ* **47**(3):193–201; 2011.
39. Excess heat flow from the earth is consistent with a young age rather than billions of years, even taking into account heat from radioactive decay. See Woodmorappe, J., 1999. Lord Kelvin revisited on the young age of the earth, *J. Creation (TJ)* **13**(1):14, 1999.
40. Radiometric Dating and the Age of the Earth
41. Carbon-14 in *coal* suggests ages of thousands of years and clearly contradict ages of millions of years.
42. Carbon-14 in oil again suggests ages of thousands, not millions, of years.
43. Carbon-14 in *fossil wood* also indicates ages of thousands, not millions, of years.
44. Carbon-14 in *diamonds* suggests ages of thousands, not billions, of years. Note that attempts to explain away carbon-14 in diamonds, coal, etc., such as by neutrons from uranium decay converting nitrogen to C-14 do not work. See: Objections.
45. Incongruent radioisotope dates *using the same technique* argue against trusting the dating methods that give millions of years.
46. Incongruent radioisotope dates using different techniques argue against trusting the dating methods that give millions of years (or billions of years for the age of the earth).
47. Demonstrably non-radiogenic 'isochrons' of radioactive and non-radioactive elements undermine the assumptions behind isochron 'dating' that gives billions of years. 'False' isochrons are common.
48. Different faces of the same zircon crystal and different zircons from the same rock giving different 'ages' undermine all 'dates' obtained from zircons.
49. Evidence of a period of rapid radioactive decay in the recent past (lead and helium concentrations and diffusion rates in zircons) point to a young earth explanation.
50. The amount of helium, a product of alpha-decay of radioactive elements, retained in zircons in granite is consistent with an age of 6,000±2000 years, not the supposed billions of years. See: Humphreys, D.R.,

Young helium diffusion age of zircons supports accelerated nuclear decay, Chapter 2 (pages 25–100) in: Vardiman, Snelling, and Chaffin (eds.), *Radioisotopes and the Age of the Earth: Results of a Young Earth Creationist Research Initiative, Volume II*, Institute for Creation Research and Creation Research Society, 2005.

51. Lead in zircons from deep drill cores vs. shallow ones. They are similar, but there should be less in the deep ones due to the higher heat causing higher diffusion rates over the usual long ages supposed. If the ages are thousands of years, there would not be expected to be much difference, which is the case (Gentry, R., *et al.*, Differential lead retention in zircons: Implications for nuclear waste containment, *Science* **216**(4543):296–298, 1982; DOI: 10.1126/science.216.4543.296).

52. Pleochroic halos produced in granite by concentrated specks of short half-life elements such as polonium suggest a period of rapid nuclear decay of the long half-life parent isotopes during the formation of the rocks and rapid formation of the rocks, both of which speak against the usual ideas of geological deep time and a vast age of the earth. See, Radio-halos: startling evidence of catastrophic geologic *processes*, *Creation* **28**(2):46–50, 2006.

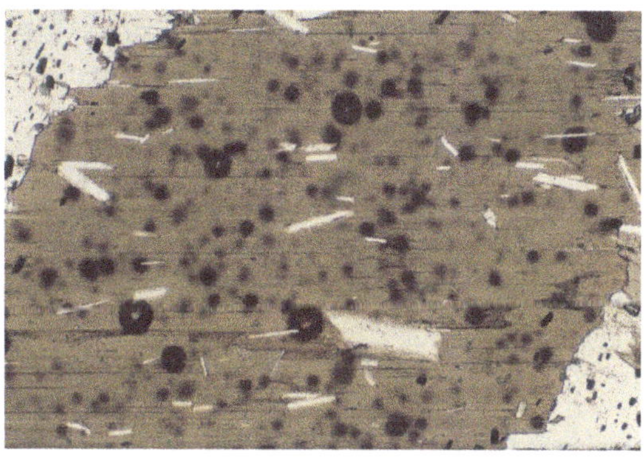

Fig.20.12: Pleochroic Halos

53. Squashed pleochroic halos (radio-halos), Figure 20.12, formed from decay of polonium, a very short half-life element, in coalified wood from several geological eras suggest rapid formation of all the

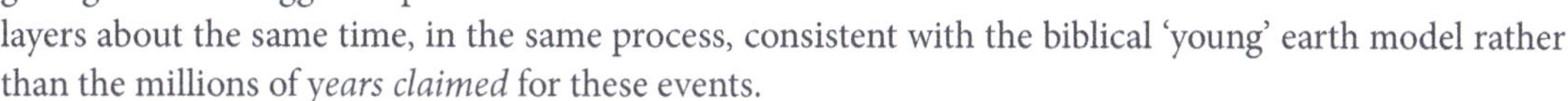

layers about the same time, in the same process, consistent with the biblical 'young' earth model rather than the millions of *years claimed* for these events.

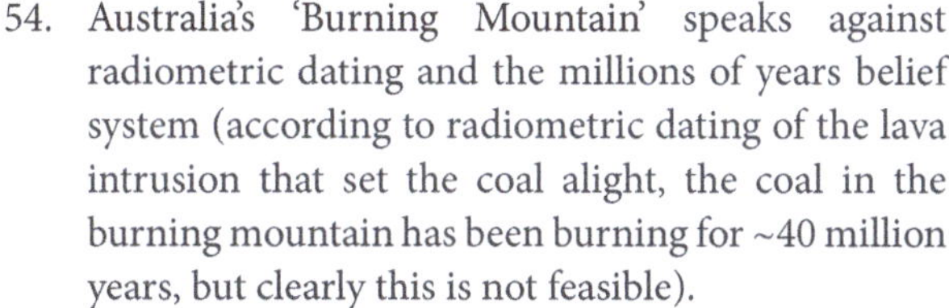

54. Australia's 'Burning Mountain' speaks against radiometric dating and the millions of years belief system (according to radiometric dating of the lava intrusion that set the coal alight, the coal in the burning mountain has been burning for ~40 million years, but clearly this is not feasible).

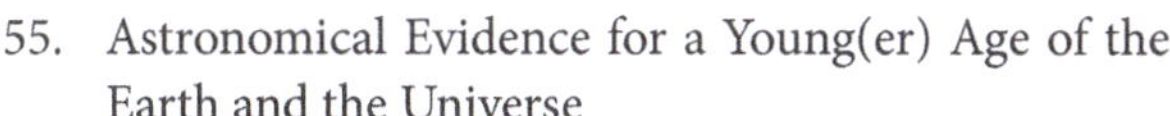

55. Astronomical Evidence for a Young(er) Age of the Earth and the Universe

Fig.20.13: Saturn's rings are increasingly recognized as being relatively short-lived rather than essentially changeless over millions of years. Curtsy Photo by NASA

56. Evidence of recent volcanic activity on Earth's moon is inconsistent with its supposed vast age because it should have long since cooled if it were billions of years old. See: Transient lunar phenomena: a permanent problem for evolutionary models of Moon formation and Walker, T., and Catchpoole, D., Lunar volcanoes rock long-age timeframe, *Creation* **31**(3):18, 2009. See further corroboration: "At Long Last, Moon's Core 'Seen'"; www.sciencemag.org/news/2011/01/long-last-moons-core-seen. Also, Saturn's rings are increasingly recognized as being relatively short-lived rather than essentially changeless over millions of years, Figure 20.13.

57. Recession of the moon from the earth. Tidal friction causes the moon to recede from the earth at 4 cm per year. It would have been greater in the past when the moon and earth were closer together. The moon and earth would have been in catastrophic proximity (Roche limit) at less than a quarter of their supposed age.

58. The moon's former magnetic field, Figure 20.14. Rocks sampled from the moon's crust have residual magnetism that indicates that the moon once had a magnetic field much stronger than earth's magnetic field today. No plausible 'dynamo' hypothesis could account for even a weak magnetic field, let alone a strong one that could leave such residual magnetism in a billions-of-years' time-frame. The evidence is much more consistent with a recent creation of the moon and its magnetic field and free decay of the magnetic field in the 6,000 years since then. Humphreys, D.R., *The moon's former magnetic field—still a huge problem for evolutionists, J. Creation* **26**(1):5–6, 2012.

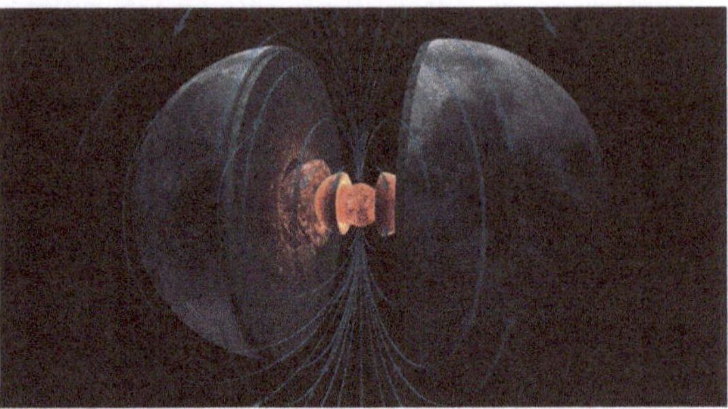

Fig.20.14: *Ancient moon's mega magnetic field*

59. Ghost craters on the moon's *maria* (singular *mare*: dark 'seas' formed from massive lava flows) are a problem for the assumed long ages. Enormous impacts evidently caused the large craters and lava flows within those craters, and this lava partly buried other, smaller impact craters within the larger craters, leaving 'ghosts.' But this means that the smaller impacts can't have been too long after the huge ones, otherwise the lava would have flowed into the larger craters before the smaller impacts. This suggests a very narrow time frame for all this cratering, and by implication the other cratered bodies of our solar system. They suggest that the cratering occurred quite quickly. See Fryman, H., Ghost craters in the sky, *Creation Matters* **4**(1):6, 1999; A biblically based cratering theory (Faulkner); Lunar volcanoes rock long-age timeframe.
60. The presence of a significant *magnetic field around Mercury* is not consistent with its supposed age of billions of years. A planet so small should have cooled down enough so any liquid core would solidify, preventing the evolutionists' 'dynamo' mechanism. See also, Humphreys, D.R., Mercury's magnetic field is young! *J. Creation* **22**(3):8–9, 2008.
61. The outer planets Uranus and Neptune have magnetic fields, but they should be long 'dead' if they are as old as claimed according to evolutionary long-age beliefs. Assuming a solar system age of thousands of years, physicist Russell Humphreys successfully predicted the strengths of the magnetic fields of Uranus and Neptune.
62. Jupiter's larger moons, *Ganymede, Io, and Europa, have magnetic fields*, which they should not have if they were billions of years old, because they have solid cores and so no dynamo could generate the magnetic fields. This is consistent with creationist Humphreys' predictions. See also, Spencer, W., Ganymede: the surprisingly magnetic moon, *J. Creation* **23**(1):8–9, 2009.
63. Volcanically active moons of Jupiter (Io) are consistent with youthfulness (Galileo mission recorded 80 active volcanoes). If Io had been erupting over 4.5 billion years at even 10% of its current rate, it would have erupted its entire mass 40 times. Io looks like a young moon and does not fit with the supposed billions of year's ages for the solar system. Gravitational tugging from Jupiter and other moons accounts for only some of the excess heat produced.

64. The surface of Jupiter's moon Europa. Studies of the few craters indicated that up to 95% of small craters, and many medium-sized ones, are formed from debris thrown up by larger impacts. This means that there have been far fewer impacts than had been thought in the solar system and the age of other objects in the solar system, derived from cratering levels, have to be reduced drastically (see Psarris, Spike, *what you aren't being told about astronomy, volume 1: Our created solar system* DVD, available from CMI).
65. Methane on Titan (Saturn's largest moon)—the methane should all be gone because of UV-induced breakdown. The products of photolysis should also have produced a huge sea of heavier hydrocarbons such as ethane. An *Astrobiology item* titled "The missing methane" cited one of the Cassini researchers, Jonathan Lunine, as saying, "If the chemistry on Titan has gone on in steady-state over the age of the solar system, then we would predict that a layer of ethane 300 to 600 meters thick should be deposited on the surface." No such sea is seen, which is consistent with Titan being a tiny fraction of the claimed age of the solar system (needless to say, Lunine does not accept the obvious young age implications of these observations, so he speculates, for example, that there must be some unknown source of methane).
66. The rate of change / disappearance of Saturn's rings is *inconsistent with their supposed vast age*; they speak of youthfulness.
67. Enceladus, a moon of Saturn, looks young. Astronomers working in the 'billions of years' mindset thought that this moon would be cold and dead, but it is a very active moon, spewing massive jets of water vapor and icy particles into space at supersonic speeds, consistent with a much younger age. Calculations show that the interior would have frozen solid after 30 million years (less than 1% of its supposed age); tidal friction from Saturn does not explain its youthful activity (Psarris, Spike, *what you aren't being told about astronomy, volume 1: Our created solar system DVD*; Walker, T., Enceladus: Saturn's sprightly moon looks young, *Creation* **31**(3):54–55, 2009).
68. Miranda, a small moon of Uranus, should have been long since dead, if billions of years old, but its extreme surface features suggest otherwise. See *Revelations in the solar system.*
69. Neptune should be long since 'cold', lacking strong wind movement if it were billions of years old, yet Voyager II in 1989 found it to be otherwise—it has the fastest winds in the entire solar system. This observation is consistent with a young age, not billions of years. See Neptune: monument to creation.
70. Neptune's rings have thick regions and thin regions. This unevenness means they cannot be billions of years old, since collisions of the ring objects would eventually make the ring very uniform. Revelations in the solar system.
71. Young surface age of Neptune's moon, Triton—less than 10 million years, even with evolutionary assumptions on rates of impacts (see Schenk, P.M., and Zahnle, K. On the negligible surface age of Triton, *Icarus* **192**(1):135–149, 2007. <doi:10.1016/j.icarus.2007.07.004>.
72. Uranus and Neptune both have magnetic fields significantly off-axis, which is an unstable situation. When this was discovered with Uranus, it was assumed by evolutionary astronomers that Uranus must have just happened to be going through a magnetic field reversal. However, when a similar thing was found with Neptune, this *AD hoc* explanation was upset. These observations are consistent with ages of thousands of years rather than billions.
73. The orbit of Pluto is chaotic on a 20-million-year time scale and affects the rest of the solar system, which would also become unstable on that time scale, suggesting that it must be much younger. (See: Rothman, T., God takes a nap, *Scientific American* **259**(4):20, 1988).
74. The existence of short-period comets (orbital period less than 200 years), e.g., Halley, which have a life of less than 20,000 years, is consistent with an age of the solar system of less than 10,000 years, Figure

20.15. *ad hoc* hypotheses have to be invented to circumvent this evidence (see Kuiper Belt). See Comets and the age of the solar system.

Fig.20.15: *Short Period Comet*

75. "Near-infrared spectra of the Kuiper Belt Object, Quaoar and the suspected Kuiper Belt Object, Charon, indicate both contain crystalline water ice and ammonia hydrate. This watery material cannot be much older than 10 million years, which is consistent with a young solar system, not one that is 5 billion years old." See: The 'waters above'.
76. Lifetime of long-period comets (orbital period greater than 200 years) that are sun-grazing comets or others like Hyakutake or Hale–Bopp means they could not have originated with the solar system 4.5 billion years ago. However, their existence is consistent with a young age for the solar system. Again an *ad hoc* Oort Cloud was invented to try to account for these comets still being present after billions of years. See, Comets and the age of the solar system, Figure 20.15.
77. The maximum expected lifetime of near-earth asteroids is of the order of one million years, after which they collide with the sun, Figure 20.16. And the Yarkovsky effect moves main belt asteroids into near-earth orbits faster than had been thought. This brings into question the origin of asteroids with the formation of the solar system (the usual scenario), or the solar system is much younger than the 4.5 billion years claimed. Henry, J., *The asteroid belt: indications of its youth, Creation Matters* **11**(2):2, 2006.

Fig.20.16: *Asteroid 2021 GT2 will fly by Earth on June 6, 2022 - Curtsy - Live Science*

78. The lifetime of binary asteroids—where a tiny asteroid 'moon' orbits a larger asteroid— in the main belt (they represent about 15–17% of the total): tidal effects limit the life of such binary systems to about 100,000 years. The difficulties in conceiving of any scenario for getting binaries to form in such numbers to keep up the population, led some astronomers to doubt their existence, but space probes confirmed it (Henry, J., The asteroid belt: indications of its youth, *Creation Matters* **11**(2):2, 2006).
79. The observed *rapid rate of change in stars* contradicts the vast ages assigned to stellar evolution. For example, Sakurai's Object in Sagittarius: in 1994, this star was most likely a white dwarf in the center of a planetary nebula; by 1997 it had grown to a bright yellow giant, about 80 times wider than the sun (*Astronomy & Astrophysics* **321**: L17, 1997). In 1998, it had expanded even further, to a red supergiant 150 times wider than the sun. But then it shrank just as quickly; by 2002 the star itself was invisible even to the most powerful optical telescopes, although it is detectable in the infrared, which shines through the dust (Muir, H., 2003, Back from the dead, *New Scientist* **177**(2384):28–31).
80. The faint young sun paradox. According to stellar evolution theory, as the sun's core transforms from hydrogen to helium by means of nuclear fusion, the mean molecular weight increases, which would compress the sun's core increasing fusion rate. The upshot is that over several billion years, the sun ought to have brightened 40% since its formation and 25% since the appearance of life on earth. For the latter, this translates into a 16–18 °C temperature increase on the earth. The current average temperature is 15 °C, so the earth ought to have had a -2 °C or so temperature when life appeared. See: Faulkner, D., The young faint Sun paradox and the age of the solar system, *J. Creation (TJ)* **15**(2):3–4, 2001. As of 2010, the faint young sun remains a problem: Kasting, J.F., Early Earth: Faint young Sun redux, *Nature* **464**:687–689, 1 April 2010; doi:10.1038/464687a; www.nature.com/nature/journal/v464/n7289/full/464687a.html
81. Evidence of (very) recent geological activity (tectonic movements) on the moon is inconsistent with its supposed age of billions of years and its hot origin. Watters, T.R., *et al.*, Evidence of Recent Thrust Faulting on the Moon Revealed by the Lunar Reconnaissance Orbiter Camera, *Science* **329**(5994):936–940, 20 August 2010; DOI: 10.1126/science.1189590 ("This detection, coupled with the very young apparent age of the faults, suggests global late-stage contraction of the Moon.") NASA pictures support biblical origin for Moon.
82. The giant gas planets Jupiter and Saturn radiate more energy than they receive from the sun, suggesting a recent origin. Jupiter radiates almost twice as much energy as it receives from the sun, indicating that it may be less than 1 % of the presumed 4.5 billion years old solar system. Saturn radiates nearly twice as much energy per unit mass as Jupiter. See The age of the Jovian planets.
83. Speedy stars are consistent with a young age for the universe. For example, many stars in the dwarf galaxies in the Local Group are moving away from each other at speeds estimated at to 10–12 km/s. At these speeds, the stars should have dispersed in 100 Ma, which, compared with the supposed 14,000 Ma age of the universe, is a short time. See Fast stars challenge big bang origin for dwarf galaxies.
84. The ageing of spiral galaxies (much less than 200 million years) is not consistent with their supposed age of many billions of years. The *discovery of extremely 'young' spiral galaxies* highlights the problem of this evidence for the evolutionary ages assumed.
85. The number of types I *supernova remnants* (SNRs) observable in our galaxy is consistent with an age of thousands of years, not millions or billions. See Davies, K., *Proc. 3prd ICC*, pp. 175–184, 1994.
86. There is a great paucity of highly expanded SNRs compared to what is expected under evolutionary cosmogony. See *supernova remnants.*
87. Human History is Consistent with a Young Age of the Earth

88. Human population growth. Less than 0.5% p.a. growth from six people 4,500 years ago would produce today's population. *Where are all the people?* if we have been here much longer?
89. 'Stone age' human skeletons and artefacts. There are not enough for 100,000 years of a human population of just one million, let alone more people (10 million?). See Where are all the people?
90. Length of recorded history. Origin of various civilizations, writing, etc., all about the same time several thousand years ago. See Evidence for a young world.
91. Languages. Similarities in languages claimed to be separated by many tens of thousands of years speaks against the supposed ages (e.g., compare some aboriginal languages in Australia with languages in south-eastern India and Sri Lanka). See The Tower of Babel account affirmed by linguistics.
92. Common cultural 'myths' speak of recent separation of peoples around the world. An example of this is the frequency of stories of an earth-destroying flood.
93. Origin of agriculture. Secular dating puts it at about 10,000 years and yet that same chronology says that modern man has supposedly been around for at least 200,000 years. Surely someone would have worked out much sooner how to sow seeds of plants to produce food. See: Evidence for a young world.

EVIDENCE FOR A YOUNG UNIVERSE

Fig.20.17: Here are a dozen natural phenomena which conflict with the evolutionary idea that the universe is billions of years old.

The numbers I list below in bold print (often millions of years) are **maximum possible** ages set by each process, not the actual ages. The numbers in italics are the ages *required by evolutionary theory* for each item.

The point is that the maximum possible ages are always much less than the required evolutionary ages, while the *Biblical age* (6,000 to 10,000 years) always fits comfortably within the maximum possible ages. Thus, the following items are evidence against the evolutionary time scale and for the Biblical time scale.

Much more young-world evidence exists, Figure 20.17, but I have chosen these items for brevity and simplicity. Some of the items on this list can be reconciled with an old universe only by making a series of improbable and unproven assumptions; others can fit in only with a young universe. The list starts with distant astronomic phenomena and works its way down to Earth, ending with everyday facts.

1. Galaxies Wind themselves up too Fast

The stars of our own galaxy, the Milky Way, rotate about the galactic center with different speeds, the inner ones rotating faster than the outer ones. The observed rotation speeds are so fast that if our galaxy were more than **a few hundred million years** old, it would be a featureless disc of stars instead of its present spiral shape, Figure 20.18.

Fig.20.18: Spiral galaxy NGC 1232 in the constellation of Eridanus (courtesy of the European Southern Observatory).

Yet our galaxy is supposed to be at least *10 billion years* old. Evolutionists call this 'the winding-up dilemma,' which they have known about for fifty years. They have devised many theories to try to explain it, each one failing after a brief period of popularity. The same 'winding-up' dilemma also applies to other galaxies.

For the last few decades, the favored attempt to resolve the dilemma has been a complex theory called 'density waves.' The theory has conceptual problems, has to be arbitrarily and very finely tuned, and lately has been called into serious question by the Hubble Space Telescope's discovery of very detailed spiral structure in the central hub of the 'Whirlpool' galaxy, M51.

2. Comets Disintegrate too Quickly

According to evolutionary theory, comets are supposed to be the same age as the solar system, about *5 billion years*. Yet each time a comet orbits close to the sun; it loses so much of its material that it could not survive much longer than about **100,000 years.** Many comets have typical ages of **10,000 years**, Figure 20.19.

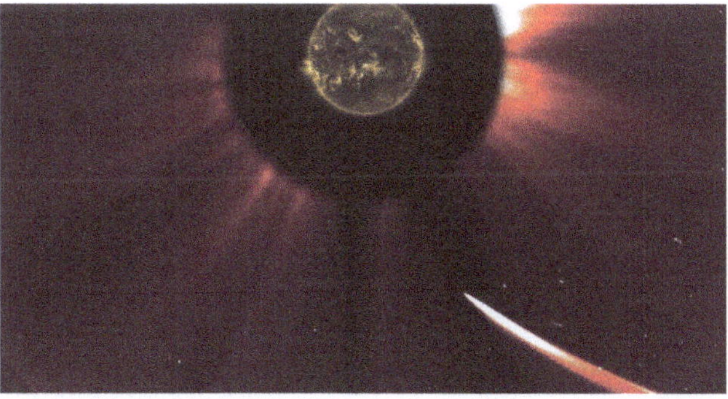

Fig.20.19: Comet ISON Disintegrates in Sun Flyby, Disappointing Scientists – Curtsy ABC News

Evolutionists explain this discrepancy by assuming that

a. comets come from an unobserved spherical 'Oort cloud' well beyond the orbit of Pluto,
b. improbable gravitational interactions with infrequently passing stars often knock comets into the solar system, and
c. other improbable interactions with planets slow down the incoming comets often enough to account for the hundreds of comets observed.

So far, none of these assumptions has been substantiated either by observations or realistic calculations.

Lately, there has been much talk of the 'Kuiper Belt,' a disc of supposed comet sources lying in the plane of the solar system just outside the orbit of Pluto. Even if some bodies of ice exist in that location, they would not really solve the evolutionists' problem, since according to evolutionary theory the Kuiper Belt would quickly become exhausted if there were no Oort cloud to supply it.

3. Not Enough Mud on the Sea Floor

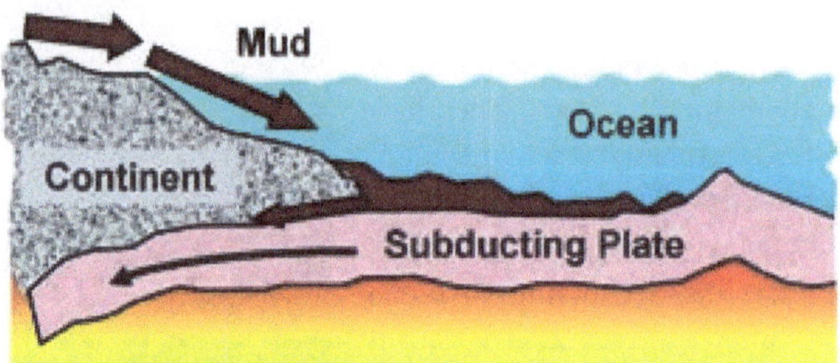

Fig.20.20: Not Enough Mud in the Sea Floor

Each year, water and winds erode about 25 billion tons of dirt and rock from the continents and deposit it in the ocean, Figure 20.20. This material accumulates as loose sediment (i.e., mud) on the hard basaltic (lava-formed) rock of the ocean floor. The average depth of all the mud in the whole ocean, including the continental shelves, is less than 400 meters.

The main way known to remove the mud from the ocean floor is by plate tectonic subduction. That is, sea floor slides slowly (a few cm/year) beneath the continents, taking some sediment with it. According to secular scientific literature, that process presently removes only 1 billion tons per year. As far as anyone knows, the other 24 billion tons per year simply accumulate. At that rate, erosion would deposit the present amount of sediment in less than **12 million years**.

Yet according to evolutionary theory, erosion and plate subduction have been going on as long as the oceans have existed, an alleged *3 billion years*. If that were so, the rates above imply that the oceans would be massively choked with mud dozens of kilometers deep. An alternative (creationist) explanation is that erosion from the waters of the Genesis flood running off the continents deposited the present amount of mud within a short time about 5000 years ago.

4. Not Enough Sodium in the Sea

Every year, river and other sources dump over 450 million tons of sodium into the ocean. Only 27% of this sodium manages to get back out of the sea each year. As far as anyone knows, the remainder simply accumulates in the ocean. If the sea had no sodium to start with, it would have accumulated its present amount in less than 42 million years at today's input and output rates. This is much less than the evolutionary age of the ocean, *3 billion years*. The usual reply to this discrepancy is that past sodium inputs must have been less and outputs greater. However, calculations which are as generous as possible to evolutionary scenarios still give a maximum age of only **62 million years**. Calculations for many other sea water elements give much younger ages for the ocean.

5. The Earth's Magnetic Field is Decaying too Fast

The total energy stored in the Earth's magnetic field has steadily decreased by a factor of 2.7 over the past 1000 years. Evolutionary theories explaining this rapid decrease, as well as how the Earth could have maintained its magnetic field for *billions of years*, are very complex and inadequate, Figure 20.21.

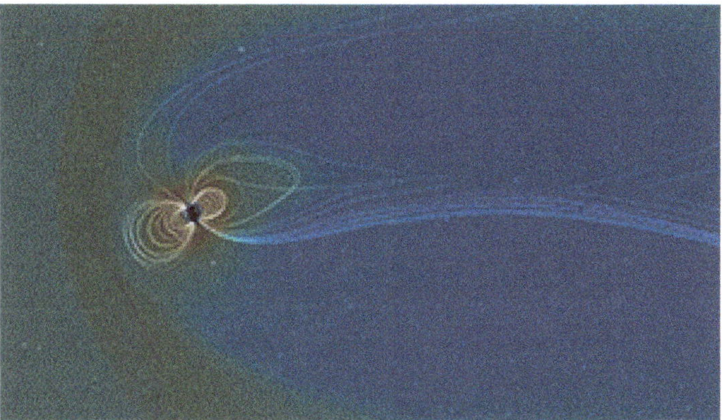

Fig.20.21: The Earth's Magnetic Field is Decaying too Fast

A much better creationist theory exists. It is straightforward, based on sound physics, and explains many features of the field: its creation, rapid *reversals* during the Genesis flood, surface intensity decreases and *increases* until the time of Christ, and a steady decay since then. This theory matches paleomagnetic, historic, and present data. The main result is that the field's total energy (not surface intensity) has always decayed at least as fast as now. At that rate the field could not be more than **10,000 years** old.

6. Many Strata are too Tightly Bent

In many mountainous areas, strata thousands of feet thick are bent and folded into hairpin shapes. The conventional geologic time scale says these formations were deeply buried and solidified for *hundreds of millions of years before* they were bent. Yet the folding occurred without cracking, with radii so small that the entire formation had to be still wet and unsolidified when the bending occurred. This implies that the folding occurred **less than thousands of years** after deposition, Figure 20.22.

Fig.20.22: Folded rocks: testimony to Noah's Flood

7. Injected Sandstone Shortens Geologic 'Ages'

Strong geologic evidence exists that the Cambrian Sawatch sandstone—formed an alleged 500 million years ago—of the Ute Pass fault west of Colorado Springs was still unsolidified when it was extruded up to the surface during the uplift of the Rocky Mountains, allegedly 70 million years ago. It is very unlikely that the sandstone would not solidify during the supposed *430 million years* it was underground., Figure 20.23 Instead, it is likely that the two geologic events were **less than hundreds of years** apart, thus greatly shortening the geologic time scale.

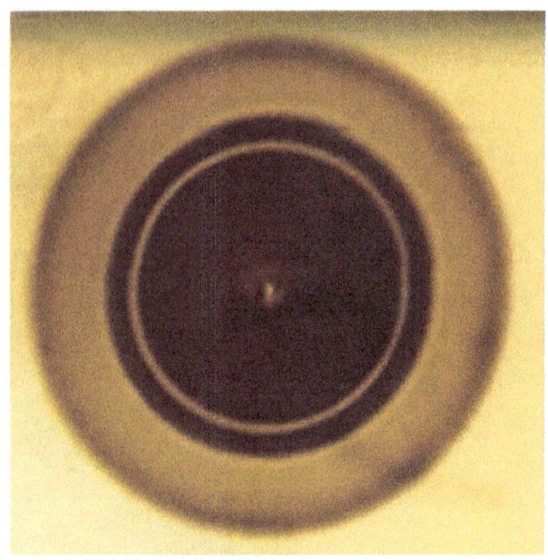

Fig.20.23: Radio-halos Curtsy - (Mark Armitage)

8. Fossil Radioactivity Shortens Geologic 'Ages' to a Few Years

Radio-halos are rings of color formed around microscopic bits of radioactive minerals in rock crystals. They are fossil evidence of radioactive decay. 'Squashed' Polonium-210 radio-halos indicate that Jurassic, Triassic, and Eocene formations in the Colorado Plateau were deposited within months of one another, not hundreds of millions of years apart as required by the conventional time scale. 'Orphan' Polonium-218 radio-halos, having no evidence of their mother elements, imply either instant creation or drastic changes in radioactivity decay rates.

9. Helium in the Wrong Places

All naturally-occurring families of radioactive elements generate helium as they decay. If such decay took place for billions of years, as alleged by evolutionists, much helium should have found its way into the Earth's atmosphere. The rate of loss of helium from the atmosphere into space is calculable and small. Taking that loss into account, the atmosphere today has only 0.05% of the amount of helium it would have accumulated in 5 billion years. This means the atmosphere is much younger than the alleged evolutionary age. A study published in the *Journal of Geophysical Research* shows that helium produced by radioactive decay in deep, hot rocks has not had time to escape. Though the rocks are supposed to be over *one billion years* old, their large helium retention suggests an age of only **thousands of years**.

10. Not Enough Stone Age Skeletons

Evolutionary anthropologists say that the stone age lasted for at least *100,000 years*, during which time the world population of Neanderthal and Cro-Magnon men was roughly constant, between 1 and 10 million. All that time they were burying their dead with artefacts. By this scenario, they would have buried at least 4 billion bodies. If the evolutionary time scale is correct, buried bones should be able to last for much longer than 100,000 years, so many of the supposed 4 billion stone age skeletons should still be around (and certainly the buried artefacts). Yet only a few thousand have been found. This implies that the stone age was much shorter than evolutionists think, **a few hundred years** in many areas.

11. Agriculture is Too Recent

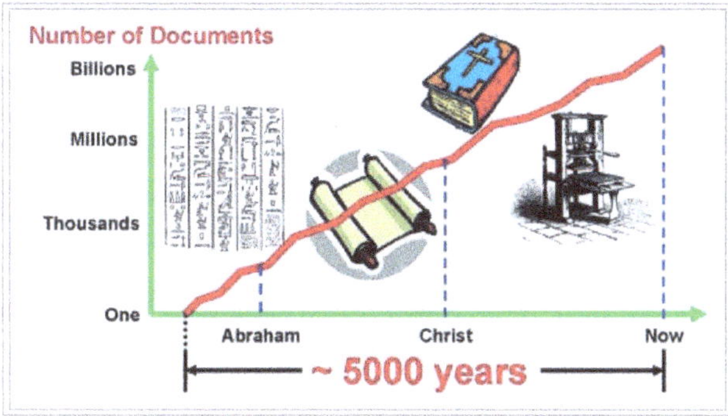

Fig.20.24: Agriculture is Too Recent

The usual evolutionary picture has men existing as hunters and gatherers for *100,000 years* during the stone age before discovering agriculture less than 10,000 years ago. Yet the archaeological evidence shows that stone age men were as intelligent as we are, Figure 20.24. It is very improbable that none of the 4 billion people mentioned in item 10 should discover that plants grow from seeds. It is more likely that men were without agriculture **less than a few hundred years** after the flood, if at all.

12. History is Too Short

According to evolutionists, stone age man existed for *100,000 years* before beginning to make written records about **4000 to 5000 years** ago. Prehistoric man built megalithic monuments, made beautiful cave paintings, and kept records of lunar phases. Why would he wait a thousand centuries before using the same skills to record history? The Biblical time scale is much more likely.

SALTY SEAS

Evidence for a young earth

Our planet, Earth, is the only place in the universe known to have liquid water. In fact, astronauts looking at Earth's surface from outer space see mainly water, Figure 20.25. The ocean covers 71% of the total area, and contains enough water to cover the whole planet to a depth of 2.7 km (1.7 miles) if the surface were completely flat.

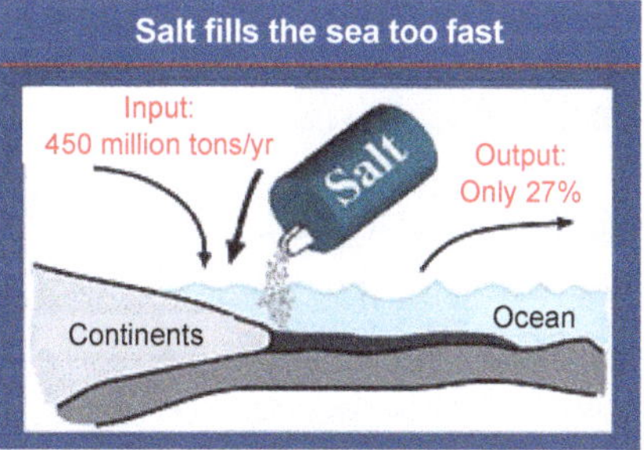

Fig.20.25: Not Enough Salt – Young Earth

Salinity

The ocean is essential for life on Earth, and also helps make the climate fairly moderate. However, although the ocean contains 1,370 million cubic kilometers (334 million cubic miles) of water, humans can't survive by drinking from it—it is too salty.

To a chemist, 'salt' refers to a wide range of chemicals where a metal is combined with a non-metal. Ordinary common salt is a compound formed when the metal sodium combines with the non-metal chlorine—sodium chloride. This contains electrically charged atoms, called *ions*, that attract each other, resulting in a fairly hard crystal. When salt dissolves, these ions separate. Sodium and chloride ions are the main ions in seawater, but not the only ones. The salty seas benefit man, because the ocean provides many useful minerals for our industries.

The Age of the Sea

Many processes stated below bring salts into the sea, while they don't leave the sea easily. So, the saltiness is increasing steadily. Since we can work out how much salt there is in the sea, as well as the rates that salts go into and out of the sea, we should be able to calculate a maximum age for the sea.

In fact, this method was first proposed by Sir Isaac Newton's colleague, Sir Edmond Halley (1656–1742), of comet fame. More recently, the geologist, physicist, and pioneer of radiation therapy, John Joly, (1857–1933) estimated that the oceans were 80–90 million years old at the most. But this was far too young for evolutionists, who believed that life evolved in the ocean billions of years ago.

More recently, the geologist Dr Steve Austin and the physicist Dr Russell Humphreys analyzed figures from secular geoscience sources for the quantity of sodium ion (Na^+) in the ocean, and its input and output rates. *The slower the input and faster the output, the older the ocean could be.*

Every kilogram of seawater contains about 10.8 grams of dissolved Na^+ (about 1% by weight). This means that there is a total of 1.47×10^{16} (14,700 million million) tons of Na^+ in the ocean.

Sodium Input

Water on the land can dissolve salt outcrops, and can weather many minerals, especially clays and feldspars, and leach the sodium out of them. This sodium can be carried into the ocean by rivers. Some salt is supplied by water through the ground directly to the sea—called submarine groundwater discharge (SGWD). Such water is often very concentrated in minerals. Ocean floor sediments release much sodium, as do hot springs on the ocean floor (hydrothermal vents). Volcanic dust also contributes some sodium.

Austin and Humphreys calculated that about 457 million tons of sodium now comes into the sea every year. The minimum possible rate in the past, even if the most generous assumptions are granted to evolutionists, is 356 million tons/year.

Actually, a more recent study shows that salt is entering the oceans even faster than Austin and Humphreys thought. Previously, the amount of SGWD was thought to be a small fraction (0.01–10%) of the water from surface runoff, mainly rivers. But this new study, measuring the radioactivity of radium in coastal water, shows that the amount of SGWD is as much as 40% of the river flow. This means that the maximum possible age of the ocean is even smaller.

Sodium Output

People who live near the sea often have problems with rust in cars. This is due to salt spray—small droplets of seawater escape from the ocean, the water evaporates, leaving behind tiny salt crystals. This is a major process that removes sodium from the sea. Another major process is called ion exchange—clays can absorb sodium ions and exchange them for calcium ions, which are released into the ocean. Some sodium is lost from the ocean when water is trapped in pores in sediments on the ocean floor. Certain minerals with large cavities in their crystal structure, called zeolites, can absorb sodium from the ocean.

However, the rate of all of this sodium output is far less than the input. Austin and Humphreys calculated that about 122 million tons of sodium leaves the sea every year. The maximum possible rate in the past, even if the most generous assumptions are granted to evolutionists, is 206 million tons/year.

Estimating the Age of the Ocean

Granting the most generous assumptions to evolutionists, Austin and Humphreys calculated that the ocean must be *less* than 62 million years old. It's important to stress that this is *not* the *actual* age, but a *maximum* age. That is, this evidence is consistent with any age up to 62 million years, including the biblical age of about 6,000 years.

The Austin and Humphreys calculation assumes the lowest plausible input rates and fastest plausible output rates. Another assumption is that there was no dissolved salt to start with. If we assume more realistic conditions in the past, the calculated maximum age is much less.

For one thing, God probably created the oceans with some saltiness, so that saltwater fish could live comfortably in it. Noah's Flood would have dissolved large amounts of sodium from land rocks. This would have found its way into the oceans when the Flood waters retreated. Finally, the larger-than-expected SGWD would further reduce the maximum age.

The salinity of the oceans is strong evidence that they, and the Earth itself, are far younger than the billions of years required for evolution, and is consistent with the biblical age of about 6,000 years. It is also far younger than the evolutionists' 'dates' for many marine creatures. In short, the sea is not salty enough to suit the taste of evolutionists! Of course, all such calculations depend on assumptions about the past, like the starting conditions and constant rates of processes. They can never *prove* the age of something. For that, we need an eye-witness (*cf.* Job 38:4). The point of such calculations is to demonstrate *that even under the evolutionists' own assumptions about the past*, the Earth is far younger than is usually claimed and does not contradict the Bible.

MISSING POPULATION

Fig.20.26: *Eight billion people live on planet Earth. That sounds like a lot of people. Curtsy stock.xchng*

Well, I would not want to invite them all to a barbecue at my house! However, they could all fit into an area the size of England, with more than 20 square meters each. Many of us live in cities, so we have the impression that the world is bursting with people. However, much of the world is sparsely populated, Figure 20.26.

Nevertheless, many wonder at how the population could have grown to eight billion from Noah's family who survived the Flood that wiped out everyone else about 4,500 years ago.

When you do the figures, it confirms the biblical truth that everyone on Earth today is a descendant of Noah's sons and daughters-in-law.

Not only that, but if people have been here for much longer, and there was no global Flood of Noah's day, there should be a lot more people than there are—or there should be a lot more human remains!

Many people have problems understanding growth *rates* of things. When the population doubles from 16 to 32, it does not seem like much, but when it doubles from three billion to six billion it seems like a lot more. However, it is exactly the same *rate* of growth. Given enough generations, the number of people being added with each generation becomes astronomical. It's like compound interest on an investment—eventually the amount being added each year becomes very great.

To illustrate this, think of the story of the inventor of chess. His king offered him a reward, but instead of gold he asked for one grain of rice doubled for each successive square on a chessboard. The number of grains would have been 1, 2, 4, 8, 16, 32 etc. The 10th square would have 512; the 20th, 524 *thousand;* the 30th, 537 *million*. The amount of rice on the last square would have been a number so great—vastly in excess of the total world rice harvest at present—that it would have represented wealth far exceeding that of the king. Such is the power of compounding. And population growth is compound growth—that's why so many people are now being added each year. It's not necessarily that people are having more children than they once did, or that fewer people are dying.

What Causes Population Explosion

The population grows when more people are born than die. The current growth rate of the world population is about 1.7% per year. In other words, for every 100 million people, 1.7 million are added every year; i.e., births net of deaths.

Many assume that modern medicine accounts for the world's population growth. However, 'third world' countries contribute most of the population growth, suggesting that modern medicine is not as important as many think.

Population growth in a number of South American and African countries exceeds 3% per year. In many industrialized countries with modern medical facilities, the population growth is less than 0.5%. Some relatively wealthy countries are actually declining in population.

The move from agriculture to manufacturing/technology has been a big factor in slowing population growth in industrialized countries. Farmers needed to have sons to help with the farm work. This was particularly necessary before mechanization. In the early- to mid-1800s in Australia, couples commonly had 8–10 surviving children. One couple had 16! And this was before the discovery of the germ basis of disease, aseptic surgery, vaccines and antibiotics. Opportunity to expand, combined with biology, saw growth in population of 4% or more, *plus* increases due to immigration. High rates of population growth were also seen in Quebec, Canada, from 1760 to 1790, following the British conquest of Canada in 1759, and well before the impact of modern medical knowledge.

In industrialized countries, the advent of social security pensions and retirement plans (superannuation) has probably been another major factor in the decline of population growth. These schemes mean that people do not see the need to have children for security in their old age. Furthermore, people can now easily choose how many children they have because of modern birth control methods, such as the contraceptive tablets.

What Growth Rate is Needed to Reach Eight Billion Since the Flood?

It is relatively easy to calculate the growth rate needed to reach today's population from Noah's three sons and their wives, after the Flood. With the Flood at about 4,500 years ago, it needs less than 0.5% per year growth. That's not very much.

Of course, population growth has not been constant. There is reasonably good evidence that growth has been slow at times—such as in the Middle Ages in Europe. However, data from the Bible (*Genesis 10,11*) shows that the population grew quite quickly in the years immediately after the Flood.

- Shem had five sons,
- Ham had four, and
- Japheth had seven.

If we assume that they had the same number of daughters, then they averaged 10.7 children per couple. In the next generation,

- Shem had 14 grandsons,
- Ham, 28 and
- Japheth, 23, or

130 children in total. That is an average of 8.1 per couple. These figures are consistent with God's command to 'be fruitful and multiply and fill the earth' (*Genesis 9:1*).

Let us take the average of all births in the first two post-Flood generations as 8.53 children per couple. The average age at which the first son was born in the seven post-Flood generations in Shem's line ranged from 35 to 29 years (*Genesis 11:10–24*), with an average of 31 years, so a generation time of 40 years is reasonable.

Hence, just four generations after the Flood would see a total population of over **3,000** people (remembering that the longevity of people was such that Noah, Shem, Ham, Japheth, etc., were still alive at that time). This represents a population growth rate of 3.7% per year, or a doubling time of about 19 years.

If there were 300 million people in the world at the time of Christ's Resurrection, this requires a population growth rate of only 0.75% since the Flood, or a doubling time of 92 years—much less than the documented population growth rate in the years following the Flood.

A REMARKABLE COINCIDENCE

The Jews are descendants of Jacob (also called Israel). The number of Jews in the world in 1930, before the Nazi Holocaust, was estimated at 18 million. This represents a doubling in population, on average, every 156 years, or 0.44% growth per year since Jacob. Since the Flood, after which the world population was eight, the world population has doubled every 155 years, or grown at an average of 0.45% per year. There is agreement between the growth rates for the two populations. Is this just a lucky coincidence?

Hardly! The figures agree because the real history of the world is recorded in the Bible.

What if People had Been Around for one Million Years?

Evolutionists claim that mankind evolved from apes about a million years ago. If the population had grown at just 0.01% per year since then (doubling only every 7,000 years), there could be $\mathbf{10^{43}}$ people today—that's a number with 43 zeros after it. This number is so big that not even the Texans have a word for it!

To try to put this number of people in context, say each individual is given 'standing room only' of about one square meter per person. However, the land surface area of the whole Earth is 'only' $\mathbf{1.5 \times 10^{14}}$ **square meters**. If every one of those square meters were made into a world just like this one, all these worlds put together would still 'only' have a surface area able to fit **10^{28} people** in this way. This is only a tiny fraction of 10^{43} (10^{29} is 10 times as much as 10^{28}, 10^{30} is 100 times, and so on). Those who adhere to the evolutionary story argue that disease, famine and war kept the numbers almost constant for most of this period, which means that mankind was on the brink of extinction for most of this supposed history. This stretches credulity to the limits.

Where are all the Bodies?

Evolutionists also claim there was a 'Stone Age' of about 100,000 years when between one million and 10 million people lived on Earth. Fossil evidence shows that people buried their dead, often with artefacts—cremation was not practiced until relatively recent times (in evolutionary thinking). If there were just one million people alive during that time, with an average generation time of 25 years, they should have buried 4 billion bodies, and many artefacts. If there were 10 million people, it would mean 40 billion bodies buried in the earth. If the evolutionary timescale were correct, then we would expect the skeletons of the buried bodies to be largely still present after 100,000 years, because many ordinary bones claimed to be much older have been found. However, even if the bodies had disintegrated, lots of artefacts should still be found.

Now the number of human fossils found is nothing like one would expect if this 'Stone Age' scenario were correct. The number found is more consistent with a 'Stone Age' of a few hundred years, which would have occurred after Babel. Many people groups could have used stone tools as they moved out from Babel (*Genesis 11*), having lost the technologies of metal smelting (*Genesis 4:22*) due to the Flood and the confusion of languages at Babel.

Immigrant peoples, when they settled in a new area, would have had an initial phase where they would shelter in caves, or have rudimentary housing. They would have made use of stone tools, for example, while they developed agricultural techniques appropriate to the local soils and climate, found sources of ores, and rediscovered how to manufacture tools, etc.

Groups that descended into animism might never emerge from this 'stone age' of their development, because of the stifling effects of such things as taboos, and fear of evil spirits. One tribal group in the Philippines, for example, had a taboo against water, causing rampant disease due to lack of hygiene—before the Gospel of Jesus Christ rescued them from superstition.

Fig.20.27: Australian Aborigines

Australian Aborigines—How Long have They Been in Australia?

When Europeans came to settle in Australia in 1788, it was estimated that there were perhaps only 300,000 Aboriginal people. And, Figure 20.27 yet today we are told that the people have been here for 60,000 years or more. Now there is no way that a mere 300,000 people had exhausted the plenty of this large country so as to account for a long period of very low population growth. If we allow for one-third of the land area as desert, it means that there was only one person for every 18 square kilometers (7 square miles) of habitable land area—hardly overpopulated, even for a subsistence existence.

If 20 people had come to settle sometime after the Flood, say 3,500 years ago, it would have needed a population growth of a mere 0.28% per year to produce 300,000 people. Such a minimal rate operating over 60,000 years could produce more people than there are atoms in the *Milky Way Galaxy!*

The real history of the world is recorded in the Bible, the Word of the Creator-God who was there in the beginning. This record shows that the world was deluged and destroyed (*Genesis 6–9, 2 Peter 3*) so that all people living today came from those who survived aboard Noah's Ark. A study of population growth clearly supports this biblical record.

THE TOWER OF BABEL

Secular linguists are puzzled by the existence of twenty or so language families in the world today. The languages within each family (and the people that speak them) have been shown to be genetically related, but few genetic links have been observed between families, Figure 20.28. This is a problem for secular linguists.

Fig.20.28: The Tower of Babel – *Curtsy – Public domain, via Wikimedia Commons*

If, as they believe, man evolved from an ape-like ancestor, man would at some point have gained the ability to speak. This process of change would actually be superbly dangerous, as they admit. But still, if speech did evolve somewhere, somehow, we would expect to find that all languages *are* genetically related. They clearly are not.

Some have therefore suggested that man evolved speech simultaneously in more than one place. This suggestion is beyond belief, considering the dangers involved in the supposed evolution of speech. So how did the language families come into existence

In the conventional view, language developed from a single 'protolanguage' and diverged into different languages as time progressed, Figure 20.29.

In the Biblical model, man was created with language. This was supernaturally changed at Babel where God 'confused' the languages. Scripture does not directly state how many languages arose at Babel.

Only Genesis provides a credible explanation. It records how God gave the people new languages to speak. Groups speaking the same language moved away together. The languages they spoke then, have slowly evolved into the six thousand-or-so languages we find today, but the distinctions between the groups of languages are still observable, as detailed below.

Determining whether or not languages share a common ancestor is not easy. A Dutch student learning Hindi might not realize that Hindi is related to Dutch. Yet, both languages have been shown to be part of the Indo-European language family.

Steel has previously covered in detail the development of the Indo-European languages, clearly refuting claims that this paralleled biological evolution. Apparently, all languages in this family have developed from a 'parent language', which no longer exists.

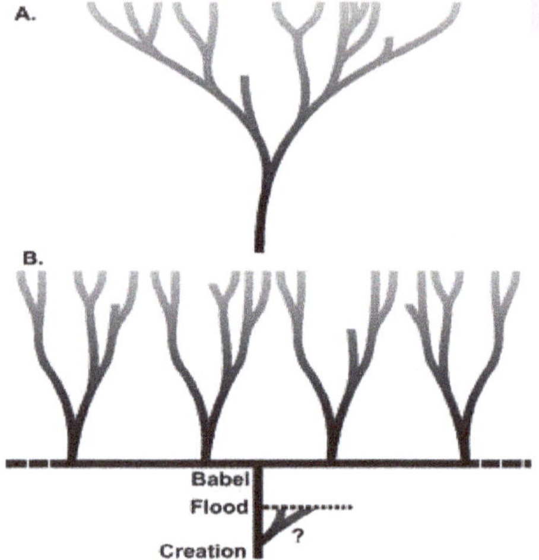

*Fig.20.29: The conventional view (**A**) and Biblical view (**B**) of language development.*

This idea was unknown in the late 18th century, until Sir William Jones suggested that *Greek, Latin and Sanskrit* had independently 'sprung from some common source, which, perhaps, no longer exists.' He also suggested that other groups of languages, such as the *Celtic and Germanic* languages, though quite different, might also be related in the same way. Few question his findings today. Comparative and Historical Linguistics have more or less carried on what Jones began. Two centuries have revealed much, and the findings are encouraging for Creation scientists, who believe the account of the Tower of Babel in Genesis 11 to be a true, historical account of events.

In this essay we shall be looking at some of the evidence for Babel, and examine two rival theories as well. Before we do that, let us have a brief look at the methods involved.

LANGUAGE CLASSIFICATION

The Indo-European language family is not the only language family in the world. There are others, which are more difficult to examine. We have many writings of some European languages, covering more than 2,500 years of development. For many other languages, however, there are no writings at all. That makes the study of their development more complicated.

The traditional way of comparing languages was to compare the history and grammar structures of two languages, while keeping in mind physical and cultural similarities between the tribes. This method was useful in Europe, but it was time-consuming and proved difficult in Africa. Several decades of hard work at the beginning of this century had uncovered only the tip of the iceberg, as far as all languages in Africa were concerned.

A dramatic breakthrough came in the person of Joseph Greenberg in the middle of last century. Greenberg came up with a new method. He collected lists of words from many African languages, and compared them with each other. He noticed clear patterns. Several languages had similar sounding words for similar things, and Greenberg concluded that these languages must therefore be related. His method has become the norm in comparative linguistics.

Greenberg's method is one of two major ways of classifying languages. Typological classification looks at grammatical structures and classifies languages accordingly. However, there may not be a genetic relationship between languages with a similar typological makeup. Since we are interested in genetic relationship, we will now have a brief look at the second method, Genetic Qualification, and consider its findings in relation to our essay question.

GENETIC QUALIFICATION

Core Vocabulary

Genetic qualification prefers to use only 'core vocabulary', i.e., words which are said to change little over time. The method aims to see how many of these words are similar in different languages, while keeping in mind how words usually change in pronunciation.

The core vocabulary includes, amongst others, words for body parts, numbers, and personal pronouns. When clear patterns of similarities between languages are observed, then those languages are said to be related.

Cognate Words

The word 'patterns' in the previous paragraph was carefully chosen, because the core vocabulary between related languages is never identical, but similar, or 'cognate'. Words are cognate when they are shown to be consistent to the pattern of phonetical change that has taken place in the past. For example, the word *tahi* in Tongan might not look like *kai* in Hawaiian, even though they both mean 'sea'. But, if you also compare Tongan *tapu* to Hawaiian *kapu* (both meaning 'forbidden') and Tongan *tanata* to Hawaiian *kanaka* (meaning 'man') you begin to see a pattern: Where Tongan has an initial 'T' Hawaiian has an initial 'K', and one begins to see that the words might be related. They are cognate.

Common Phonetic Changes

Deciding which words are cognate and which words are not is never easy. Different scholars have made different judgements when comparing the same lists. There is no general agreement in all cases. There are, however, a few rules to go by, as certain phonetical changes are more likely to occur than others. Stronger sounds, for example, may become weaker.

Equally, words may lose initial or final letters, or merge two consonants into one. These changes are fairly common. The opposites of these examples may also happen, but are less common. Words easily lose sound; they rarely gain it.

Findings after Many Decades of Observation

Comparative linguistics has come a long way since Greenberg began his radical technique. His method has been used extensively across the world, and has led to the systematical genetic categorization of most languages in the world. There is no agreement in detail, but the following groupings, with several variations, are common.

European and Asian Families

The Indo-European family covers most of Europe plus a part of south west Asia. In northern Europe we find the Uralic Family, which includes Finnish and Hungarian. In north-east Asia we find the Chukchi-Kamchatkan family. Central Asia and the rest of northern Asia host the Altaic family, which also contains Turkish. Southern Asia hosts the Sino-Tibetan, Dravidian, Daic and Austroasiatic families. Finally, the Caucasus may host two further families.

Pacific Families

The Pacific is host to three or four families. The languages of the Australian Aborigines are usually grouped as one family, as are the languages spoken on mainland Papua. There is no agreement on the treatment of Tasmanian, which is now extinct. The Austronesian family includes languages spoken on Madagascar, the Southern part of the Malaysian Peninsula, the Indonesian Islands, the Philippines, and the Māori languages.

African Families

The Afro-Asiatic (Arabic) family is found in North Africa, the Nilo-Saharan languages are spoken in the center of Africa; the Niger-Kongo family, which includes Swahili, is found in west and east Africa and the Khosian languages are spoken in the south-west of Africa.

American Families

The Americas host three major families, with many sub-groupings. The Aleut-Eskimo is found in northern Canada, from the eastern part of Alaska to Greenland. The Na-Dene group is found in north-eastern Canada and Alaska, and also includes some languages spoken in the south west of the United States. Finally, the Amerind family covers the rest of the Americas.

Picture Incomplete

In this classification we count some twenty major families. However, this classification is far from complete. Several languages seem unrelated to any other language, and are treated by some as separate families. Moreover, new discoveries are made regularly, which may show two families to be related. This, in turn, may cause two families to merge into one. Ruhlen, for example, found many similarities recently between the isolated language Ket (spoken in Siberia) and some of the Na-Dene languages, which suggests they may be related. These discoveries do not surprise all linguists. Some believe that all families ultimately go back to one single language, which came into existence when humans first developed speech. Others argue that human speech developed independently in different places, thus resulting in several language families, while Creation scientists argue that the Tower of Babel Account in Genesis 11 explains the existence of the variety of families observed today. Let us examine what evidence there is for or against each side.

EVIDENCE FOR ONE SINGLE PROTO LANGUAGE

'The ultimate question, is', says Ruhlen, 'whether all human languages are genetically related,' but the evidence for this is scarce. There are a few words which, he says, are similar in all languages. However, the words he gives in his example do not have the same meaning in every language. The meanings vary from 'one' to 'finger' and 'hand.' There are similarities between them, but this is not convincing evidence of genetic relationship between language families.

It must be pointed out, though, that we cannot go back too far in time. Core vocabulary is stable, but does change. In some languages this change has been measured for more than 2,000 years. The result shows that 19.5% of the core vocabulary changes every 1,000 years. If this is the same for all languages, it means that statistically all words in a language should be replaced within a period of about 10,000 years. That would make any research beyond that period of time impossible. This, in turn, makes it impossible to prove that all language families are ultimately related.

EVIDENCE FOR THE EVOLUTION OF SPEECH

Trask shows that humans differ from their 'closest relatives, the apes' in that their vocal tracts are much longer and differently shaped, thus making speech possible. However, the shape is also dangerous, as it could lead to choking. 'The idea is', says Trask, 'that speech and language proved to be so beneficial to the species that we became specialized for it even at the cost of losing a number of fellows to death by choking every year.' However, Trask remains unsure as to how and when this change occurred.

O'Grady and Dobrovolsky, similarly, despite describing in some detail how the brain processes speech, admit ignorance as to how and when speech developed. 'We know considerably less about the evolutionary specialization for non-vocal aspects of language … and the interpretation of meaning.' Again, there is no evidence to back their view that speech evolved.

It seems clear from their writings that they take the Evolution Theory for granted. Ruhlen admits that 'scholars supporting monogenesis or the relatability of all languages run the risk of being branded Creationists and of therefore having their work disregarded by colleagues.'

EVIDENCE POINTING TO BABEL

We have seen that the history of languages cannot be traced back for more than 10,000 years. We have also seen lack of knowledge regarding the evolution of human speech. It seems that there is little evidence to support the view that all languages evolved from one or more proto-languages. There is, however, another explanation for the existence of the language families in the world today. This explanation is found in Genesis. We will now examine the evidence supporting the Babel account found in Genesis 11. We will focus in particular on three areas where the findings of historical and comparative linguistics back this account.

LANGUAGE FAMILIES

We are unsure how many languages spread out from Babel. The Bible teaches that everyone at Babel spoke the same language; it says 'the whole world had one language and a common speech.' Clearly not enough time had passed for other fully fetched languages to develop since Noah and his family left the Ark, especially since all the people were in one place—though slightly different dialects might have developed in the period between Noah leaving the Ark and the tower of Babel.

In any case, the conclusions reached in this writing that Genesis adequately explains the findings of historical and comparative linguistics would be the same. The exact location of Babel is unknown. It is possible that one of the Ziggurats unearthed in modern Iraq is the remains of the infamous tower. As the number of people alive at the time would not have been great, about a dozen or so languages would probably have been plenty.

Keeping in mind how languages change, we would expect, as Wieland suggests, to find several distinct language families today.

'… it should be possible to group [languages] together into "families" like the Indo-European family of languages. But there should be no links between one "family" and another. That is because, in this model, each distinct language family is the offshoot of an original Babel "stem language" which did not arise by chance from a previous ancestral language.'

The Babel account suggests that several languages came into existence on that day. It is presented as a miraculous intervention by God. It is unlikely to have been acceleration of normal language changes (i.e., they did speak the same language, but dialects began to form) as people in the same area generally speak the same dialect.

We have already seen that at present about twenty language families are being distinguished, with yet some 'isolated' languages unaccounted for. There are indications that further research will reveal that some of these families may be related to each other, resulting in fewer families.

If the percentage of word loss described above is correct, we expect to see many similarities, still, between all languages in each family after only 5,000 years since Babel. However, the absence of any contact between languages in one family, over such a period, could result in greater word loss, so that comparing those languages today would indicate no relation. Also, some languages have a much quicker word loss, for cultural reasons. These languages, therefore, change much quicker.

The findings of some twenty language families, then, with several 'isolated' languages unaccounted for, is consistent with our expectations, as outlined above. Clearly distinguishable families however, continue to puzzle secular linguists, who generally believe languages evolved naturally, as these distinctions are not consistent with the expectations their hypotheses demand.

TIMING CONSISTENT

Even though he seems convinced that all languages stem from a single Proto Language, Robins talks of a 'unitary state' of the Indo-European language family, which is 'as far as one can at present go by comparative and historical inference.' He adds, 'Whatever date may be ascribed *(and 3,000 BC has been suggested)* eons of linguistic history lie behind it [emphasis mine].' However, it seems odd to believe in those 'eons of linguistic history', without any evidence, unless one takes the evolution theory for granted, as he does. The evidence indicates otherwise. We *do* observe an original language, or at least, traces of it, from which the Indo-European languages have derived. The 3,000 years BC, which Robins mentions, is significant, as such a time span is consistent with the Biblical account.

Moreover, recent findings, as we have seen, suggest that Ket is related to the Na-Dene languages. This suggests that the tribes are related, but that they separated when the American tribes moved from Asia across the Bering Strait into America. Wieland points out that 'to have such close correlation's still existing makes little sense if the migrations were as much as 11,000 years ago, as is commonly believed. From the biblical record, they would have been less than some 4,000 years ago.' Again, the evidence backs the Genesis account.

LANGUAGE DESIGN AND CHANGE

'All languages are something of a ruin,' a quote Crowley attributes to Dutch linguist van der Tuuk. 'What he meant, was that as a result of changes having taken place, some "residual" forms are often left to suggest what the original state of affairs might have been.' Crowley carries on to share how languages can change from sophisticated to simpler versions, and from simpler to more complex systems. He distinguishes between, 'isolating,' 'agglutinating' and 'inflecting' languages and shows how languages change in circular patterns.

Isolating languages, he says, are those where every word has only one meaning, i.e., no endings. They tend to become agglutinating when free form grammatical markers, i.e., prepositions, are phonologically reduced to endings or suffixes. Agglutinating languages thus look 'as if the bits of the language were simply "glued" together to make up larger words.' Subsequent morphological reduction make the original grammatical markers unrecognizable, but the endings remain functional.

The language has become an inflecting language, in which 'there are many morphemes included within a single word, but the boundaries between one morpheme and another are not clear,' Finally, morphological reduction makes the language lose its 'cases' and the language returns to being an isolating language.

In the case of Greek, this change can be seen very clearly in history. Classical Greek was a highly inflected language; it used five cases, as well as Active, Middle and Passive voice. Koine Greek was almost reduced to four cases, and the Middle voice was used rather inconsistently. Modern Greek distinguishes only three cases, but many endings have disappeared. It is a good example of van der Tuuk's Ruin, as it is slowly becoming an isolated language.

Crowley's model shows that languages can change from inflecting languages, with 'endings', to isolating languages. These may appear to be easier in structure, but are in fact equally complex, as the lack of subtle nuances, which the endings and prefixes often provide, leads to ambiguity.

Crowley does show that isolating languages can change further and pick up complicated case systems after they have lost them. However, this model cannot be used to explain the *origin* of highly sophisticated language systems like Sanskrit and Greek.

History shows that when a language changes, it tends to become more user-friendly. It likes to be flexible. When it has rid itself of cases, it is free to make them up again. However, as these changes are spontaneous, unplanned, and often unnoticed, it seems impossible that a language as sophisticated and regular as the Indo-European 'parent' was made up from a simpler form. Language change, as Crowley's model shows, would be unlikely to produce *consistent* endings for the whole of the Inflecting Language.

Steel has shown how modern Indo-European languages have reduced the number of noun inflections for different case, gender and number; and different verbs inflection for tense, voice, number and person. He also showed how English has also lost 65–85% of the Old English vocabulary, and many Classical Latin words have also been lost from its descendants, the Romance languages (Spanish, French, Italian, etc.).

Steel also pointed out that most of the changes were *not* random, but the result of *intelligence*. For example: forming compound words by joining simple words and derivations by adding prefixes and suffixes, modification of meaning, and borrowing words from other languages including calques (a borrowed compound word where each component is translated and then joined). There are also unconscious but definitely non-random changes such as systematic sound shifts, for example those described by Grimm's Law (which relates many Germanic words to Latin and Greek words).

The fact remains that the Greek/Sanskrit parent was utterly consistent, and highly sophisticated. If chance, then, did not make this Proto Language, where did it get its consistency from? The only credible explanation is found in Genesis 11. It suggests a Designer.

In Babel one of the groups was given the sophisticated, and utterly consistent, Proto Indo-European language. Sadly, as people in a fallen world began to use this language, it slowly began to lose some its consistency, as grammatical mistakes became fashionable. Today, some 5,000 years later, some of the languages which descended from it, still have some traces left of the original case system. Some have completely rid themselves of it, and have become isolating languages, while others have made up entirely new case systems, but less consistent and less sophisticated than the original.

The observation of language structure and language change, therefore, is also more consistent with the account of the *Tower of Babel* than rival theories.

The facts we observe today are consistent with the Tower of Babel account in Genesis 11, but this does not prove the correctness of the account. Since the history of languages cannot be reconstructed beyond 10,000 years, evidence for (and against) alternative views is limited.

However, if we take an objective look at the facts at our disposal, we cannot but draw the conclusion that the Bible account has far more going for it than the alternatives, for which there is little, if any, evidence. We therefore wholeheartedly believe that the findings of historical and comparative linguistics have served indeed to affirm the Tower of Babel account recorded in Genesis 11, beyond reasonable doubt. As always, Scripture cannot be 'proven to be right' by man's findings. We believe Scripture *is* right, as it originates from an *infallible* God. However, where man's findings are objectively interpreted, they usually affirm the accounts in Scripture, rather than deny them. This writing has applied the findings of historical and comparative linguistics to the Genesis account, and found that the facts to our disposal are affirmative indeed.

Believing this account, however, requires believing in God, and the denial of the evolution theory, which suggests that all animals, humans, and even human language, arose by chance. For many, this might prove too big a price to pay, despite the evidence.

FISH SURVIVE THE FLOOD

How did saltwater fish survive dilution of the sea water with fresh water?

How did freshwater types survive in salt water?

How did plants survive?

If the whole Earth were covered by water in the Flood, then there would have been a mixing of fresh and salt waters. Many of today's fish species are specialized and do not survive in water of radically different saltiness to their usual habitat.

So how did they survive the Flood? Please observe that the Bible tells us that only land-dwelling, air-breathing animals and birds were on the Ark (Gen. 7:14–15, 21–23).

We do not know how salty the sea was before the Flood. The Flood was initiated by the breaking up of "the fountains of the great deep" (Gen. 7:11). Whatever "the fountains of the great deep" were, the Flood must have been associated with massive earth movements, because of the weight of the water alone, which would have resulted in great volcanic activity.

Volcanoes emit huge amounts of steam, and underwater lava creates hot water/steam, which dissolves minerals, adding salt to the water.

Furthermore, erosion accompanying the movement of water off the continents after the Flood would have added salt to the oceans. In other words, we would expect the pre-Flood ocean waters to be less salty than they were after the Flood. The problem for fish coping with saltiness is this: fish in fresh water tend to absorb water, because the saltiness of their body fluids draws in the water (by osmosis). Fish in saltwater tend to lose water from their bodies because the surrounding water is saltier than their body fluids.

SALTWATER/FRESHWATER ADAPTATION IN FISH TODAY

Many of today's marine organisms, especially estuarine and tidepool species, are able to survive large changes in salinity. For example, starfish will tolerate as low as 16–18% of the normal concentration of sea salt indefinitely. Barnacles can withstand exposure to less than one-tenth the usual salt concentration of sea water.

There are migratory species of fish that travel between salt and fresh water. For example, salmon, striped bass and Atlantic sturgeon spawn in fresh water and mature in salt water. Eels reproduce in salt water and grow to maturity in freshwater streams and lakes. So, many of today's species of fish are able to adjust to both fresh water and salt water, Figure 20.30.

Fig.20.30: *Eels, like many sea creatures, can move between salt and fresh water*

There is also evidence of post-Flood specialization within a kind of fish. For example, the Atlantic sturgeon is a migratory salt/freshwater species but the Siberian sturgeon (a different species of the same kind) lives only in fresh water.

Many families of fish contain both fresh- and saltwater species. These include the families of toadfish, garpike, bowfin, sturgeon, herring/anchovy, salmon/trout/pike, catfish, clingfish, stickleback, scorpionfish, and flatfish. Indeed, most of the families alive today have both fresh- and saltwater representatives. This suggests that the ability to tolerate large changes in salinity was present in most fish at the time of the Flood. Specialization, through natural selection, may have resulted in the loss of this ability in many species since then.

Hybrids of wild trout (freshwater) and farmed salmon (migratory species) have been discovered in Scotland, suggesting that the differences between freshwater and marine types may be quite minor.

Indeed, the differences in physiology seem to be largely differences in degree rather than kind.

The kidneys of freshwater species excrete excess water (the urine has low salt concentration) and those of marine species excrete excess salt (the urine has high salt concentration). Saltwater sharks have high concentrations of urea in the blood to retain water in the saltwater environment whereas freshwater sharks have low concentrations of urea to avoid accumulating water.

When sawfish move from salt water to fresh water, they increase their urine output twenty-fold, and their blood urea concentration decreases to less than one-third.

Major public aquariums use the ability of fish to adapt to water of different salinity from their normal habitat to exhibit freshwater and saltwater species together. The fish can adapt if the salinity is changed slowly enough.

So, many fish species today have the capacity to adapt to both fresh and salt water within their own lifetimes.

Aquatic air-breathing mammals such as whales and dolphins would have been better placed than many fish to survive the Flood, not being dependent on clean water to obtain their oxygen.

Many marine creatures would have been killed in the Flood because of the turbidity of the water, changes in temperature, etc. The fossil record testifies to the massive destruction of marine life, with marine creatures accounting for 95% of the fossil record. Some, such as trilobites and ichthyosaurs, probably became extinct at that time. This is consistent with the Bible account of the Flood beginning with the breaking up of the "fountains of the great deep" (i.e., beginning in the sea; 'the great deep' means the oceans).

Fig.20.31: *Freshwater trout can hybridize with (saltwater) salmon.*

There is also a possibility that stable fresh and saltwater layers developed and persisted in some parts of the ocean, Figure 20.31. Fresh water can sit on top of salt water for extended periods of time. Turbulence may have been sufficiently low at high latitudes for such layering to persist and allow the survival of both freshwater and saltwater species in those areas.

SURVIVAL OF PLANTS

Many terrestrial seeds can survive long periods of soaking in various concentrations of salt water. Indeed, salt water impedes the germination of some species so that the seed lasts better in salt water than fresh water. Other plants could have survived in floating vegetation masses, or on pumice from the volcanic activity. Pieces of many plants are capable of asexual sprouting.

Many plants could have survived as planned food stores on the Ark, or accidental inclusions in such food stores. Many seeds have devices for attaching themselves to animals, and some could have survived the Flood by this means. Others could have survived in the stomachs of the bloated, floating carcasses of dead herbivores.

The olive leaf brought back to Noah by the dove (Gen. 8:11) shows that plants were regenerating well before Noah and company left the Ark, Figure 20.32.

Fig.20.32: The olive leaf brought back to Noah by the dove (Gen. 8:11)

There are many simple, plausible explanations for how fresh- and saltwater fish and plants could have survived the Flood. There is no reason to doubt the reality of the Flood as described in the Bible.

A WITNESS AT THE "ENDS OF THE EARTH"

The knowledge of the one true God—preserved for millennia in Polynesian culture.

Since all people groups come from Noah and his family, it should be no surprise that most have memories of the one true Creator God. The Māori people of New Zealand are no exception. Their traditions speak of such a God they call "Io Matua Kore" (Io the Parentless). Christian missionaries have found that this ancient tradition has been most significant in their Christian outreach to the Māori and their Pacific Island relatives, much like Paul's teaching of the "unknown God" to the Athenians on Mars Hill (Acts 17).

Fig.20.33: New Zealand Māori's Retain Knowledge of the Creator, Io, who has the Same Attributes as the LORD God of the Bible. This Knowledge Pre-dates any Missionary Influence. Curtsy - Image iStockphoto/HultonArchive

It was from the missionary's outreach experience that came upon the significance of the Io knowledge to the Māori people and the Hebrew Scriptures. This knowledge had travelled with the ancestors of the Māori in their odyssey from the Middle East across the planet all the way to Hawaii and from there to New Zealand, Figure 20.33—to the ends of the earth.

BIBLICAL HISTORY CONTAINED IN MĀORI FOLKLORE

The missionaries were first alerted that the Māori culture preserved the knowledge of God by Daniel Kikawa's book about Io belief in Hawaii. Recalling the Māori migration from Hawaii, the missionaries researched New Zealand Io knowledge and found it to be faithful to that of Hawaii, despite 1,000 years of oral transmission. When they compared the names of Io with the names and attributes of Almighty God in the Bible, he found they agreed precisely—there were no contradictions.

The missionaries stated, ".. that was enough to know that here was a remarkable treasure residing in Māori culture."

Sometimes at a Māori "marae" (meeting house) the speaker will begin, "*Hawaiki, Hawaiki Roa, Hawaiki Nui, Hawaiki Pamamao.*" Missionaries explained, "The names in this recitation retrace their ancestors' journey from the west, beginning in the Middle East, through the Indo-Malay regions and then out into the wide expanses of the Pacific. To this day, enclaves of Polynesian peoples remain in places like the Naga Hills in Assam, the Karen Hills of Burma, the remote parts of Tibet and the islands of Indonesia and Melanesia.

Another remarkable confirmation of Genesis biblical history is found in the folklore of Polynesian groups dotted around the Pacific. In his book, *Io Origins*, Missionaries detail the common denominators in these stories, including:

A man known as Lua-Nu'u, (the second Noah) who left his country, settling far to the south.

- By command of his God, Lua-Nu'u introduced circumcision for all his descendants.
- He had two sons, one by a slave girl and the other by a "chieftainess".
- He went up a mountain to sacrifice to his God.

Not only did missionaries realize that Lua-Nu'u is the Abraham of Scripture, but they explained that further details of the folklore delineate Jacob. Even Joseph is enshrined in these stories, which predate any missionary contact.

A LONG JOURNEY OF DISCOVERY

One of the missionaries' parents and grandparents were dairy farmers in New Zealand—pioneers who broke in the land. His great, great grandfather came from Scotland and other ancestors from Ireland. "Actually, the missionary is a fourth generation Kiwi", he said. "I grew up with Māori in my home district and knew at about age nine that I would marry a Māori, and did so." However, four generations in New Zealand is not such a big deal. His wife's ancestors are from three of the many Māori tribes: the Ngati Kahu, the Ngapuhi, and the Ngati Whatua, and they can boast of a much longer sojourn in New Zealand, Figure 20.34.

Fig.20.34: *Maori chiefs and priests of some tribes have stories of events recorded in the Bible from over 4,000 years ago, before they immigrated from the Middle East some time after Noah's Flood. Curtsy - (Image stockxpert)*

As well as dairy farmers, the missionary's parents were active Christians, and for nearly 50 years ran the district Sunday school, collecting children in the family car. The missionary vividly remembers when he was about age 11 a visitor came and preached and he was deeply convicted of sin. "I refused to yield, being a first-class, self-righteous sinner! It took two miserable weeks before I became ashamed that I was defying God and surrendered my life to Christ."

Such dramatic encounters with God, including God's removal of his irksome stutter, shaped his character and gave him direction for his life-long work. With his wife, he shared a powerful call to preach the Gospel to the Pacific nations. They spent eight years working in the South Sea islands with Samoans, Tongans, Fijians and Indians, pastoring cross-cultural churches and working at church planting and evangelism. Over many years, they have been senior pastors of a large church and Bible school in Whangarei, a city in the far north of New Zealand.

The couple deep interest in the indigenous "God factor" they had observed in South Pacific cultures came slowly. They had noticed how the modern university culture intentionally sidelined and even aggressively opposed any recognition of God in the teaching of science. He could see that the evolutionary underpinnings for science that these academics insisted upon were nothing to do with operational science but were virtually a religious belief code.

"True science should examine all the evidence without a closed mind to the 'God factor'," The missionaries said. "The apostle Paul speaks about those who suppress the evidence about God even when the evidence is before them (Romans 1:18–19). In fact, Paul demonstrated open mindedness by acknowledging the God factor in Greek culture (Acts 17:22–32)."

The missionaries were also influenced by Don Richardson's landmark book *Eternity in Their Hearts*. "That book vividly outlined the 'God factor' in many diverse cultures," Graham said, "preserving knowledge of the one true God and predating any missionary influence.

THE RICH LEGACY OF MĀORI FOLKLORE

"Māori Christians long for accurate spiritual information about their past," The missionary couple said. "Evolutionary-based anthropology has had its own agenda and has actually suppressed the truth about the indigenous knowledge of the Almighty God—knowledge that is enshrined in their ancient beliefs about Io."

Io Origins describes how this knowledge travelled with the first Māori immigrants from their previous home in Hawaii to New Zealand. Belief in the creator God, "Io", was once pervasive in Hawaii and the society was peaceful and benevolent. Their tradition speaks of them travelling from the west and being known as the Menehune people.

This Hawaii culture was then overrun by enemies from Tahiti who installed their own bloodthirsty religion that elevated a cruel system of tapu (sacredness upheld by harsh penalties). Io priests were silenced by death or compelled to cease imparting Io knowledge.

Others fled from Hawaii carrying Io knowledge with them, some arriving in Aotearoa. Later immigrants did not carry Io knowledge, which is why some tribes in New Zealand have Io knowledge and others do not. At all times Io knowledge was tapu (sacred) and taught only to chiefs and tohunga (priests). That is why the early missionaries in both Hawaii and New Zealand were initially unaware of Io.

The missionaries outline some of the rich legacy of the traditional Māori names of the Almighty—names that are still in use after millennia of oral transmission. All the names are biblically sound. The missionaries have brought together traditions from both sides of the vast Pacific, New Zealand and Hawaii, and show that the people who migrated from the Middle East carried accurate information about Almighty God, right around the globe.

PATHWAY OF HOPE AND CHANGE

For the missionaries, the information about Io provides an effective pathway to cultural evangelism to much of Polynesia. The knowledge of Io links directly to their own history yet it is biblically accurate. They want to see it widely disseminated.

The missionaries have a deep love for the Māori people of New Zealand. They see the Māori as being under siege today from traditionalists who falsely suggest Christianity has robbed them of their culture, from modernists who deny the rich heritage of belief in Io, and from activists who are intent on using Māori culture as a weapon against Christianity.

Deeply concerned about the welfare of the people of New Zealand and the islands of the Pacific, The missionaries grieve over the way seminaries are churning out graduates without confidence in the inspired truth of Scripture. Their desire is for seminaries to reclaim their confidence in the historicity of the events of Genesis, events such as the global Flood and the Tower of Babel, which link so powerfully with Māori culture. And they expressed hope for uplift and change.

"We need reformers. We need uncompromising proclaimers of the good news of the Gospel. There are hopeful signs with some very able Māori pastors with sizable congregations. But our greatest need is for spiritual revival flowing from a confident proclamation of the Word of God, beginning in Genesis and turning the people back to the one true, Creator God—known traditionally in the Pacific region as Io."

APPENDIX

a. If r = % rate of growth per year, and the number of years of growth = n, then after n years, the population produced by the eight survivors of the Flood = $8(1+r/100)^n$. For a more comprehensive formula that takes into account longevity, number of children born and generation time, see Morris, H.M., World population and Bible chronology, *Creation Research Society Quarterly* **3**(3):7–10, 1966.

b. It is possible that the births mentioned are not the firstborn; they could just be the sons leading to Abraham. This would shorten the generation times and make the population growth even greater.

c. This answers a common skeptical objection regarding the population at the time of Babel about 100 years after the Flood. This dating assumes that Peleg was named because of this event (Genesis 10:25)—see In the Days of Peleg. However, his naming could have been prophetic, like Methuselah, who died in the year of the Flood and whose name means 'When he dies, it shall be sent'. If this is true, then Babel could have been some time after Peleg's birth, but during his lifetime.

d. The 'rule of 72' states that dividing 72 by the annual growth (in %) gives the years to double the population. This is an approximation that makes the calculations easy. A figure of 69.3 is more accurate (100 x ln2 = 69.3).

e. Europa, one of Jupiter's moons, is suspected to have liquid water under an icy crust, but this is not known for certain.

f. It could possibly make such patterns for an agglutinating language, but not for an inflecting language, as the phonological reduction would not be consistent.

REFERENCES

1. Scheffler, H. and H. Elsasser, *Physics of the Galaxy and Interstellar Matter*, Springer-Verlag, 1987, Berlin, pp. 352–353, 401–413.

2. D. Zaritsky *et al.*, *Nature*, 22 July 1993. *Sky & Telescope*, December 1993, p. 10.
3. Steidl, P.F., Planets, comets, and asteroids, *Design and Origins in Astronomy*, pp. 73–106, G. Mulfinger, ed., Creation Research Society Books, 1983, 5093 Williamsport Dr., Norcross, GA 30092.
4. Whipple, F.L., Background of modern comet theory, *Nature* 263:15–19, 2 September 1976.
5. Gordeyev, V.V. *et al.*, The average chemical composition of suspensions in the world's rivers and the supply of sediments to the ocean by streams, *Doklady Akademii Nauk SSSR* 238:150, 1980.
6. Hay, W.W., *et al.*, Mass/age distribution and composition of sediments on the ocean floor and the global rate of subduction, *Journal of Geophysical Research* 93(B12):14,933–14,940, 10 December 1988.
7. Maybeck, M., Concentrations des eaux fluviales en elements majeurs et apports en solution aux oceans, *Revue de Geologie Dynamique et de Geographie Physique* 21:215, 1979.
8. Sayles, F.L. and P.C. Mangelsdorf, Cation-exchange characteristics of Amazon River suspended sediment and its reaction with seawater, *Geochimica et Cosmochimica Acta* 41:767–779, 1979.
9. Austin, S.A. and D.R. Humphreys, The sea's missing salt: a dilemma for evolutionists, *Proceedings of the 2nd International Conference on Creationism*, Vol. II, Creation Science Fellowship, 1991, icr.org/article/sea-missing-salt.
10. Austin, S.A., Evolution: The ocean says NO! *ICR Impact* No. **8**, October 1973, Institute for Creation Research.
11. Merrill, R.T. and M. W. McElhinney, *The Earth's Magnetic Field*, Academic Press, London, 1983, pp. 101–106.
12. Humphreys, D.R., Reversals of the earth's magnetic field during the Genesis Flood, *Proceedings of the 1st International Conference on Creationism*, August 1986, Pittsburgh, Creation Science Fellowship, 1987, 362 Ashland Ave., Pittsburgh, PA 15228, Vol. II, pp. 113–126, creationicc.org/abstract.php?pk=22.
13. Coe, R.S., M. Prévot, and P. Camps, New evidence for extraordinarily rapid change of the geomagnetic field during a reversal, *Nature* 374:687–692, 20 April 1995.
14. Humphreys, D.R., Physical mechanism for reversals of the earth's magnetic field during the Flood, *Proceedings of the 2nd International Conference on Creationism*, Vol. II, Creation Science Fellowship, 1991, icr.org/article/reversals-earths-magnetic-field-flood.
15. Austin, S.A. and J.D. Morris, Tight folds and clastic dikes as evidence for rapid deposition and deformation of two very thick stratigraphic sequences, *Proceedings of the 1st International Conference on Creationism* Vol. II, Creation Science Fellowship, 1986, pp.3–15, creationicc.org/abstract.php?pk=16.
16. Gentry, R.V., Radioactive halos, *Annual Review of Nuclear Science* 23:347–362, 1973.
17. Gentry, R.V. *et al.*, Radio-halos in coalified wood: new evidence relating to time of uranium introduction and coalification, *Science* 194:315–318, 15 October 1976.
18. Gentry, R. V., Radio-halos in a Radio-chronological and cosmological perspective, *Science* 184:62–66, 5 April 1974.
19. Gentry, R. V., *Creation's Tiny Mystery*, Earth Science Associates, 1986, P.O. Box 12067, Knoxville, TN 37912-0067, pp. 23–37, 51–59, 61–62.
20. Vardiman, L., *The Age of the Earth's Atmosphere: A Study of the Helium Flux through the Atmosphere*, Institute for Creation Research, 1990, P.O.Box 2667, El Cajon, CA 92021.
21. Gentry, R. V. *et al.*, Differential helium retention in zircons: implications for nuclear waste management, *Geophysical Research Letters* 9:1129–1130, October 1982. See also ref. 20, pp. 169–170.
22. Deevey, E.S., The human population, *Scientific American* 203:194–204, September 1960.
23. Marshak, A., Exploring the mind of Ice Age man, *National Geographic* 147:64–89, January 1975.

24. Dritt, J. O., Man's earliest beginnings: discrepancies in the evolutionary timetable, *Proceedings of the 2nd International Conference on Creationism*, Vol. I., Creation Science Fellowship, 1990, pp. 73–78, creationicc.org. Return to text.
25. For the nth square, the number of rice grains = $2^{n-1} = 2^{63}$ for the last square, or about 10^{19} grains!
26. *Encyclopædia Britannica CD 2000*, Trends in world population. Return to text.
27. Proven/developed by the creationist scientist Louis Pasteur (see Louis Pasteur (1822–1895), *Creation* 14(1):16–19).
28. Pioneered by another great creationist scientist, Joseph Lister (see Joseph Lister: father of modern surgery, *Creation* 14(2):48–51).
29. Armstrong, H.L., More on growth of a population, *Creation Research Society Quarterly* 22(1):47,1985, citing Lower, A.R.M., *Canadians in the Making*, Longmans, Green and Co., Toronto, p. 113, 1958. There was little immigration in this period.
30. Even if the population were a million, the low reproductive rate would not be sufficient to eliminate harmful mutations. The mutational load alone would have ensured extinction. For details, see ReMine, W., *The Biotic Message*, St Paul Science, St Paul, Minnesota, 1993.
31. Such as dinosaur bones in Montana, claimed to be over 65 million years old, but so 'fresh' that blood cells and hemoglobin are still present. See Sensational dinosaur blood report! *Creation* 19(4):42–3, 1997.
32. Osgood, A.J.M., A better model of the Stone Age, *Journal of Creation* 2(1):88–102, 1986 and Part 2, *Journal of Creation* 3(1):73–95, 1988.
33. *The Australian Encyclopædia*, 5th Edition, 1988, The Australian Geographic Society, Sydney, 1:230, 1988. There has been a tendency to revise this estimate upwards, possibly driven by the obvious inconsistency of the 300,000 figures with the belief in the antiquity of the Aboriginal population.
34. How long have Aborigines been in Australia? *Creation* 15(3):48–50, 1993.
35. Steel, A.K., The development of languages is nothing like biological evolution, *Journal of Creation* 14(2):31–40, 2000.
36. Cited in: Crowley, T., *An Introduction to Historical Linguistics*, Oxford University Press, Oxford, p. 24, 1992.
37. Certain languages, like Basque, seem to have little in common with other languages. They are either not classified, or treated as a separate language family.
38. Ruhlen, M., *A guide to the World's Languages*, Edward Arnold, London, 1987.
39. Ruhlen, M., *Proceedings of the National Academy of Sciences*, 95:13994–13996, 1998, *as cited* in: Wieland, C., Siberian Links for Amerindians, *Creation* 21(3):9, 1999.
40. Trask, R.L., *Language, the Basics*, Routledge, London, p. 18, 1999.
41. O'Grady, M. and Dobrovolsky, M., *Contemporary Linguistics*, St. Martin's Press, New York, p. 10, 1989.
42. Wieland, C., Towering change, *Creation* 22(1):22–26, citing p. 26, 1999.
43. Robins, R.H., *General Linguistics: An Introductory Survey*, Longmans, London, p. 229.
44. Wieland, C., Siberian links for Amerindians, *Creation* 21(3):9, 1999.
45. Halley, E., A short account of the cause of the saltness [*sic*] of the ocean, and of the several lakes that emit no rivers; with a proposal, by help thereof, to discover the age of the world, *Philos. Trans. R. Soc. Lond., B, Biol. Sci.*, 29:296–300, 1715; cited in Ref. 4.
46. Joly,J., An estimate of the geological age of the earth, *Scientific Transactions of the Royal Dublin Society*, New Series 7(3), 1899; reprinted in *Annual Report of the Smithsonian Institution*, June 30, 1899, pp. 247–288; cited in Ref. 4.

47. Austin S.A. and Humphreys, D.R., The sea's missing salt: a dilemma for evolutionists, *Proceedings of the Second International Conference on Creationism*, Vol. II, pp. 17–33, 1990. This paper should be consulted for more detail than is possible in this article.
48. Moore, W.S., Large groundwater inputs to coastal waters revealed by ^{226}Ra enrichments, *Nature* 380(6575):612–614, April 1996 | doi:10.1038/380612a0; perspective by T.M. Church, An underground route for the water cycle, same issue, pp.579–580 | doi:10.1038/380579a0.
49. Church, YT.M., p. 580, comments: "The conclusion that large quantities of SGWD are entering the coastal ocean has the potential to radically alter our understanding of oceanic chemical mass balance."
50. Kikawa D., *Perpetuated in Righteousness from Hawaii*, Kikawa, 1994.
51. Richardson D., *Eternity in Their Hearts*, Regal Books, California, 1984.

21

SATURN, URANUS, NEPTUNE AND JUPITER ARE YOUNG PLANETS

THE YOUNG AGE OF SATURN

When Saturn, Figure 21.1, was the only known ringed planet, the rings were believed to be as old as the solar system—4.6 billion years in the conventional chronology. The existence of the rings to the present day was taken as evidence of this chronology. In the 1970s and 1980s, other planets were found to have short-lived, rapidly dissipating rings with life times of the order of millennia. Subsequently, the view of the age of Saturn's rings began to change. They are now viewed conventionally as no more than hundreds of millions of years old, and a former prop of the conventional chronology has now vanished. Furthermore, an examination of ring observations and data unconstrained by conventional chronology indicates that the actual life time of Saturn's rings may be of the order of tens of thousands of years, and possibly less. This age fits in perfectly with the biblical Creation/Fall/Flood model, and opens up possibilities for effectively explaining their origin within a biblical framework.

Fig.21.1: This enhanced-color image was created by combining three images taken through ultraviolet, violet and green filters on July 12, 1981. - Curtsy NASA Several changes were apparent in Saturn's atmosphere since Voyager 1's November 1980

A puzzle for evolutionary chronology began with the Voyager 1, Figure 21.2, flyby past Saturn's rings in 1980. Before then, Earth-bound telescopes provided little ring detail, and planetary rings were assumed to have endured virtually changeless since the emergence of the solar system from the solar nebula—a vast cloud of gas and dust—some 4.6 billion years ago.

'Everyone had expected that collisions between particles in Saturn's rings would make the rings perfectly uniform.'

Fig.21.2: Voyager 1 Forty-Five Years Later - - Curtsy NASA

For example, Jeffreys (1919,1947 – p 267), had claimed that 'the frequency of collision [of ring particles] is very great, and ... on account of the loss of relative motion at every collision, the rings must long ago have reached a state in which all the particles are moving in very accurate circles, all in the same plane.'

This view arose from belief in the rings' great age, but Voyager 1 showed that the rings are highly structured and probably young, as there is more structure than can be expected to persist over 4.6 billion years. Efforts to locate sufficient binding forces have failed, and a 'growing number [of astronomers] believe that the rings of Saturn are constantly ... changing due to fragmentation of moonlets, Figure 12.3, and input of new ring particles.'

Fig.21.3: *Mystery of Saturn's Moonlets*

However, there remains a reluctance to associate ring change with ring dissipation, since this could imply a young solar system. This reluctance did not exist before the ascendancy of evolutionary chronology, as in James Maxwell's day, Saturn's rings were acknowledged to be rapidly changing and possibly dissipating.

Fig.21.4: *Saturn's Amazing Rings - Curtsy NASA*

Today, space probes have rediscovered rapid ring change and dissipation, as is evident from statements by many astronomers. For Jupiter, ring 'particles should last only a very short time—perhaps only a few thousand years ...'

Of Saturn's rings, Figure 12.4, and planetary rings generally, 'it now appears that the length of time for planetary rings to dissipate is relatively short.'

URANUS' RINGS

'The thin outer atmosphere of Uranus extends into the rings, so it should slow down very tiny dust particles and cause them to sink into the inner atmosphere in a few thousand years or less ... Collisions between ring particles ... slowly [make] the ring wider.' Figure 21.5.

Saturn's rings have little matter, 'only about a millionth of the mass of our moon,' similar to that of smaller asteroids such as *243 Ida* or *253 Mathilde*. Their small mass suggests that the rings could 'empty out' fairly quickly. Indeed, Jupiter's rings are thought to be, in part, the product of the dissolution of *two moons, Adrastea and Metis*, both with masses comparable to the mass of Saturn's rings.

Fig.21.5: *Uranus' Tilt essentially has the planet orbiting the Sun on its side, the axis of its spin is nearly pointing at the Sun. Curtsy - NASA and Erich Karkoschka, U. of Arizona*

Saturn's Rings have been Widening Rapidly

In the 1960s, Alexander (1962, 1980 – p 320) documented 350 years of widening in Saturn's A and B rings. One of his sources was Otto Struve, who in the 1850s assessed observations from the previous two centuries, which indicated ring spreading into Saturn at a rate of about 100 km per year. Unfortunately, however, the reigning hypothetical assumptions from the popular nebular hypothesis (which claims a naturalistic origin, an old age and little change in the solar system presently) caused many to question Struve's analysis. So strong had belief in the nebular hypothesis become that Taylor inconsistently claimed ring spreading was compatible with it.

However, Maxwell (1895, 1965 - pp. 353, 373–374) had shown that Saturn's rings are particulate rather than rigid disks or liquid, and considered Struve's (p 17–20, 1883) analysis to be consistent with his theory, the predictions of which have been confirmed by observation.

Nevertheless, Struve failed to measure continued ring spreading, and in 1895 Lewis (1895 p 385) concluded that ring observations were not in agreement ('accordant') because of 'the great difficulty in making these 'measures.' But he then dogmatically stated that Saturn's rings were 'certainly' not undergoing long-term change, even though his data showed C-ring spreading. Lewis thus laid the groundwork for Jeffreys' concept of very old rings.

SATURN'S C RING FORMED RECENTLY

Saturn's most prominent rings are the A, B, and C rings. However, the C ring was not visible until the 1800s:

'William Herschel, the foremost astronomical observer of his time (1738–1822), makes no mention of the [C ring] in any of his writings, and it is inferred that it was not then a conspicuous object. If this inference be correct, we must conclude that this ring is rapidly growing, and that the rings of Saturn are probably comparatively recent introductions to the solar system.' Figure 21.6.

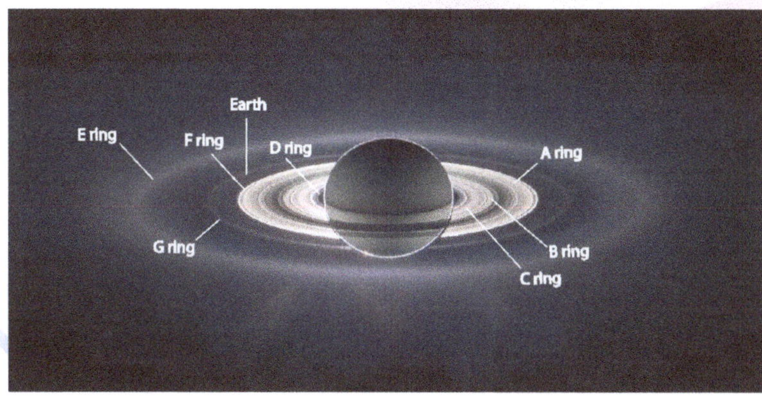

Fig.21.6: *Saturn's Rings – Curtsy Hubble Heritage*

Today the C ring can be seen 'with telescopes of moderate size.' Since Herschel's telescopes were among the best of his day, with Saturn a 'favorite object of study,' one is led to conclude that he missed the C ring because it was absent. The first recorded observation of the C ring was in 1848. Thus, one of the three prominent rings of Saturn has evidently developed since the early 1800s. The inner edge of the C ring is approaching the planet, and Napier and Clube calculated the rate of approach as 100 km per year. Figure 21.7.

Fig.21.7: *NASA's Voyager 2 turns 45; The Voyager 2 probe was launched on 20 August 1977. Curtsy NASA spacecraft | Space*

The history of C ring observations implies rapid ring spreading and dissipation. The inner edge of the B ring is now 91,975 km from the center of Saturn and the inner edge of the C ring is at 74,658 km.44 Thus the width of the C ring is 17,317 km, or about 15,000 km, a width which developed since about 1850. This implies an infill of ring particles in agreement with the computation of Napier and Clube (1979 p 457).

Like Jupiter's and Uranus's rings, Saturn's rings appear to be decaying in a millennial time-frame. Ring dissipation does not require millions of years. When planetary rings were thought to be old, they were taken as evidence for an old solar system. Intimation of their youth therefore obliterates a prop of the conventional chronology.

NEW SATURNIAN RINGS IS FORMED

In 1954, Baum reported 'dusky nebulous matter in the form of an additional ring' beyond ring A, with 'a diffuse fringe [extending] the ring system beyond its normal limits.' Baum (1954 p 194) may have been seeing one or more of the now-recognized tenuous outer rings (the F, G, and E rings). On the other hand, he may have been seeing dissipation of A ring material outward, and if ring particles 'reach the outer edge of the rings, they leave the ring system.'

16 In 1967 Feibelman likewise reported 'an extension or at least a gradual tapering of the outer edge of the A ring.' Thus, it appears that the A ring is losing particles to the outer F, G, and E rings, and eventually to space beyond. How trustworthy are such ground-based observations? Dismissing them as subjective phenomena would be premature. In fact, existence of the F ring had been theorized before the Voyager flybys, though in characteristic fashion Jeffreys discounted this prediction.

Furthermore, inside the C ring, 'the possibility of a faint ring ... was raised some time ago [from ground-based observations], and this D ring was actually found.' Ground-based discovery of the D ring before its Voyager detection implies validity for ground-based ring-spreading observations. Like the outer F, G, and E rings, the D ring seems to be composed of small particles. These particles are spiraling into Saturn:

Ring particles of Jupiter and Uranus also show this behavior. To sum up, particles in outer rings dissipate into space; those in innermost rings fall toward the planet.

FAILING EFFORTS TO SAVE LONG AGES

The Uranian and Jovian ring systems were discovered shortly before the Voyager views of Saturn's rings and, according to NASA, appeared too young to exist in an old solar system:

'The theory that explained how Saturn's rings could persist through 4.6 billion years of solar system evolution also explained why Saturn was the only planet that could have a ring. Then those theories had to be revised to account for the rings of Uranus. The revisions implied that Jupiter would not have a ring. Now Jupiter has been found to have a ring and we have to invent a theory to explain it.' Figure 21.8.

Fig.21.8: *Jovian Planets and Their Rings – Curtsy NASA*

The older 'unworkable' theory was *the orbital resonance hypothesis*. When Saturn was the only known ringed planet, orbital resonances, due to moons of Saturn gravitationally acting on ring particles, could account for the limited ring structure visible from Earth. The resonance hypothesis 'had been worked out with fewer than a half-dozen rings [of Saturn] known. The ring structure the Voyagers discovered is too complex to ... explain thousands of rings.' 'A thousand rings seemed a monumental problem for theorists. They had run out of resonances long ago.' NASA's conclusion: 'No theory has yet been developed that explains how all three of these planets could have rings for so long', i.e., 4.6 billion years.

The '*shepherd moon hypotheses* were subsequently proposed to give planetary rings a long lifetime. As originally conceived, shepherd moons were supposed to corral ring particles, keeping entire ring systems together over eons, Figure 21.9. The shepherd moon theory was, therefore, once used to account for all ring structures of Saturn, Jupiter, and Uranus.

Fig.21.9: *Found this rare Gas Giant with two moons inside its rings. It's called shepherd moons! : Curtsy - r/EliteDangerous*

After the Voyager 2 flyby of Uranus' rings in 1986, NASA scientist Bradford Smith stated, 'We are assuming [the existence of shepherds], because we don't know any other way to do it [i.e., preserve the rings].'60 Since then, conventional opinion on the antiquity of planetary rings has changed due to difficulties in the shepherd moon theory. Rings are no longer viewed as debris from the solar nebula with an age of billions of years. Instead, the rings have formed by the fracturing of one or more moons, and therefore, must have formed 'recently'. 'Recently,' however, is a relative term, and may signify millions of years.

Nevertheless, shepherd moons continue to be presented as the reason planetary rings exist. Though ring decay occurs, it is still not acceptable to allow this fact to imply a young solar system, and shepherds are invoked to extend a ring's chronology. Therefore, rings must be simultaneously decaying, yet confined by shepherds:

'[Planetary rings] tend to spread ... Sometimes planetary rings are kept in place by the gravitational force of shepherd moons. Saturn has a very intricate ring system with lots of moons helping to keep its rings together.'

- This is false— *'lots' of shepherds have not been found.*
- Another false claim is that the '"shepherding" effect has been found to confine a number of rings in the solar system'.

Out of hundreds of thousands of ringlets in planetary ring systems, only a few have been found with nearby moonlets interpreted as shepherds. Most notable are the F ring of Saturn, Jupiter's ring system, and Uranus's thick ring. As mentioned above, the last two are now viewed primarily as rapidly decaying, despite putative shepherding effects.

THE SHEPHERD MOONS

'Shepherd moons' such as *Prometheus and Pandora* (moons of Saturn near the F ring) have been photographed, but mere existence does not confirm they are acting as shepherds. Further, moons once described as 'shepherds' seem to be disintegrating into the ring structure, as is acknowledged for Jupiter and Uranus.

During the 1995 Saturn ring plane crossing, the Hubble Space Telescope looked for new satellites. Two were announced as new in a press release and were designated *1995S1 and 1995S2*. They turned out to be the already known moons *Atlas and Prometheus*. Even more interesting, five other bodies, *1993S3 to S7*, were observed, but were later 'hypothesized to be shattered moonlets' in the F ring. The obvious conclusion is that bodies perceived as 'shepherd' moons of Saturn are undergoing disintegration within the ring structure.

Fig.21.10: *Saturn's Rings are Increasingly Recognized as being relatively short-lived rather than essentially changeless over millions of years.*

Discussing these fragmented satellites, Philip Nicholson of Cornell University said:

'[O]ne scenario for the origin of Saturn's ring system is that it is made up of countless fragments from several pulverized moons. ... the new objects orbit Saturn near the narrow F ring, which is a dynamic transition zone between the main rings and the larger satellites. [Fragmented moons would eventually] spread around the moon's orbit to form a new ring.'

Showalter surmised that Saturn's narrow G ring, thought to be composed of very fine dust, may in fact be the 'decaying corpse' of a moon destroyed by meteoroid impact. Since the F ring is a 'dynamic transition zone' where satellite fragmentation is likely to occur, what is the possibility that the so-called '*shepherds,' Prometheus and Pandora*, could be undergoing the same type of dissolution?

A stunning observation answered this question. The reason the previously mentioned satellite 1995S2 was not initially recognized as *Prometheus* is that its location did not match the position expected. *Prometheus* had 'slipped in its orbit by 20 degrees from the predicted position ... a consequence of a "collision" of *Prometheus* with the F ring, which is believed to have occurred in early 1993.' Thus, *Prometheus* is not so much 'shepherding' the F ring as mutually interacting with it, sometimes colliding with it and likely disintegrating as a result.

It is doubtful that the so-called shepherds of the F ring ever fulfilled that function. In 1980, Voyager 1 detected a twisting or 'braiding' in the F ring attributed to *Prometheus* and Pandora, but Voyager 2 in 1981 detected 'no signs of braiding in the F ring.' Thus, the 'shepherds' *Prometheus* and *Pandora* are not shepherds after all. Instead, *Prometheus and Pandora* are fragments of larger bodies en-route to further disintegration, the same process thought to have produced the moonlets 199S3 to S7. *Prometheus and Pandora* are not spherical and have an irregular shape. They seem either to be captured asteroids or fragments of a larger moon. The F ring itself is expected to widen over time, eventually dissipating altogether.

The Voyager missions demolished the belief that planetary rings must be old. The Cassini probe began orbiting Saturn in 2004. Preliminary Cassini data confirm that at least some of the ring structure is the 'crumbled remains of an ancient Saturnian moon' destroyed possibly by meteorite impact.

During the ring-plane crossings of 1995 and 1996, Earth-based fluorescence measurements indicated that 'the Saturnian ring system must be losing about 3 tons of water per second, Figure 21.10. That's too much to be explained by the impact of interplanetary dust alone.' Despite this evidence of ring dissipation, total replenishment of the ring system by meteoritic impact was modeled as a way of preserving the rings. However, preliminary Cassini data indicate that oxygen is given off by the rings at about 4 times the expected rate, again confirming high collision rates and dissipation of material from the rings.

Longevity estimates for Saturn's rings have undergone steady downward revision since the 1970s. The resonance theory was invoked to prevent such downward revision, but failed to counter indications that planetary rings are much younger than the conventional age of the solar system. The shepherd moon theory continues to be employed to minimize the downward revision. Despite widespread belief that shepherd moons such as *Prometheus and Pandora* have preserved Saturn's rings for possibly hundreds of millions of years, the putative shepherd moons appear to be pulverized and dissipating along with the ring structure.

The origin of Saturn's rings seems to be the 'destruction' of once-existing moons and appear to be a short-lived phenomenon which will have dissipated in a timeframe of the order of tens of thousands of years at *most—possibly only thousands of years*. This demonstrates amazing *consistency with the biblical model* because it shows the rings are young and only recently formed. It also opens up plausible explanations within the model of how the rings formed, e.g., on the fourth day of creation with the creation of the planets or a ***meteorite collision near Saturn around the time of the Flood***.

Distant Stars in a Young Universe

A. If the universe is young and it takes millions of years for light to get to us from many stars, how can we see them?

B. Did God create light in transit?

C. Was the speed of light faster in the past?

D. Does this have anything to do with the big bang?

E. What about Relativity?

SOME galaxies are billions of light-years away. Since a light-year is the distance light would travel over the time period of one year, and we can see such galaxies, does this mean that the universe1 is very old?

Despite all the biblical and scientific evidence for a young earth/universe, this has long been a seemingly intractable problem. However, any scientific understanding of origins will always have opportunities for research—problems that need to be solved. We can never have complete knowledge and so there will always be things to learn. Figure 21.11

Fig.21.11: The Planets Are Young: Uranus and Neptune

TRAVEL PROBLEM THE BIG BANG LIGHT

It's important to remember that the most widely held cosmology, the standard secular big bang theory, has a problem of its own with time and light travel, called the horizon problem.

According to the big bang, the universe began in a fireball from which all matter in the universe is ultimately derived. For galaxies to have any hope of forming at all during the expansion process, the fireball must have begun with an uneven distribution of temperatures. However, we see radiation coming from the cosmos, in all directions of the sky that has a very uniform temperature. This is the cosmic microwave background (CMB) radiation and its temperature has been measured to be uniform to one part in 100,000.

If the regions started at uneven temperatures, and are now almost at the same temperature, then energy must have been transferred from hot regions to cooler ones. The fastest way that energy can be transferred is by radiation, at the speed of light. Consider, then, a region of space 10 billion light years (a light year is the distance light travels in a year) away from earth in the north sky, and the other 10 billion light years in the south. They are 20 billion light years apart. However, since the big bang was allegedly only 13.7 billion years ago, this is not enough time for light to have travelled from one region to the other. Yet the background temperature is almost identical.

However, the problem for the big bang is even more severe than this. The CMB radiation is alleged to be the radiation that appeared when the temperature of the initial fireball cooled enough for it to become transparent to radiation. This is alleged to have happened about 300,000 years after the initial fireball appearance. Consequently, only those regions within about 300,000 light-years of each other could have become uniform in temperature during this time. Yet we have regions separated by at least 20 billion light-years that are at essentially the same temperature.

This horizon problem gave rise to hypothetical fudge factors such as faster-than-light 'inflation' of space being added to the big bang— expanding by a factor of 1050 in 10^{-33} seconds. However, there is no known mechanism to start or stop the process in a smooth fashion—it is effectively a naturalistic 'miracle'. Even New Scientist asked whether inflation was "just wishful thinking." Dr Paul Steinhardt, winner of the 2002 Dirac Medal for his contributions to inflation theory, wrote an article, featured on the cover of Scientific American as "Quantum Gaps in the big bang: Why our best explanation of how the universe evolved must be fixed—or replaced." Steinhardt identified four ways in which inflationary theory fails.

Other big bang cosmologists have even suggested that the speed of light (radiation) may have been much faster in the past (see also. So, no-one can rightly claim this issue as a reason not to believe the Bible, because the standard secular big bang cosmology has a similar problem.

At this point we could just say, 'The big bang has miracles without any miracle worker, so surely we Christians can have miracles with a miracle worker!' Creation Week was, after all, a miraculous event.

LIGHT SPREAD

A few decades ago, perhaps the most common explanation from biblical creationists was that God said 'Let There be Light' - 'on its way,' so that Adam could see the stars immediately without having to wait years for the light from even the closest ones to reach the earth. While we should not limit the power of God, this has some immense difficulties.

It would mean that whenever we look at a very distant object, what we apparently see happening never really happened at all. For instance, say we see an object a million light-years away that appears to be rotating; that is, the light we receive in our telescopes carries this information, 'recording' this behavior. However, according to the 'created in transit' explanation, the light we are now receiving did not come from the star, but was created 'en route.'

This would mean, for a, say, 10,000-year-old universe, that anything we see happening beyond about 10,000 light-years is actually part of a gigantic picture-show of things that have not actually happened, showing us objects that may not even exist.

To explain this problem further, consider an exploding star (supernova) at, say, an accurately measured distance of 100,000 lightyears. (Remember we are using this explanation in a 10,000-year-old universe.) As the astronomer on Earth watches this exploding star, he is not just receiving a beam of light. If that were all, then it would be no problem at all to say that God could have created a whole chain of photons (light particles) already on their way. However, what the astronomer receives is also a particular, very specific pattern of variation within the light, showing the changes that one would expect to accompany such an explosion—a predictable sequence of events involving neutrinos, visible light, X rays and gamma-rays. For example, because most neutrinos pass through solid matter as if it were not there, while light is slowed down, we can detect a massive neutrino burst before the light reaches us.

The light and neutrino burst carry information recording an apparently real event. The astronomer is perfectly justified in interpreting this 'message' as representing actual reality—that there really was such an object, which exploded according to the laws of physics, brightened, emitted X-rays, dimmed, and so on, all in accord with the expected outcomes of known physical laws.

Everything the astronomer sees is consistent with this, including the spectral patterns in the light from the star, giving us a chemical signature of the elements contained in it. Yet the 'light created en-route' explanation would mean that this recorded message of events, transmitted through space, had to be contained within the light beam from the moment of its creation, or planted into the light beam at a later date, without ever having originated from that distant point. (If it had started from the star—assuming that there really was such a star—the light beam would still be 90,000 light-years away from Earth, if the universe was 10,000 years old and the speed of light constant.)

To create such a detailed series of signals in light beams reaching Earth, signals which seem to have come from a series of real events but in fact did not, has no conceivable purpose. Worse, it is like saying that God created fossils in rocks to fool us, or even test our faith, and that they don't represent anything real (a real animal or plant that lived and died in the past). This would be a strange deception for a holy God to engage in.

LIGHT ALWAYS TRAVEL AT THE SAME SPEED

An obvious solution would seem to be a higher speed of light in the past, allowing the light to cover the same distance in less time. This seems at first glance a too-convenient ad hoc explanation. Some years ago, Barry Setterfield raised such a possibility to a high profile by showing that there seemed to be a decreasing trend in the historical observations of the speed of light (c) over the past 300 years or so. Setterfield (and his later co-author, Trevor Norman) produced evidence in favor of their 'cdk' theory.8 They believed that it would have affected radiometric dating results, and even have caused the red-shifting of light from distant galaxies, although this idea was later overturned, and other modifications were made also.

Many attacked the idea on the fallacious grounds that Einstein's Special Relativity said that the speed of light could not change. It actually just says that the speed of light measured by observers will be invariant regardless of the speed of the source or observer.

Much debate raged to and fro among capable people within creationist circles about whether the statistical evidence really supported cdk or not. The biggest difficulty, however, is with certain physical consequences of the theory. If c had declined the way Setterfield proposed, these consequences should still be discernible in the light from distant galaxies, but they are apparently not. High-precision tests of Einstein's Theory of General Relativity, in our galaxy, using co-orbiting pairs of neutron stars, where at least one is a pulsar, within thousands of light-years distance, indicate the same value for c as we measure locally. In short, none of the theory's defenders have been able to answer all the problems raised. Interestingly, big bang defenders treated the idea of cdk with contempt, but then one of their own, João Magueijo, proposed a similar idea to rescue the big bang from its own light-travel (horizon) problem!

COSMOLOGIES OF NEW CREATIONIST

Nevertheless, the cdk theory stimulated much thinking about the issues. For example, creationist physicist Dr Russell Humphreys says that he spent a year, on and off, trying to get the cdk theory to work consistently, but without success. However, the thinking inspired him to develop ideas for a new creationist cosmology as an alternative to big bang theory.

This sort of development, in which one creationist theory, cdk, is overtaken by another, is a healthy aspect of science. The basic biblical framework, because it comes from the Creator, is non-negotiable, as opposed to the changing views and models of fallible people seeking to understand the data within that framework (evolutionists also often change their ideas on exactly how things have made themselves, but never whether they did; that materialistic framework remains non-negotiable).

AN EVIDENCE

Consider that the time taken for something to travel a given distance is the distance divided by the speed it is travelling.

That is,

Time = Distance (divided by) Speed = Distance / Speed

When this is applied to light from distant stars, the time calculates out to be billions of years. Some have sought to challenge the distances, but they are very unlikely to be substantially wrong.

Astronomers use many different methods to measure the distances, and no informed creationist astronomer would claim that errors would be so vast that billions of light-years could be reduced to several thousand, for example. Even our own Milky Way Galaxy is about 100,000 light years across.

If the speed of light (c) has not changed, the only thing left in the equation is time itself. In fact, Einstein's Relativity Theory has been telling the world for a hundred years that time is not an absolute. Scientists may not know what time is but they do know how to measure it. Nowadays very precise and exact atomic clocks measure the rate or flow of time and it has been measured to vary from place to place. In fact, two things have been observed to distort the flow of time— one is speed and the other is gravity. Einstein's general theory, the best theory of gravity we have at present, indicates that gravity distorts time.

This effect has been measured experimentally, many times. Clocks at the top of tall buildings, where gravity is slightly less, run slightly faster than those at the bottom, just as predicted by the equations of General Relativity (GR).

Gravity distorts time so that a clock on the top of a mountain will run faster than a clock on the plains, Figure 21.12.

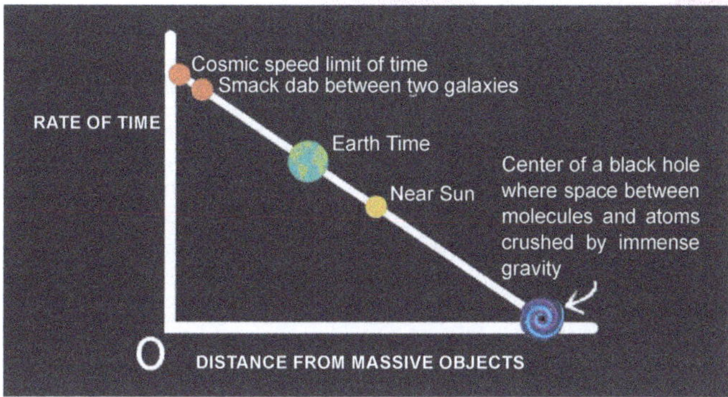

Fig.21.12: Gravity distorts time

EARTH - THE CENTER OF THE UNIVERSE

Most people think of the universe as having a center and an edge. This means that if you were to travel into space, you would eventually come to a place beyond which there was no more matter. In this understanding, Earth is near the center, as it appears to be as we look out into space.

This might sound like common sense, as indeed it is, but all modern secular cosmologies deny this. That is, they make the assumption that the universe has no boundary—no edge and no center—dubbed the 'cosmological principle.' In this assumed universe, every galaxy would be surrounded by galaxies spread evenly in all directions, Figure 21.12. In such a universe, all net gravitational forces cancel out and there is no preferred direction, so there are also no net effects of movement of astronomical objects.

This is a philosophical assumption; that is, religious. And it is made to remove Earth from its apparently privileged position near the center of the universe (because that's what the Bible implies—that Earth is the focus of God's attention in creating the universe). Note the views of respected cosmologist George Ellis, once a colleague of the famous Stephen Hawking; as reported by Scientific American:

"People need to be aware that there is a range of models that could explain the observations" Ellis argues. "For instance, I can construct you a spherically symmetrical universe with Earth at its center, and you cannot disprove it based on observations." Ellis has published a paper on this. "You can only exclude it on philosophical grounds. In my view there is absolutely nothing wrong in that. What I want to bring into the open is the fact that we are using philosophical criteria in choosing our models. A lot of cosmology tries to hide that."

Not only can you have such an understanding of the universe, but it actually fits the evidence better than the no-center, boundless universe assumed by secularists. There is now observational evidence that the universe has a center. For example, galaxies appear to have a large-scale structure centered near our galaxy. These observations do not fit the materialists' no-center, unbounded, randomly generated universe, but are consistent with a universe designed by a creator.

The big bang has many other problems, so much so that even many secularists are calling for a radical rethink:

"Big bang theory relies on a growing number of hypothetical entities—things that we have never observed. Inflation, dark matter and dark energy are the most prominent. Without them there would be fatal contradictions between the observations made by astronomers and the predictions of the big bang theory."

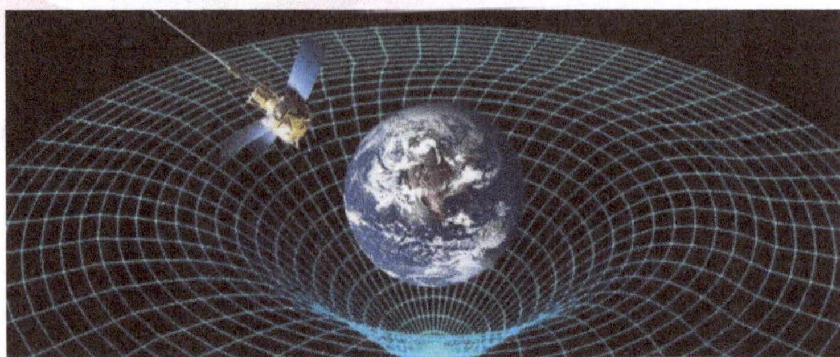

Fig.21.13: A 3-D spherical ball of space and matter has a center and thus a net gravitational force. In the big bang model, the matter of our universe is imagined to be spread over the surface of a 4-dimensional or higher dimensional space, which has no center (balloon analogy).

According to G R, if the universe has a boundary and center, then there can be net gravitational effects on a cosmological scale and these can affect the flow of time during its history. Depending on the how the universe was created, clocks could have run at different rates on Earth compared to other parts of the universe, Figure 21.13. In other words, it is no longer enough to say God made the universe in six days. He certainly did (Exodus 20:11 and Genesis 1), but six days as measured by which clocks? (If we say 'God's time' we miss the point that He created the flow of time as we now experience it; He is outside of time, seeing the end from the beginning. Equally seriously, God inspired Scripture to instruct us (2 Timothy 2:15– 17). This entails that words and logical inferences must be the same for God and man, otherwise Scripture would not be able to equip us with truth He reveals.)

NEW APPROACHES

We now have two creationist cosmologies that could explain how God created everything in six earth days and Adam and Eve could see distant starlight. Both these concepts are rather mind-stretching, but we should not be surprised that when we are trying to get a glimpse of the miracle of creation it is not easy to understand (God's ways are higher than our ways!).

1. Mistakenly - Dr Russ Humphreys

Dr Humphreys had an earlier model, as explained in the book, Starlight and Time, but it failed to account for observations in relation to nearby galaxies. He has developed a new explanation of light-transit-times, to explain how light travelled from the distant cosmos and reached Earth, all during one ordinary-length day on Earth, the fourth day of creation week. This understanding depends on the effect of gravity on time (gravitational time dilation). Humphreys takes the "waters that are above the heavens" (Psalm 148:4 cf. Genesis 1:6–10), to mean that God created the universe with a massive layer of water that encircles the universe. If the mass of this water were very large, it would have a large effect on the flow of time throughout the universe. And then there is the effect of God's creating the stars during the fourth day of creation week as well (Isaiah 40:26). He also takes it that God ***'stretching out the heavens'***, mentioned in various places in Scripture, refers to the expansion of the universe, especially during the fourth day. This expansion could have started on Day 2, when God created the 'expanse' (Hebrew raqia, KJV "firmament", Genesis 1:7).

The model indicates that early on the fourth day, Earth plunged into a zone of timelessness. In this zone all physical processes, including clocks, come to a complete stop. The spherical zone of timelessness expands out from the earth at the speed of light, engulfing the newly-created stars and galaxies. After reaching the most distant galaxies, the timeless zone reverses direction and begins shrinking back toward the earth at the speed of light. As it does so, it uncovers the new galaxies, so that the light can be seen on Earth. Dr Humphreys: "When the sphere reaches zero radius and disappears, Earth emerges, and immediately the light that has been following the sphere will reach Earth, even light that started billions of light-years away. On the fourth day, an observer on the night side of the earth would see a black sky one instant, and a sky filled with stars the next instant."

A universe with a center and an edge, plus Humphreys' concept of the waters above, provided an explanation for the 'Pioneer anomaly', which is a small but strange deceleration of four outgoing spacecraft: Galileo, Ulysses, and Pioneers 10 and 11.

2. Dr John Hartnett

Dr John Hartnett has taken a different approach, which uses a different aspect of Einstein's relativity theory. His cosmology applies a concept developed by Israeli cosmologist Dr Moshe Carmeli (1933–2007) called 'cosmological relativity.' Carmeli argued that to adequately describe the large-scale structure of the universe, in addition to length, breadth, depth, time (four dimensions), another measure, or dimension, was needed: *the velocity of the expansion of space*. This dimension has an effect on gravity and time—hence 'cosmological general relativity.' Carmeli's ideas have been successful in explaining long-standing astronomical puzzles, such as high redshift supernovas, galactic rotation observations, spheroidal galaxy anomalous dispersion, and expansion of the large-scale universe. A great strength of Carmelian relativity is that it does away with hypothetical unobserved entities such as dark matter and dark energy, both of which are needed for big bang cosmology.

Carmeli developed his cosmology with the assumption of the cosmological principle (no center and no edge to the universe), but Hartnett realized that these ideas also worked with a universe with a center and an edge. Furthermore, with this approach, an acceleration (increasing velocity) of the expansion of space, such as could be expected on the fourth day of the creation week, would have profound implications for time during that period. Time dilation results, but not due to a net gravitational effect—it is due to the enormous accelerated stretching of the fabric of space. This means that on Day 4, the clocks in the outer reaches of the expanding universe were running very fast compared to clocks on Earth. This allows time for distant starlight from the galaxies being created on the fourth day to travel to Earth and be visible to Adam and Eve. Again, it's the fourth day as measured by Earth clocks, the clocks the Bible uses.

What if no-one had ever thought of the possibility of time dilation? Many might have felt forced to agree with those scientists (including some Christians) who have asserted that there was no possible solution— vast ages for Earth are a fact because we can see distant stars, and the Bible must be 'reinterpreted' (massaged) or rejected. Many have urged Christians to abandon the Bible's clear teaching of a recent creation because of these 'undeniable facts.' However, this reinterpretation of Scripture would also mean that Earth is old and the rocks containing fossils under our feet are old. Accordingly, this also entails (if it is logically thought through) accepting that there were billions of years of death, disease, and bloodshed before Adam, thus eroding the Creation/Fall/Restoration historical framework presented in the Bible—the framework in which the Gospel makes sense, and upon which western civilization has been built, with all its many benefits.

However, even without the new ideas that seem to solve the problem, such an approach would still have been wrong-headed. The authority of the Bible should never be compromised by mankind's 'scientific' proposals. One little previously unknown fact, or one change in a starting assumption, can drastically alter the whole picture so that what was 'fact' is no longer so.

This is worth remembering when dealing with other areas of difficulty which, despite the substantial evidence for Genesis creation, still remain. As shown, this particular area of difficulty is shared by the big bang theory, and creationists should point this out. Only God possesses infinite knowledge. By basing our scientific research on the assumption that His Word is true (instead of the assumption that it is wrong or irrelevant at points where today's 'science' cannot explain it) our scientific theories are much more likely, in the long run, to come to represent reality accurately. However, creation was a miraculous process and we must recognize that God is able to do things that we, in our human limitations, will struggle to understand. And big bangers invoke secular (God-less) 'miracles' to try to solve the same problems.

PLANETS ROTATIONS

Stars and planets form in the collapse of huge clouds of interstellar gas and dust. The material in these clouds is in constant motion, and the clouds themselves are in motion, orbiting in the aggregate gravity of the galaxy, Figure 21.14. As a result of this movement, the cloud will most likely have some slight rotation as seen from a point near its center. This rotation can be described as angular momentum, a conserved measure of its motion that cannot change. Conservation of angular momentum explains why an ice skater spins more rapidly as she pulls her arms in. As her arms come closer to her axis of rotation, her speed increases and her angular momentum remains the same. Similarly, her rotation slows when she extends her arms at the conclusion of the spin.

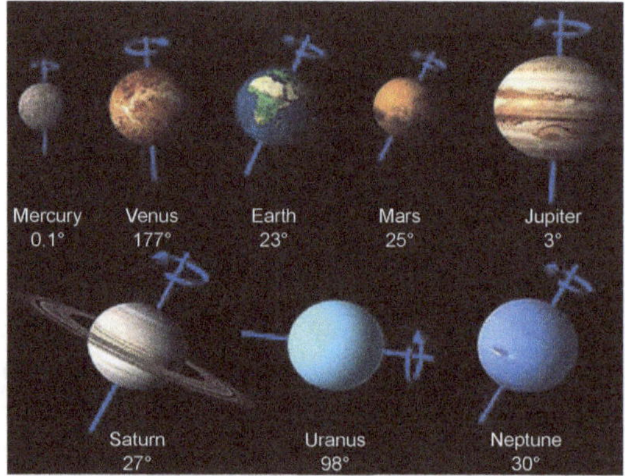

Fig.21.14: Planets Rotational Axes

As an interstellar cloud collapses, it fragments into smaller pieces, each collapsing independently and each carrying part of the original angular momentum. The rotating clouds flatten into proto-stellar disks, out of which individual stars and their planets form. By a mechanism not fully understood, but believed to be associated with the strong magnetic fields associated with a young star, most of the angular momentum is transferred into the remnant accretion disk. Planets form from material in this disk, through accretion of smaller particles.

In our solar system, the giant gas planets (Jupiter, Saturn, Uranus, and Neptune) spin more rapidly on their axes than the inner planets do and possess most of the system's angular momentum. The *sun itself rotates slowly*, only once a month. The planets all revolve around the sun in the same direction and in virtually the same plane. In addition, they all rotate in the same general direction, with the exceptions of ***Venus and Uranus***.

Our solar system is actually flat, with most of its planets orbiting within three degrees of the plane of the Earth's orbit around the sun, called the ecliptic. This flatness extends to the asteroid belt between Mars and Jupiter, though some members of the region of icy objects past Neptune called the Kuiper belt are more extreme, with inclinations up to 30 degrees.

This relative flatness, which is not an unusual feature of solar systems, results from how stars and planetary systems typically form. The process begins with a slowly rotating, roughly spherical cloud of gas and dust, about one light year across. Eventually, a portion of this material collapses toward the center, forming a star, and the spinning cloud begins to flatten into a disk due to its rotation. It's out of this rotating protoplanetary disk of gas and dust that planets are then spun out, resulting in a relatively flat solar system. Eventually, when most of the gas has settled onto the star or planets or has dissipated, the system is left with a debris disk of planetary leftovers, like our own asteroid-strewn Kuiper belt.

Some astronomers at Penn State study protoplanetary and debris disks to get a better idea of how planetary systems form. But not all stars and planets form in exactly the same manner — and not all planetary systems are flat.

"It's an exciting time, because so many planets have been discovered in other solar systems, for example by NASA's Kepler space telescope and Transiting Exoplanet Survey Satellite (TESS), and a lot of them look very different from the planets in our solar system," said Rebekah Dawson, Shaffer Career Development Professor in Science and assistant professor of astronomy and astrophysics. "So, we have to come up with new ways of thinking about planet formation that can account for the diversity of planets we now know about." Figure 21.15.

Fig.21.15: *A newly formed star is surrounded by a rotating disk of gas and dust, called a protoplanetary disk. This disk, illustrated here around a brown dwarf, provides the materials for planet formation. Curtsy - NASA/JPL-Caltech*

In addition to studying disks, researchers like Dawson study the exceptions to the norm, unusual stars and planets that could support or make us rethink current theories. Together, these investigations are helping scientists improve our understanding of how and where different kinds of stars and planets form, and what makes a planet habitable.

CATCHING PLANETARY FORMATION IN THE ACT

While some researchers study mature systems and infer aspects of the planet formation process, Assistant Professor of Astronomy and Astrophysics Ian Czekala tries to catch planetary formation in the act.

"I study the protoplanetary disks that surround young stars for the first 10 million years of their lives," he said. "That may sound like a long time period, but it's actually very small compared to a star's total lifetime.

While Kepler and other survey missions have found thousands of mature solar systems, there are fewer nearby proto-planetary systems that easily lend themselves to detailed study. To investigate these early systems, Czekala uses the Atacama Large Millimeter/submillimeter Array (ALMA), one of the most complex astronomical observatories ever built. Located in Chile, ALMA uses a network of high-precision antennas working together to provide a high-resolution look at the universe, using wavelengths of light between the infrared and radio regions of the electromagnetic spectrum.

VENUS AND URANUS SPIN THE WRONG WAY

The cosmos offers us many mysteries that astronomers are still struggling to solve, but there is one that is in our own solar system and remains unexplained – why do Venus and Uranus rotate in different directions from the other planets in the solar system?

Venus rotates from east to west, while Uranus is so inclined that it practically rotates on its side. Every other planet, including Earth, rotates from west to east, and scientists have not yet determined why, Figure 21.16.

But Venus and Uranus are exceptions and have retrograde motion, which is opposite of the direction of rotation of the Sun? Why is this so?

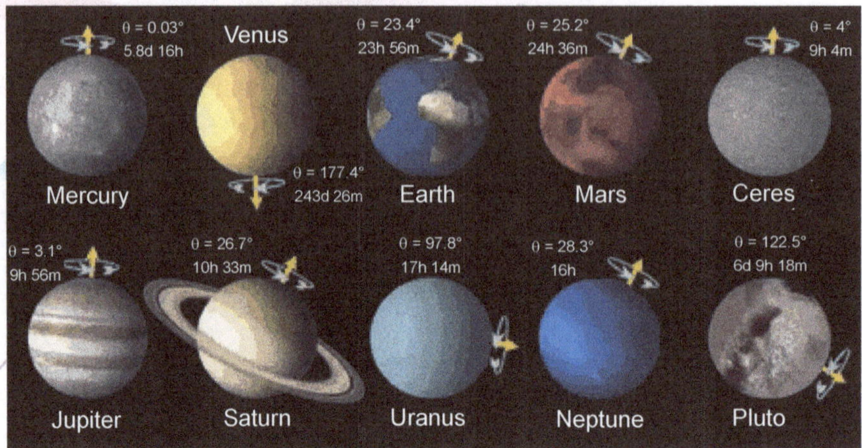

Fig.21.16: *Spinning Direction of Planets*

Venus' hypotheses are that it began to rotate counterclockwise, then slowed down at rest, and then began to rotate clockwise.

This may explain the very low rotation speed of the planet today – on Venus the days are longer than the year and the Sun rises from the West.

A planet in a locked orbit to its star doesn't rotate with reference to its star; it rotates once (in essence its day) in the same period as its year (orbital period) from an external viewpoint.

To get a planet that doesn't rotate in respect to the rest of the universe is far harder. It would have to rotate in the opposite direction to its star orbit with a day period exactly equal to its year.

We have two counter rotating planets in our own solar system, but what then are the chances of the planet's day being exactly equal to its year? There are a lot of planets in the universe, so maybe one fits this concept.

With such long days and nights, temperature extremes probably preclude life on such a planet. The bible was exceptionally explicit to make the earth particularly very special to be inhabited. Accordingly, the earth was miraculously fined tuned for human beings, animals, birds and plants. Speculating of other life in other planets would be wats of time. It is immaculate view that the sun, and other planets move very slowly across their sky while the stars at night are stationary.

Planets normally spin but, given that the universe contains billions of stars, a few must stop spinning.

A star or solar system is formed from a collapsing cloud of gas. In the highly unlikely event that this cloud has no angular momentum and therefore no spin, the result would be a non-spinning star without any orbiting planets. Even if almost imperceptible, a molecular cloud normally has some angular momentum.

Conservation of angular momentum means that this cloud spins faster as it collapses, much in the same way that ice skaters spin faster as they pull in their arms. Some of this gas collapses into a proto-stellar disc, a ring of material that acts as the nursery for planets, which then naturally spin and orbit.

For example, viewed from above the ecliptic (an imaginary plane that corresponds to what was the proto-stellar disc for the solar system), ***Earth*** spins anticlockwise and this is the same direction that it orbits the sun.

Venus is unusual in that it spins clockwise, which is opposite to its orbital direction.

Venus will eventually become tidally locked to the sun, as the moon is to Earth. This means that its spin and orbital periods will become the same.

Eventually Venus will show the same face to the sun at all times and a day on the planet will equal a Venusian year. However, because its spin and orbital directions are opposite, Venus would be quasi-non-spinning with respect to the "fixed stars" or the cosmic background radiation.

From the surface of a non-spinning planet, its sun would appear to move across the sky, but the other stars would be stationary. However, this would mean that Venus would still be spinning a small amount, because our solar system orbits the center of the Milky Way in that time.

Also, some rogue planets that have been knocked out of their solar systems and wander across the cosmos might have lost their spin due to chance interactions and collisions with other objects.

However, Planets originate as dust particles that are attracted to each other mainly by static electricity. Once all those accumulated particles build up enough mass, then gravity will attract other masses.

As particles hit the developing planetoid, they are likely to hit offset to the center of gravity. The momentum of the particles is then converted into angular momentum as they contribute to the planetoid's mass. Each particle, pebble, rock asteroid and comet that strikes provides more angular momentum as they hit at an angle and contribute to the growing planet.

It is unlikely that all contributing debris would hit the planetoid directly on the center of gravity and therefore, it is likely that all planets will rotate.

In fact, there are two planets that spin on their axes from east to west. The other is Uranus.

The rotation of Venus is very slow, taking a little more than 243 Earth days to make a complete turn, while the planet's orbit around the sun takes just over 224 days.

An explanation for the backward, or retrograde, rotation is not certain. A long-held theory is that Venus once rotated as the other planets do, but was struck by a planet-size object. The impact and its aftermath caused the rotation to change directions or flipped the planetary axis.

Another theory is that tidal effects from core-mantle friction and a thick atmosphere might have accounted for the change in direction.

A more recent idea, put forth in the journal Nature in 2001, suggests that an initial counterclockwise rotation was an unstable state for various complex reasons and that the planet slowed down and slipped into a more stable state of clockwise rotation. As for the initial direction of rotation for all of the sun's planets, it is assumed to be a legacy of their formation through a gradual accretion of the whirling original material of the solar system.

SIGNIFICANCE

The Sun's equator lines up with the orbits of the planets. This fact supports the theory that stars and their planets inherit their angular momentum from the same source: the gravitational collapse of a molecular cloud. Most astronomers expected spin-orbit alignment to be a universal feature of planetary systems.

This proved false: many drastic misalignments are known, and many possible reasons have been offered. In one theory, a distant companion star upsets the alignment at an early stage, while the star is still surrounded by a protoplanetary disk. Here, the *K2-290* system is shown to be the best-known candidate for such a primordial misalignment. The star rotates backward, and a companion star with suitable properties has been identified.

It is widely assumed that a star and its protoplanetary disk are initially aligned, with the stellar equator parallel to the disk plane. When observations reveal a misalignment between stellar rotation and the orbital motion of a planet, the usual interpretation is that the initial alignment was upset by gravitational perturbations that took place after planet formation.

Most of the previously known misalignments involve isolated hot Jupiter's, for which planet–planet scattering or secular effects from a wider-orbiting planet are the leading explanations. In theory, star/disk misalignments can result from turbulence during star formation or the gravitational torque of a wide-orbiting companion star, but no definite examples of this scenario are known.

An ideal example would combine a coplanar system of multiple planets—ruling out planet–planet scattering or other disruptive post-formation events—with a backward-rotating star, a condition that is easier to obtain from a primordial misalignment than from post-formation perturbations. There are two previously known examples of a misaligned star in a coplanar multi-planet system, but in neither case has a suitable companion star been identified, nor is the stellar rotation known to be retrograde.

A BACKWARD-SPINNING

It is widely assumed that a star and its proto-planetary disk are initially aligned, with the stellar equator parallel to the disk plane. When observations reveal a misalignment between stellar rotation and the orbital motion of a planet, the usual interpretation is that the initial alignment was upset by gravitational perturbations that took place after planet formation. Most of the previously known misalignments involve isolated hot Jupiter, for which planet-planet scattering or secular effects from a wider-orbiting planet are the leading explanations. In theory, star/disk misalignments can result from turbulence during star formation or the gravitational torque of a wide-orbiting companion star, but no definite examples of this scenario are known.

An ideal example would combine a coplanar system of multiple planets—ruling out planet-planet scattering or other disruptive post-formation events—with a backward-rotating star, a condition that is easier to obtain from a primordial misalignment than from post-formation perturbations. There are two previously known examples of a misaligned star in a coplanar multi-planet system, but in neither case has a suitable companion star been identified, nor is the stellar rotation known to be retrograde. Here, we show that the star *K2-290* A is tilted by 124±6 degrees compared to the orbits of both of its known planets, and has a wide-orbiting stellar companion that is capable of having tilted the proto-planetary disk. The system provides the clearest demonstration that stars and protoplanetary disks can become grossly misaligned due to the gravitational torque from a neighboring star.

VENUS AND URANUS SPINNING IN THE WRONG DIRECTION

Space offers plenty of mysteries for astronomers to solve, and there's one in our own Solar System that's been unexplained for decades: why are Venus and Uranus spinning in different directions to the other planets around the Sun?

Venus spins on its axis from east to west, while Uranus is tilted so far over, it's virtually spinning on its side. Every other planet, including our own, Figure 21.17, spins from west to east, and scientists haven't figured out why.

The planets should really all be spinning the same way: our Solar System was formed by a collapsing and rotating cloud of gas, and it›s thought that the spin direction of most planets (like Earth) has been carried over from that ancient rotation.

But Venus and Uranus are the exceptions: they have what's known as retrograde rotation, spinning counter to the rotation of the Sun. But how is this possible?

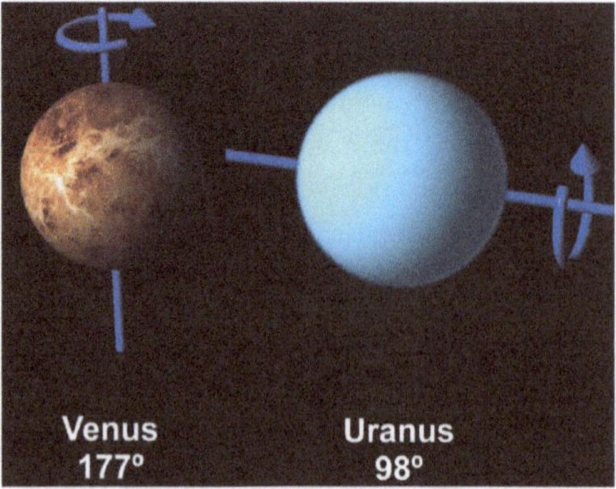

Fig.21.17: Venus and Uranus Spin The Wrong Way

One of the most long-standing hypotheses is that Venus and Uranus originally rotated counter-clockwise – like Earth and the other planets still do – but were struck at some point by massive objects (perhaps other planets) that sent them spinning in different directions.

In recent years, astronomers have looked for other explanations, examining Venus and Uranus independently.

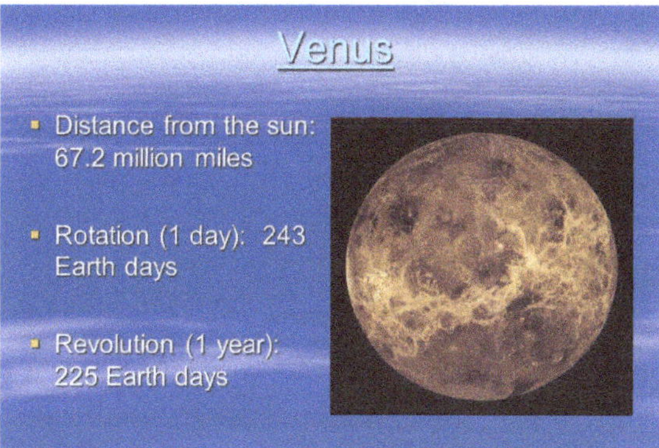

Fig.21.18: *Venus – A Day is 243 Earth Days*

In 2011, simulations suggested that a number of smaller collisions, rather than one big impact, knocked Uranus' spin to an angle of 98 degrees. This could also explain why the planet's moons rotate at the same angle – something that would be unlikely if there were just one massive hit.

An alternative explanation put forward by astronomers in 2009 is that Uranus once had a large moon, the gravitational pull of which caused the planet to fall on its side. Eventually, the moon could have been knocked out of orbit by another planet, a bit like a game of cosmic pinball.

As for Venus, our closest neighbor, scientists have suggested that it started off rotating counter-clockwise, then slowed down to be almost static, before starting to spin clockwise like it does now.

This might explain the planet's very slow rotation speed today – it takes Venus 243 Earth days to rotate fully, Figure 21.18, but only 225 Earth days to orbit the Sun. So, if you lived on Venus, your days would be longer than your years (and the Sun would rise in the west).

How does that happen to a planet? Astronomers think that the Sun›s strong gravitational pull on the dense atmosphere of Venus; the atmospheric tides that would create; and the tidal pulls from other planets, could all have combined to reverse the planet›s spin.

This idea of tidal torques – where the dense atmosphere on the warm, Sun-drenched side of a planet is pulled away from the cold side – is one of the most well-established explanations for Venus› retrograde rotation, along with a planetary collision.

For now, though, no one's 100 percent sure what makes Venus and Uranus the odd ones out in our Solar System's family of planets.

Our next close look at Venus should come from a flyby with the BepiColombo probe, which is eventually headed for Mercury and launching in 2018. That mission might give us new data to help solve the mystery.

MYSTERIES OF URANUS' ODDITIES EXPLAINED

All the planets in our Solar System revolve around the Sun in the same direction and plane, and rotate in the same direction with their poles oriented perpendicular to their orbit. All but one, that is, Uranus, Figure 21.19.

Fig.21.19: *Uranus, its moons, and rings are all "tipped", suggesting they formed during a cataclysmic impact early in its history. Curtsy – NASA*

Uranus has puzzled scientists for decades, rotating essentially on its side with its pole pointing almost directly at the Sun. Like Saturn, Uranus also has a ring system and 27 moons, all of which orbit the planet's equator and are "tipped over" too. Many astronomers have theorized the reason for this unique and unusual orientation, but now a research team led by Professor Shigeru Ida from the Earth-Life Science Institute (ELSI) at Tokyo Institute of Technology suggests that early in the history of our Solar System, Uranus was struck by a small icy planet roughly 1-3 times the mass of the Earth.

According to the study, the team came to this conclusion while constructing a novel computer simulation of moon formation around icy planets. According to the researchers, Uranus' ring and moons provided the evidence they needed to unravel the planet's oddities.

Moons are not uncommon in our Solar System, and are present in a range of sizes, orbits, compositions, and other properties, which have helped scientists determine how they formed. For example, our own Moon is thought to have formed when a Mars-sized planetoid collided with a smaller proto-Earth. These types of collisions were common during the formation of the Solar System, according to the authors, but Uranus must have experienced something different, which they believe has to do with its composition.

The outer-planets are largely composed of volatile elements, like frozen water and ammonia, as the environment in which they formed was colder than planets like Earth, which formed from non-volatile elements due to its proximity to the sun. The differing composition of the planets would result in different consequences after impact with another celestial body. For example, water freezes at a (relatively) low temperature, so the team believes that any impact with icy Uranus in the early Solar System would have vaporized any ice debris that might have formed — meaning this debris would remain in a gaseous state longer, resulting in smaller moons as there is less debris to collect and build them.

The team highlights that the ratio of Uranus' mass to Uranus' moons' masses is greater than the ratio of Earth's mass to its moon by a factor of more than a hundred, and their new model is able to accurately reproduce this.

According to Ida, "This model is the first to explain the configuration of Uranus' moon system, and it may help explain the configurations of other icy planets in our Solar System such as Neptune. Beyond this, astronomers have now discovered thousands of planets around other stars, so-called exoplanets, and observations suggest that many of the newly discovered planets known as super-Earths in exoplanetary systems may consist largely of water ice, and this model can be applied."

NEPTUNE - THE EIGHTH PLANET FROM THE SUN

Neptune is the farthest planet from the sun and was the first to be predicted before it was discovered, Figure 21.20.

Fig.21.20: Neptune - the Eighth Planet from the Sun

Neptune is the eighth and farthest planet from the sun but it is not the coldest.

Our solar system's blue gas giant is far larger than Earth, at more than 17 times Earth's mass and nearly 58 times Earth's volume, according to NASA. Neptune's rocky core is surrounded by a slushy fluid mix of water, ammonia and methane ice.

Astronomer Galileo Galilei was one of the first people to identify Neptune as a space object, however, he assumed it was a star based on its slow movement.

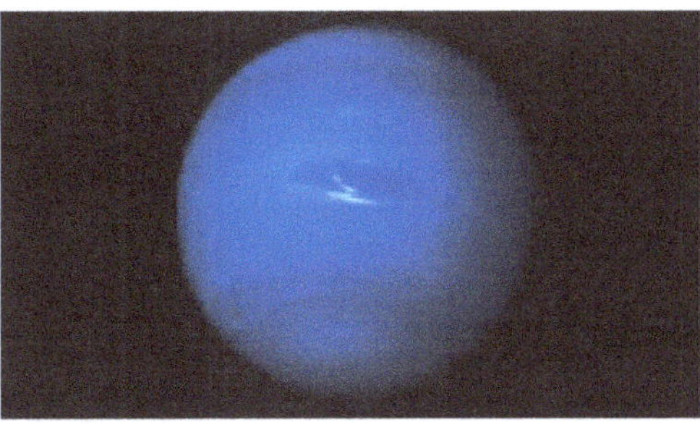

Fig.21.21: Voyager 2 snapped this close up shot of a blue windy Neptune back in 1989 – Curtsy - NASA/JPL-Caltech)

Around two hundred years later, in 1846, French astronomer Urbain Jean Joseph Le Verrier calculated the approximate location of Neptune by studying gravity-induced disturbances in the motions of Uranus, according to a synopsis written by researchers at the University of St. Andrews in Scotland.

At the same time Le Verrier was calculating the existence of Neptune, so was English astronomer John Couch Adams. The two scholars independently came up with nearly identical mathematical predictions about Neptune's existence, Figure 21.21. Le Verrier then informed his colleague, German astronomer Johann Gottfried Galle, about his calculations, and Galle and his assistant Heinrich d'Arrest confirmed Le Verrier's predictions by viewing and identifying Neptune through the telescope at his observatory in Berlin.

In accordance with all the other planets seen in the sky, and as suggested by Le Verrier, this new world was given a name from Greek and Roman mythology — Neptune, the Roman god of the sea.

Only one mission has flown by Neptune — Voyager 2 in 1989. Today, there are still many mysteries about the cool, blue planet, such as why its winds are so speedy and its magnetic field is offset. While Neptune is interesting because it is in our own solar system, astronomers are also interested in learning more about the planet to assist with exoplanet studies. Specifically, astronomers are interested in learning about the habitability of worlds that are bigger than Earth.

Like Earth, Neptune has a rocky core, but it has a much thicker atmosphere that prohibits the existence of life as we know it. Astronomers are still trying to figure out at what point a planet is so giant that it may pick up a lot of gas in the area, making it difficult or impossible for life to exist.

WHY IS NEPTUNE BLUE

The planet's cloud cover has an especially vivid blue tint that is partly due to an as-yet-unidentified compound and the result of the absorption of red light by methane in the planet's mostly hydrogen-helium atmosphere, Figure 21.22.

By studying the cloud formations on the gas giant, scientists were able to calculate that a day on Neptune lasts just under 16 hours long.

Neptune is the fourth largest planet in the solar system, with a radius of 15,599.4 miles (24,622 kilometers) — the distance between its core and the surface. However, Neptune is a spheroid shape, meaning that it bulges around its equator, making the radius of the pole slightly smaller.

Neptune's elliptical, oval-shaped orbit keeps the planet an average distance from the sun of almost 2.8 billion miles (4.5 billion kilometers), or roughly 30 times as far away as Earth, making it invisible to the naked eye. A single orbit of the sun takes Neptune 165 Earth years to complete.

Fig.21.22: Cloud streaks over Neptune's surface, taken by Voyager 2. The width of the cloud streaks ranges from 31 to 124 miles. Curtsy NASA/JPL

Despite Neptune's distance from the sun, which means it receives little sunlight to help warm and drive its atmosphere, Neptune's winds can reach up to 1,500 miles per hour (2,400 kilometers per hour), the fastest detected in the solar system so far. These winds were linked with a large dark storm that Voyager 2 tracked in Neptune's southern hemisphere in 1989.

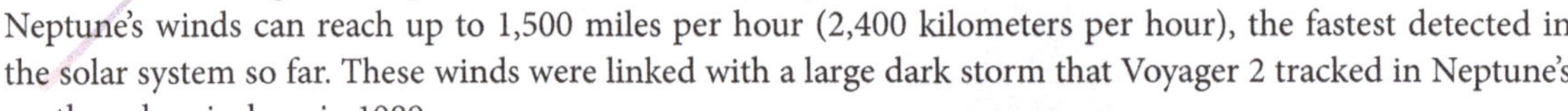

This oval-shaped, counterclockwise-spinning "Great Dark Spot" was large enough to contain the entire Earth and moved westward at nearly 750 mph (1,200 km/h).

The storm seemed to have vanished when the Hubble Space Telescope searched for it later, and since then, Hubble has witnessed the appearance and then fading of other Great Dark Spots on Neptune over the past decade, Figure 21.23.

Fig.21.23: A closeup of Neptune's Great Dark Spot taken by Voyager 2. The dark boundary of the storm and the surrounding white cirrus clouds suggests it is moving in a counterclockwise rotation. Curtsy - NASA / Jet Propulsion Lab

DIAMOND RAIN

Because of the high temperatures and pressures on Neptune and Uranus, scientists believe compressed carbon in the form of diamonds causes a "diamond rain" phenomenon on these icy giants. In 2017, researchers could simulate the conditions that would cause diamonds to form in the lab, supporting the hypothesis that diamond rain occurs on Neptune and Uranus, Figure 21.24.

Neptune's magnetic poles are tipped to the side by roughly 47 degrees compared with the poles along which it spins. As such, the planet's magnetic field, which is about 27 times more powerful than Earth's, undergoes wild swings during each rotation.

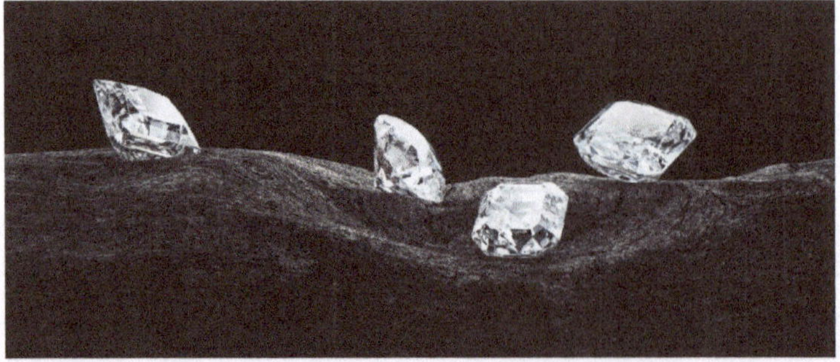

Fig.21.24: It Rains Diamonds on Neptune and Uranus – Curtsy - East Bay Times

NEPTUNE TYPE OF PLANET

Planetary scientists refer to Uranus and Neptune as 'ice giants' to emphasize that these planets are fundamentally different in bulk composition (and, consequently, formation) from the solar system's other giant planets, the 'gas giants' Jupiter and Saturn.

Based on their bulk densities (their overall masses relative to their sizes) Jupiter and Saturn must be composed mostly of the less massive ('lighter') elements, namely hydrogen and helium, even down into their deep interiors. Hence, they are called gas giants. However, in comparison, the bulk densities of Uranus and Neptune indicate that they must have significantly more heavy elements in their interior — specifically in the form of ammonia, methane, and water molecules — to explain their densities. They are, therefore, compositionally distinct, with implications for different formation processes and origins in the early solar system.

ICE GIANT

But why the term 'ice giant'? Astronomers and planetary scientists group molecules broadly by whether they are gaseous or solid at the cold temperatures found across the solar system. Hydrogen and Helium are gases at very low temperatures, whereas ammonia, methane and water condense and freeze to form ice. Hence, these heavier molecules are referred to as ices, and planets rich in these materials are classed as ice giants.

Neptune Composition

We can measure that the observable atmosphere is mostly hydrogen (more than 80%) and helium (~15%), with a small amount of methane and trace amounts of other molecules, including ethane, acetylene, and several other hydrocarbons.

However, as you go deeper into the planet, the composition must change, as the overall larger bulk density indicates that heavier elements must be present at greater depths. The bulk composition of Uranus and Neptune is roughly only 10–20% hydrogen and helium and 80–90% heavier elements, overall, by mass. From our understanding of the fundamental chemical building blocks of the solar system, we can infer that these heavier elements are likely mostly methane, ammonia and water (that form ices), plus some elements that form rocks and metals. However, the relative proportions of the elements (including the ratio of ice to rock) and how they are distributed in the deep interior are unknown.

Why is Neptune not the coldest planet in the solar system if it is the most distant from the sun?

Interestingly, the upper atmospheres of both Uranus and Neptune are warmer than what theory would predict given their composition and the amount of sunlight they receive. *This is an open mystery that has yet to be adequately explained.* It is created this way. Furthermore, compared to Neptune, Uranus appears to have much less heat escaping from its deep interior and fewer trace gas molecules (that absorb sunlight) in its upper atmosphere. This leads to Uranus being somewhat colder, but *the reasons for this are also unknown.*

Why is it important that we learn about Neptune?

It is important that we study these planets for a number of reasons. The extreme atmospheric conditions and long seasonal timescales found in the ice giants provide a unique opportunity to better understand the physics of weather and climate. More fundamentally, these planets may also help us learn more about the formation of the solar system. Understanding why the bulk densities of Uranus and Neptune differ from Jupiter and Saturn can help us to better understand how planets formed and what processes shape planetary formation in general.

From a big-picture perspective, the ice giants are seen as possible representatives of a general class of planets that appear common across the galaxy. Studies looking at the variety of exoplanets discovered in recent years show that exoplanets with sizes roughly similar to Neptune are among the most numerous. By understanding the formation and characteristics of our local ice giants, scientists hope to better understand these vastly more distant planets and the processes by which they were formed.

Fig.21.25: *The James Webb Space Telescope snapped the clearest view of Neptune›s rings in 30 years using its Near-Infrared Camera (NIRCam) on September 21, 2022. Curtsy - NASA, ESA, CSA, STScI*

Neptune is surrounded by unusual rings, which aren't uniform but possess bright thick clumps of dust called arcs.

There are at least 5 rings around Neptune — called Galle, Leverrier, Lassell, Arago, and Adams — and they are considered relatively young and short-lived. According to a writing in the journal Icarus, Earth-based observations found that Neptune's rings are apparently far more unstable than previously thought, with some dwindling away rapidly. Figure 21.25.

NEPTUNE POSSESSION

Neptune has 14 known moons, Figure 21.26, named after lesser sea gods and nymphs from Greek mythology. The largest by far is Triton, whose discovery on Oct. 10, 1846, was indirectly enabled by beer — amateur astronomer William Lassell, who discovered Triton used the funds he made as a brewer to finance his telescopes.

Triton is the only spherical moon of Neptune, Figure 21.27. The planet's other 13 moons are irregularly shaped. Triton is also unique in being the only large moon in the solar system to circle its planet in a direction opposite to its planet's rotation — this "retrograde orbit" suggests that Triton may once have been a dwarf planet that Neptune captured rather than forming in place, according to NASA, which contradicts the creationists believe that it was created this way from the beginning.

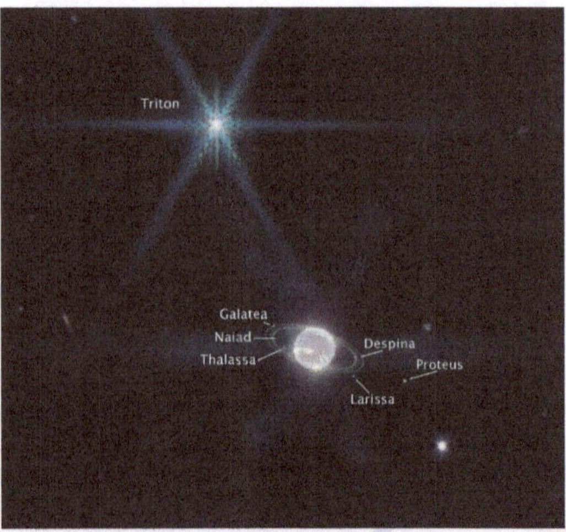

Fig.21.26: *Neptune's Moons – Curtsy Wikipedia*

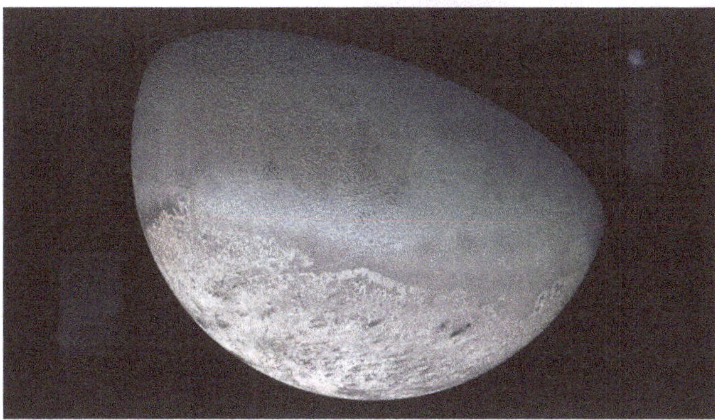

Fig.21.27: A color mosaic image of Triton taken by Voyager 2 during its flyby of the Neptune system in 1989. Curtsy - NASA / Jet Propulsion Lab / U.S. Geological Survey

Triton is extremely cold, with temperatures on its surface reaching about minus 391 degrees F (minus 235 degrees C), making it one of the coldest places in the solar system. Nevertheless, Voyager 2 detected geysers spewing icy matter upward more than 5 miles (8 km), showing its interior appears warm. Scientists are investigating the possibility of a subsurface ocean on the icy moon. In 2010, scientists discovered seasons on Triton.

In 2020, NASA announced the possibility of a new space mission to visit Triton, called Trident. "Triton has always been one of the most exciting and intriguing bodies in the solar system," Louise Prockter, director of the Lunar and Planetary Institute of the Universities Space Research Association in Houston, who leads the Trident proposal team, said in the statement.

In 2013, scientists working with SETI caught sight of Neptune's "lost" moon of Naiad using data from the Hubble Space Telescope. The 62-mile-wide (100 km) moon had remained unseen since Voyager 2 discovered it in 1989.

Also in 2013, scientists using the Hubble Space Telescope found the 14th moon, dubbed S/2004 N 1. It is Neptune's smallest moon and is just 11 miles (18 km) wide. It got its temporary name because it is the first satellite (S) of Neptune (N) to be found from images taken in 2004, according to NASA.

NEPTUNE: MONUMENT TO CREATION

Neptune Secret

Neptune is the eighth of the nine known planets in our solar system. An enormous gas giant, it is about 17 times the mass of the Earth (and 58 times its volume). It poses many problems for those who wish to deny Creation, Figure 21.28. Among other things, naturalistic theories say that Neptune shouldn't exist!

At about 30 times as far away from the Sun as Earth, as stated earlier, Neptune appears as little more than a bluish dot in all but the most powerful telescopes. Our best photographs and measurements of Neptune and its moons were taken by the Voyager II spacecraft, which

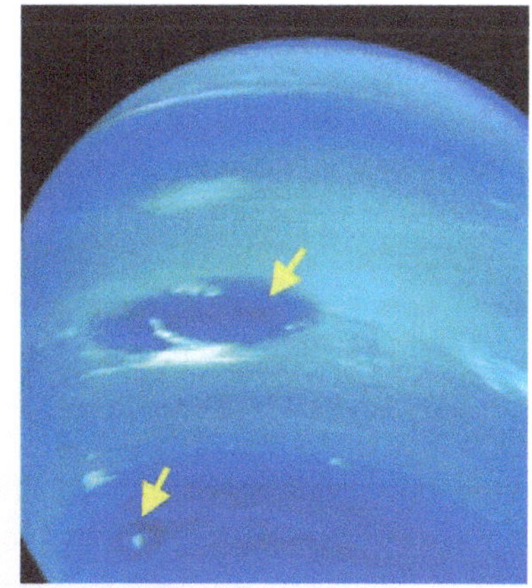

Fig.21.28: Neptune's great spot, a massive storm, which in the latest photographs has gone.

flew by the planet in August 1989. Many of these measurements greatly surprised evolutionary scientists. They had assumed Neptune would be a cold, inactive place, but it is not.

Neptune has winds that rage at almost 2,200 km/h (1,300 mph), the strongest measured in the solar system. Figure 21.28, shows two large spots thought to have been enormous atmospheric storms, the larger one approximately the same diameter as the Earth. In 1994, the Hubble Space Telescope was pointed at Neptune, which revealed that these storms are now apparently gone. However, a new one has appeared elsewhere on its surface. Neptune is a dynamic, ever-changing place.

And Neptune is nowhere near as cold as evolutionary theory predicts. Instead, it actually *generates* heat, radiating into space over twice the energy it receives from the Sun. This fits the Creation model very well, as a young Neptune could easily still be cooling off a few thousand years after its creation. However, this does not fit the evolutionary/long-age model, as many evolutionists have acknowledged.

Overall, with its raging winds, dynamic atmosphere, and heat generation, Neptune appears quite young.

The Voyager expedition measurements of Neptune's magnetic field also upset evolutionary theories. Three years earlier, Voyager had flown by the planet Uranus, discovering that Uranus's magnetic field is tilted relative to the planet's spin axis, and offset from the planet's center. Both these characteristics contradict the evolutionary 'dynamo' model of planetary magnetism (this hypothetical 'self-generating' mechanism for sustaining a magnetic field is essential for long-agers, because without some renewal such fields decay away to nothing in only a few thousand years).

So, evolutionists consoled themselves by speculating that perhaps 'Voyager had caught the field in the middle of a reversal (when the magnetic north and south poles switch places).' This is very unlikely, but not necessarily impossible. But then Voyager flew by Neptune and discovered that its field was tilted and offset, too. Scientists were forced to concede that 'it seems that the possibility of finding two planets both experiencing magnetic polarity reversals is small.'

Of course, creation scientists are not bound to dynamo theories, nor to millions of years. Creation physicist Dr Russell Humphreys was able to predict the magnetic characteristics of Uranus and Neptune (before Voyager measured them) with much more success than the evolutionists, by assuming (based on the Bible) that the planets began as masses of water (Genesis 1:2; 2 Peter 3:5) and that Creation occurred roughly 6,000 years ago.

We know from the Bible that Neptune was created on Day 4 along with the other 'lights in the heavens.' Evolutionary scientists (and 'long-age creationists') scoff at this history, believing instead that the solar system formed from an enormous cloud of gas and dust. Over postulated millions of years, the dust allegedly clumped together into rocks, these rocks clumped together into bigger rocks, and eventually there were enormous rocks ('planetesimals') flying around, which stuck together and became planets, Figure 21.29. The gas giants are supposed to have formed at the outer reaches of the solar system because it was cold enough there for ice to condense, making the growing planetoid massive enough to attract gas.

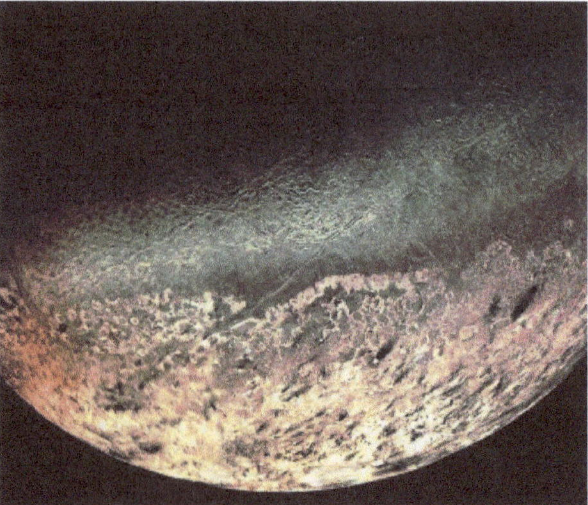

Fig.21.29: *Triton, Neptune's primary moon, is a hostile place of volcanic activity and extreme cold.*

Unfortunately for evolutionists, Neptune doesn't fit that model, Figure 21.30. An article in a (pro-evolution) astronomy magazine explained it this way:

'... *astronomers who model the formation of the solar system have kept a dirty little secret: Uranus and Neptune don't exist. Or at least computer simulations have never explained how planets as big as the two gas giants could form so far from the sun. Bodies orbited so slowly in the outer parts of the sun's protoplanetary disk that the slow process of gravitational accretion would need more time than the age of the solar system to form bodies with 14.5 and 17.1 times the mass of Earth.*'

In evolutionary models, the farther you are from the middle of the cloud (where the Sun is today), the longer the planet-formation procedure requires. Neptune and Uranus are too far out to have formed according to this process, even over the supposed 4.5-billion-year age given to the solar system. One evolutionist astronomer amusingly comments:

'*What is clear is that simple banging together of planetesimals to construct planets takes too long in this remote outer part of the solar system. The time needed exceeds the age of the solar system. We see Uranus and Neptune, but the modest requirement that these planets exist has not been met by this model.*'

How much more time is needed? Another (evolutionary) book explains:

'There have been many attempts to model the evolution of a swarm of colliding planetesimals ... Safronov calculated the characteristic time-scales for planetary growth. In the terrestrial region he found timescales of 10^7 [10,000,000] years but the time estimates increased rapidly in the outer regions of the solar system and was 10^{10} [10,000,000,000] years for Neptune—which is twice the [alleged evolutionary] age of the solar system.

It is clear that, in view of the large timescales found for the formation of the outer planets, a satisfactory theoretical model for the accretion of planets from diffuse material is not available at present.'

Therefore, even if the solar system really were 4.5 billion years old, as evolutionists believe, we would still be 5.5 billion years short of the time necessary for Uranus and Neptune to have formed by themselves. This is why the *Astronomy* magazine said that, according to evolution, '*Uranus and Neptune don't exist.*'

Safronov published these calculations in 1972. So, this problem has been recognized for at least 30 years. Why then do the textbooks and popular media so confidently proclaim that we 'know' for certain that the solar system formed by itself over thousands of millions of years? Shouldn't the fact that some of the planets 'don't exist' cause some doubt?

Of course, creationists are not alone in noticing the absurdity of this situation. Many evolutionists have been trying to come up with a solution. The *Astronomy* article mentioned above continues as follows:

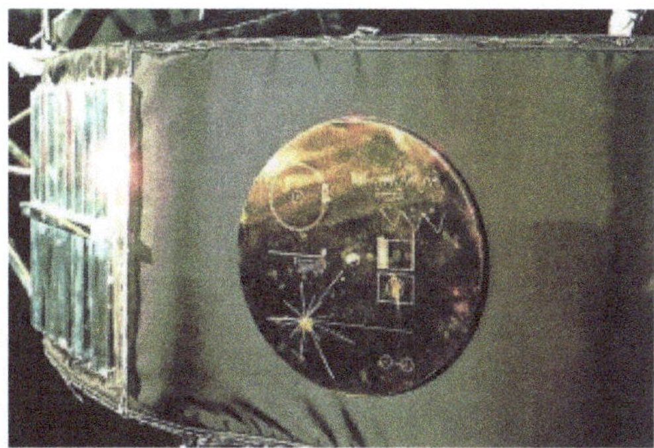

Fig.21.30: *A plaque on the Voyager spacecraft, designed to accommodate evolutionary belief by informing aliens about us.*

'... Edward Thommes and Martin Duncan of Queens University in Ontario and Hal Levison of the Southwest Research Institute in Colorado report a possible way to get around the problem. Maybe Uranus and Neptune began forming closer to the sun, where there was more material to make giant planets and timescales are much shorter ... [the article then describes their model, and how the planets migrated against the sun's gravity to their current positions, and quotes other scientists who explain why it won't work].

"'It's clear that our level of sophistication of studying planet formation is relatively primitive"? concedes Duncan. But he adds, "*So far it's been very difficult for anybody to come up with a scenario that actually produces Uranus and Neptune.*"?'

Here we see the true heart of the matter. The ultimate goal of the evolutionist is to 'come up with a scenario' of *how the universe formed by itself,* without a Creator. Sadly, they often seem to believe that the mere act of making up such a story proves that it all actually happened that way. It doesn't even have to be a *good* story; in each issue of *Creation,* we see the wildly implausible scenarios that evolutionists have to accept in order to maintain their belief system.

Indeed, in this writing we've seen that, instead of acknowledging their Creator, evolutionists would rather cling to a story that *denies the very objects that it's supposed to explain!*

Ultimately, it really doesn't matter if somebody eventually is able to 'come up with a scenario' about the formation of Neptune. Our outlook on life should not depend on whether or not someone has been able to make up a good story.

For thousands of years, sinful man has been shaking his fist at God, and inventing fables about how we all got here, without a Creator. Today, the stories are more sophisticated, and often backed up with impressive-seeming computer simulations, but it's really the same thing as before.

The 'gas and dust' story is about as good as evolutionary models get—it's been around in various forms for hundreds of years, hundreds of very intelligent people have worked on various aspects of it, and almost all evolutionary astronomers today believe it. Yet even though this 'well-proven' model pretends to explain the origin of the planets, it (embarrassingly enough) still predicts that some of those planets can't exist.

Why then would we want to put our faith in these sorts of fables, invented by sinful man? Far better to place our faith in the living Word of God, the Bible. Its historicity, accuracy and reliability are above reproach!

SOLAR SYSTEM ORIGIN

Nebular Hypothesis

According to the eye-witness account in Genesis, God created the earth on Day 1, and the sun and moon on the Fourth Day, most likely along with the planets. However, evolutionists reject a Creator *a priori,* so need to come up with another explanation.

The leading candidate is called the *nebular hypothesis.* This proposes that the sun, the earth and the rest of the solar system formed from a *nebula,* or cloud of dust and gas. The best-known pioneer of this was French atheistic mathematician Pierre-Simon Laplace (1749–1827). However, despite the dogmatic support by evolutionary astronomers, it has a number of huge problems.

Origin of Stars

First of all, if the collapsing cloud theory can't even explain the sun alone, then it is doomed from the start. To form the sun, or any star, a cloud must be dense enough to collapse and compress the interior so that it becomes hot enough for nuclear fusion to start. But most gas clouds have a tendency to expand rather than contract.

The British mathematician and astrophysicist James Jeans (1877–1946) calculated how massive a cloud must be so that gravity can overcome the tendency for gas to expand.

The main points are: high density favors collapse, and high temperature favors expansion. The minimum mass he calculated relates to both of these, and is now called the *Jeans Mass* (M_J).

But according to the big bang theory, at the time the first stars were formed, the temperature was so high that the required Jeans Mass would be about 100,000 suns. This is about the same mass as a globular cluster, i.e., no cloud less massive than this could have collapsed into a star, thus no star could have formed this way. Abraham Loeb, of Harvard's Center for Astrophysics, says, "The truth is that we don't understand star formation at a fundamental level."

Origin of Planets

According, stars alone can't be explained by such naturalistic conjectures. However, the planets are even more problematic, with several additional problems.

Angular Momentum

One major problem can be shown by accomplished skaters spinning on ice. As skaters pull their arms in, they spin faster. This effect is due to what physicists call the *Law of Conservation of Angular Momentum.*

Angular momentum = mass × velocity × distance from the center of mass,

and always stays constant in an isolated system. When the skaters pull their arms in, the distance from the center decreases, so they spin faster or else angular momentum would not stay constant.

In the formation of our sun from a nebula in space, the same effect would have occurred as the gases allegedly contracted into the center to form the sun. This would have caused the sun to spin very rapidly. But our sun spins very slowly, while the planets move very rapidly around the sun. In fact, although the sun has over 99% of the mass of the solar system, it has only 2% of the angular momentum.

This pattern is directly *opposite* to the pattern predicted for the nebular hypothesis. Evolutionists have tried to solve this problem, but a well-known solar system scientist, Dr Stuart Ross Taylor, admitted when discussing the angular momentum problem that "a predictive theory of nebular evolution is still lacking."

SUN'S AXIAL TILT

If the sun and the planets were formed by a collapsing nebula, then the sun should be spinning in the same plane as the planets. However, its axis is tilted 7.167° away from the ecliptic, which is defined by Earth's orbit. A better comparison would be Jupiter's orbital plane, since it has most of the planetary mass and angular momentum of the solar system. Jupiter's orbital inclination is 1.308° from the ecliptic, so this still leaves almost 6° difference.

The anomalous tilts of the planets are usually explained by invoking collisions, but this would not apply to the sun.

ROCKY PLANETS

Evolutionary astronomers believe that the planets arose from collisions of dust particles which melted and stuck together to form larger blobs of molten rock. These blobs further accreted to form larger and larger blobs till the inner planets were formed: Mercury, Venus, Earth and Mars. However, research has shown that the rocks would not melt, but most likely "simply zoom past each other or collide and recoil like snooker balls."

GAS GIANTS

The huge planets Jupiter and Saturn are meant to have formed far enough away from the sun so that ice could condense. This would mean extra mass, thus strong enough gravity to suck in gas from the nebula. But Jupiter's core turns out to be too small to do this. And simulations indicate that the solar nebula would have dissipated before the core had a chance to grow big enough. Furthermore, the nebula would be so unstable that the planets would spiral into the sun.

When it comes to the 'Ice Giants', Uranus and Neptune, problems are even more acute, as one evolutionary astronomer admitted:

" ... astronomers who model the formation of the solar system have kept a dirty little secret: Uranus and Neptune don't exist. Or at least computer simulations have never explained how planets as big as the two gas giants could form so far from the sun. Bodies orbited so slowly in the outer parts of the solar system that the slow process of gravitational accretion would need more time than the age of the solar system to form bodies with 14.5 and 17.1 times the mass of Earth."

RETROGRADE MOTION

The nebular hypothesis predicts that as the nebula spiraled inwards, all the resulting planets and comets would rotate and orbit in the same direction (*prograde*). But Venus rotates in the opposite direction, called *retrograde*. Furthermore, a comet was discovered with a retrograde *orbit*, and a recently discovered extra-solar stellar system has planets in retrograde orbits, opposite from the star's spin. One secular article stated:

That finding is inconsistent with the view that planets are formed by the condensation of dust from a disk surrounding a newly formed star. Some other planets were found to have highly tilted orbits that are also at odds with conventional theory.

Although the nebular hypothesis is accepted uncritically by many evolutionists, there are severe problems with forming both the sun and the planets from a collapsing cloud. The best explanation is still, "**By the word of the LORD the heavens were made, and by the breath of his mouth all their host**" (*Psalm 33:6*).

A Lesson from Pluto

Because of perceived irregularities in the motion of Uranus, Percival Lowell (1855–1916), the founder of the observatory, believed in the existence of a ninth planet. He dubbed it Planet X and calculated that it would be six times more massive than Earth. He even specified its location. Lowell searched for the planet without success from 1906 until he died.

Fig.21.31: Pluto orbits 40 times further from the sun than Earth, and for over 70 years it has been regarded as the ninth planet of our solar system. Clyde Tombaugh (1906–1997), Figure 21.32, discovered Pluto in 1930 by comparing photographs of stars taken two weeks apart at the Lowell Observatory in Arizona. Image NASA, Figure 21.31.

Tombaugh was hired by the observatory in 1929 and discovered the planet near where Lowell suggested. This apparently vindicated Lowell's predictions so the discovery was appropriately announced on Lowell's birthday (13th March) and the first two letters of Pluto's name are his initials.

Pluto is so faint that it can only be seen with a telescope 30 cm (12 in) or larger, and astronomers were unable to determine its size and mass. Early estimates could rely only on the deviations of the orbits of Neptune and Uranus. The size was quickly revised down from Lowell's estimate, and eventually astronomers settled on a mass about three quarters that of Earth.

All this changed around 1978, nearly 50 years after Pluto's initial discovery. The key evidence was found by James Christy of the US Naval Observatory when he realized that Pluto has a moon. He noticed that some of the images from their 1.5-meter telescope showed Pluto slightly elongated but the stars in the same photographs were not. From those images he was able to estimate the diameter of the moon's orbit and its orbital period. As a result, astronomers could calculate the mass of Pluto with far more certainty. It is now accepted that Pluto is only 1/500th the mass of the earth. Ongoing observations confirmed Pluto's moon, and the International Astronomical Union gave it official status in 1985 and named it Charon.

With such a tiny mass, Pluto could not possibly have affected the orbits of the gas giants Uranus or Neptune. In 1983 astronomers searched the entire sky by the Infrared Astronomical Satellite but no hidden planet was found. It is now generally believed that the perturbations to the orbits of Uranus and Neptune were imaginary, that Lowell's calculations were wrong, and Tombaugh's discovery was a coincidence, Figure 21.32.

How could so many scientists be so wrong for so long about the mass of Pluto—by a factor of 400? A similar question is often asked when creationists speak of the earth being only 6,000 years old instead of the generally accepted age of 4,600 million years.

The mass of Pluto, like the age of the earth, has not been measured directly. It is calculated from scientific models that are all based on *assumptions*.

All the scientists got the same wrong answers because they all used the same models and the same assumptions. However, ongoing observations of the behavior of Pluto led to more information that enabled an entirely different approach to the problem, overturning the previous assumptions and coming up with a radically new and soundly-based estimate.

Fig.21.32: *FClyde W. Tombaugh at the door of the Pluto discovery telescope, Lowell Observatory, Arizona. Lowell Observatory Archives*

There is another big difference. The mass of Pluto is *operational* science, where we can continue to make observations in the present using newer and better instruments and technology. But the age of the earth is *historical* science. We cannot travel back in time to make observations of things that only happened in the past. For information about the past, we need reliable reports from eyewitnesses.

PLUTO CONTRADICTS THE NEBULAR HYPOTHESIS

Pluto belongs to a class of objects that orbit the sun beyond Neptune, called TNOs (Trans Neptunian Objects). Astronomers regard these as material left over from the gas and dust nebula from which the solar system supposedly formed, supposedly 4.6 billion years ago.

But Pluto is a problem for the nebular hypothesis.

- First, it does not orbit in the same plane as the other planets (i.e., the ecliptic) but at an angle of 17°. Why not?
- Second, its axis of rotation is not perpendicular to its orbital plane but tilted so that it points almost directly at the sun at present. How come?
- Third, Pluto's orbit is not circular but highly elliptical.

In fact, it occasionally comes closer to the sun than Neptune. Why? These features of Pluto contradict the predictions of the nebular hypothesis, so astronomers have had to invent *ad hoc* secondary stories to explain them. So much for the nebular hypothesis.

Pluto and its moons don't support the idea of billions of years either. Analysis of light from Charon suggests that its surface is covered with active volcanoes of ammonia-rich water spewing out of the moon's deep interior. Similar conclusions have been reached for many TNO's. This means there must be a source of internal heat within these objects. But if they are billions of years old, they should have been cold and dead billions of years ago.

Venus: Cauldron of Fire

Venus is often called the 'morning star' and 'evening star,' since it is the brightest natural object in the sky after the Moon, but is visible at night within only about three hours of sunrise or sunset, Figure 21.33.

The ancients named it after the Latin goddess of Love. Venus is the second planet from the Sun, has the most circular orbit of all planets, and is the planet that approaches closest to Earth—42 million km (26 million miles). It is so similar to Earth in many ways that it could be considered our sister planet. But as will be seen, there are also huge differences—Earth is designed for life, while Venus is about the nearest, we have to the medieval descriptions of Hell, so ancient identification of Venus with Lucifer seems inadvertently apt.

Fig.21.33: Radar beams penetrated through Venus's thick cloud layers to reveal these surface images of both sides.

Venus's Origin According to an Eye Witness

For the truth about the origin of anything, it helps to have a reliable eye-witness record. Such a record always outweighs any circumstantial evidence that might be interpreted in another way. Genesis claims to be a witness of One who was there—the Creator. Genesis 1:14–19:

'And God said, Let there be lights in the expanse of the heavens to divide the day from the night; and let them be for signs, and for seasons, and for days, and years:

'And let them be for lights in the expanse of the heavens to give light upon the earth: and it was so.

'And God made two great lights; the greater light to rule the day, and the lesser light to rule the night: he made the stars also …

'And the evening and the morning were the fourth day.'

The Hebrew word for star, כוכב (*kokab*), refers to any bright object in the sky, so includes objects that become 'shooting stars' (meteors), planets of our solar system and, by extension, any planets around other stars.

NEBULAR HYPOTHESIS

Evolutionists believe that the solar system condensed out of a cloud of gas and dust called a *nebula*, hence the *nebular hypothesis*. There is no way to reconcile this with the Biblically revealed order of events, and there are many scientific problems with the theory as well.

Venus provides a major problem—the nebular hypothesis predicts that as the nebula spiraled inwards, all the resulting planets would rotate in the same direction (*prograde*). But Venus rotates in the opposite direction, called *retrograde*. Evolutionists once tried to explain this away by proposing that Venus rotated prograde at first, but it had a bulge on which gravitational tidal forces on Earth could act, and turn the rotation around. Aside from the weakness of tidal forces, which decrease with the *cube* of the distance, it is now known that Venus is even rounder than Earth so there is no bulge on which to act.

Venus's chemistry is also very different from Earth's. For example, the ratio of the isotopes of the inert gas argon (used to fill light bulbs), i.e., ^{36}Ar to ^{40}Ar, is 300 times greater than on Earth. If the nebular hypothesis were correct, then it would mean a vastly different composition in the region of Venus, and such differences are implausible in a relatively small region of a nebula.

Another problem for evolutionary theories about the planets is their magnetic fields. As has been shown elsewhere, Earth's magnetic field is a good example of design, and the field's decay (as well as the evidence for rapid magnetic field reversals) is excellent evidence for a young Earth. But no spacecraft has detected any magnetic field on Venus, and the sensitivity of the instruments places an upper limit on any magnetic field of 25,000 times weaker than Earth's.

Evolutionists believe that planetary magnetic fields are explained by self-sustaining dynamos, so they explain Venus's weak field by its much slower rotation (243 Earth days for one rotation). But Mercury also rotates slowly (58.82 days), yet its field is about five times stronger than Venus's, while Mars rotates almost as fast as Earth (1.03 days), but its field is less than 1/10,000th of Earth's.

However, the theory of the creationist physicist Dr Russell Humphreys explains all the data. When all the planets' cores were formed, they started off with a magnetic field produced by a decaying electrical current. The smaller the core and the poorer electrical conductor its material was, the faster the field would decay. It is thought that Venus has a smaller and less conductive core than Earth, so the field is now very weak, just as the Humphreys model suggests.

Venus provides yet another problem for billions-of-years beliefs: its surface, as shown from radar images from the Magellan satellite, seems very fresh. There are high mountains including Mt Maxwell (11,000 m or 36,000 ft above the mean surface level), rift valleys including one 9,000 km (5,600 miles) long, shield volcanoes, steep slopes, large rocks and smooth plains. There is no evidence for millions of years of erosion, although the thick atmosphere and huge atmospheric temperature differences would be expected to whip up huge sand and dust storms. There are also circular structures thought to be impact craters, but the mystery is that there are many fewer—only 935—than predicted by evolutionary theories. They are also fairly uniformly distributed. So, evolutionists propose that the *whole surface was recycled* due to volcanic and tectonic activity. They claim that the resurfacing ceased 800-300 million years ago, yet 84% of the craters show no sign of modification. Rather, the evidence seems best explained by recent cratering episodes during Creation Week or the Flood, as proposed by creationist astronomers Dr Danny Faulkner and Wayne Spencer.

CHANGING SHAPE

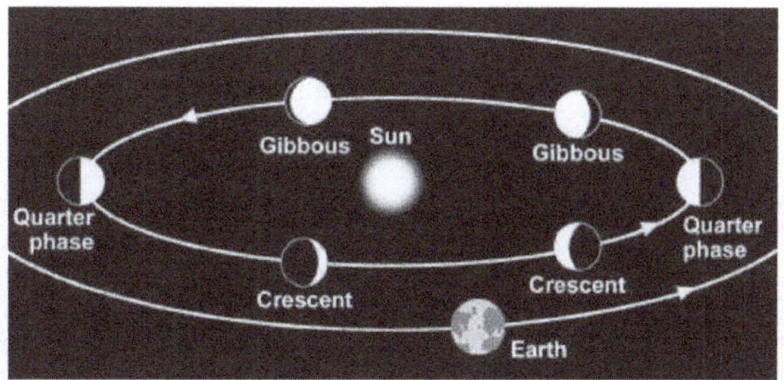

Fig.21.34: The phases of Venus as seen from the Earth.

When Galileo (1564–1642) looked through the newly-invented telescope, he discovered that Venus has phases like the Moon, Figure 21.34. But the Moon is always brightest when it is full, where we see light reflected from a whole hemisphere, and dimmer when it's in a crescent phase where we see light reflected only from a tiny fraction of the surface.

However, the brightness of Venus doesn't change that much. Galileo realized that, unlike the Moon, which orbits the Earth, Venus orbits the Sun. Thus the 'full Venus' occurs when Venus is on the opposite side of the Sun from the Earth; the extra surface reflection is counter-balanced by the far greater distance. The 'crescent Venus' is actually brighter than the full Venus because it is so much closer to us.

The time taken for Venus to return to the same orientation towards Earth, e.g., between one full Venus and the next, is called the phase cycle or *synodic* period. Showing that so-called primitive cultures were nothing of the kind, the Mayans of South America calculated the synodic period at 584 Earth days, incredibly close to the modern scientific figure of 583.92 days, and they did so without telescopes and (probably) without realizing that Venus orbits the Sun. Note that this is not to be confused with the *sidereal* period, the time for a complete orbit of Venus around the Sun, relative to an observer outside the solar system, which is 224.7 days. This is consistent with the purpose of the heavenly bodies given in Genesis 1:14.

SIMILAR, YET DIFFERENT

As shown in 'Venus Facts,' **Table 21.1**, Venus and Earth are almost the same in size, density and gravity, and Venus is almost 3/4 Earth's distance from the Sun. Yet this seemingly small difference in distance makes a huge difference to the temperature.

Earth is at an ideal distance from the Sun, and also has an atmosphere that provides a greenhouse effect, mainly due to water vapor (0.15%) but also to carbon dioxide (CO_2, 0.03%). This means that most of the Earth is in the narrow temperature range that allows water to be *liquid*. It is the only place in the universe *known* to have liquid water, aside from the *possibility* that Jupiter's moon Europa has an underground ocean.

Conversely, Venus, only a little closer to the Sun, was hot enough to drive out CO_2 from carbonate rocks (e.g., limestone) and so has had a *runaway* greenhouse effect, of a type that could never happen on Earth. It has a thick atmosphere of CO_2, and a pressure 90 times that of our own atmosphere, or as much pressure as at a kilometer's depth in Earth's ocean. This results in a surface so hot that it would melt lead and glow red. None of the large molecules required for life would stand a chance.

Also, the thick clouds reflect 76% of sunlight, which is why it is so bright. They appear slightly yellow because of some chemicals that absorb a little blue light. But the clouds mean that very little sunlight reaches the surface, although the Venera 9 and 10 spacecrafts were able to take photographs. However, the atmosphere is so thick that light bends markedly, so that an observer (extremely well protected!) on Venus would see some light even at night. Even more amazingly, the atmosphere would bend the light so much that an observer in any location could see the entire surface of Venus—it would seem like being at the bottom of a vast bowl.

If that weren't enough, the clouds do not comprise water droplets like Earth's, but instead contain concentrated sulfuric acid (H_2SO_4), and possibly some iron chloride ($FeCl_3$) crystals.

Table 21.1: Venus Facts

Mean distance from sun	108.2 million km or 67.6 million miles (72.1% Earth)
Eccentricity of orbit	0.007 (Earth's = 0.017)
Diameter (excluding atmosphere)	12,102.5 km or 7500 miles (94.9% Earth)
Mass	4.87 x 10^{24} kg (81.5% Earth)

Density	5.24 g/cm³ (94.5 % Earth)
Surface Temperature	472°C (882 °F)
Surface Pressure	90 bars (Earth's = 1 bar)
Sidereal orbital period (around sun, i.e. year)	224.7 Earth days, 61.6% Earth year
Synodic period (phase cycle seen from Earth)	583.92 Earth days
Orbital inclination	23° 27′ (Earth's = 0 by definition)
Rotation period (day)	-243.0 ± 0.1 Earth days (negative = retrograde)
Inclination of equator to orbital plane	3° (*cf.* Earth 23° 27′)
Atmospheric composition	>96% CO_2, 3.5% N_2 (Earth 78% N_2, 21% O_2, 0.9% Ar)
Gravitational acceleration at surface	8.869 m/s² (87.7% Earth's)

Source: 'Venus', The New Encyclopædia Britannica, ***12****:311; 'The Solar System, Venus'* ***27****:524–530, 1992; Britannica© CD 2000, 1994–1998. Encyclopædia Britannica, Inc.*

Venus is a beautiful-looking heavenly object, created as a sign and a marker of times, and provides a stumbling block against evolutionary theories. But the outward beauty hides an almost unimaginably harsh interior, and teaches us how finely God tuned the Earth's orbit to support life.

NEBULAR FAILURE

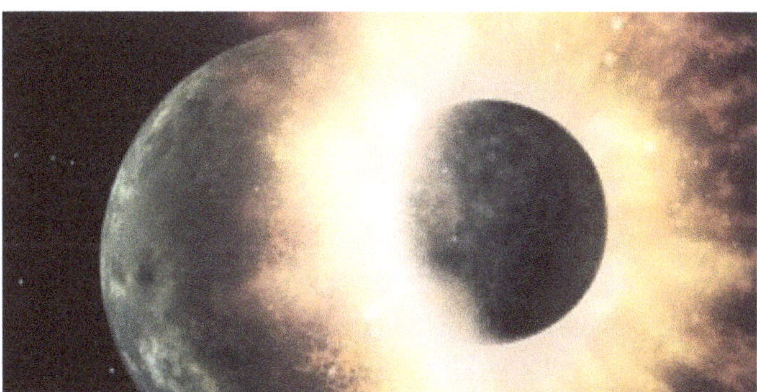

Fig.21.35: Planets crashing together - NASA/JPL-Caltech

"Planets crashing together, moons being ripped apart, heavenly bodies forming, shattering, and re-forming again, Figure 21.35. According to evolutionists, such catastrophes played a large role in the development of our solar system." This is fanciful story.

THE BIBLICAL ASTRONOMY

The Bible tells us that after creating the earth, God created the rest of the heavens during Day 4 of Creation Week. This would include all the magnificent creation we observe in our solar system: the sun, planets, moons, asteroids, comets, and other objects.

Conversely, the standard evolutionary model states our solar system formed from a cloud of gas and dust: a ***nebula***. The gas and dust condensed into rocks, then the rocks stuck together to become planets. This idea is called the '*nebular hypothesis*.'

Secular astronomers often apply the word 'evolution' to their long-ages, non-creationary model. The evolutionary model is promoted endlessly in textbooks, science magazines, television programs, and so on. However, despite the beautiful artwork and computer animations that show how it all happened, there's a glaring problem with it, as it can't produce the solar system we see today.

THE FAILURE OF THE NEBULAR HYPOTHESIS

Our solar system contradicts the nebular hypothesis in many ways. Here are a few of them, along with the typical solutions proposed by evolutionists.

1. **Mercury** is too dense: According to evolutionary theory, it must have formed at a much lower density. Therefore, reasons the evolutionist, it *did* form at a much lower density. Later, a massive asteroid smashed into it, and all the lighter material was blasted away. The material left behind is what we see today.
2. **Earth** has a moon; however, the nebular theory can't explain where it came from. Therefore, reasons the evolutionist, it didn't exist at first. Later, a massive asteroid crashed into the earth at *just* the right angle and speed. The debris sprayed into space, and some of it turned into our moon. This hypothetical asteroid, which of course doesn't exist today, has even been named—Theia.
3. **Venus** doesn't have any moons. However, if the earth got its moon in an asteroid collision, then Venus should have one too. Venus and Earth are neighbors in space, so the nebular hypothesis says they should have similar histories. Therefore, some evolutionists propose that Venus did get a moon from such a collision. Why don't we see this moon today? Because a *second* asteroid collision destroyed it.

Some evolutionists propose a catastrophe to solve yet another problem. Venus rotates *retrograde*, or backwards when compared to the other planets. Since this contradicts the nebular hypothesis, some evolutionists have proposed that Venus initially rotated in the 'correct' direction. Then a massive asteroid collision spun it around the other way.

1. **Mars** has a very thin atmosphere today. However, for various reasons, evolutionists want Mars to have had a thick atmosphere in the past. The answer, as you might have guessed by now, is that a massive asteroid collision disrupted the planet. As a result, it lost its atmosphere.
2. **Jupiter** has many 'irregular' moons. Most are retrograde, orbiting in the opposite direction of the planet's rotation. None of them could have formed in their current orbits, according to the nebular hypothesis.

Most evolutionists believe these objects formed elsewhere. Later, they were captured by gravity into their current orbits. However, such captures are extremely unlikely, and over 90 irregular moons are currently known. A favored solution is to appeal to collisions with other objects.

1. **Saturn** also has many irregular moons. These are also explained as the result of captures and collisions.
2. **Uranus** rotates on its side. Unlike the other planets, which spin like tops as they move through space, Uranus rolls along like a ball. According to the nebular hypothesis, it can't have formed this way. Therefore, reasons the evolutionist, it formed the 'correct' way. Later, a massive collision knocked it over on its side. Then it supposedly captured its moons, because their orbits are likewise sideways.

Uranus also has an unusual-looking moon named Miranda. To explain its features, some evolutionists invoke not one, not two, but *five* collisions.

1. **Neptune** has a large retrograde moon named Triton. Again, this is contrary to the nebular theory. Again, a collision is invoked to explain away the problem.

According to one version of the story, Triton used to be a moon of a planet named Amphitrite, until Neptune stole it from the smaller planet. Of course, there is no planet named Amphitrite today. There's not even a trace of it. Why? Because it allegedly collided with either Neptune or Uranus and was destroyed.

SCIENCE OR FICTION

As you can see, collisions, Figure 21.36, are invoked to explain away a long list of problems for the nebular hypothesis.

Creation scientists are frequently charged with believing in a model that is 'unscientific'. This charge is false, of course. The Bible is consistent with the physical world we live in.

On the other hand, how scientific is the evolutionary model? Science is supposed to be based on evidence. But the only 'evidence' for most of these collisions is that if they hadn't occurred, the nebular hypothesis would be disproved!

Observe also that evolutionists go so far as inventing (and even naming) planets which don't exist, and for which there is no evidence whatsoever. At the same time, they must explain why certain planets (Jupiter, Saturn, Uranus, and Neptune) *do* exist, when the evolutionary model says they cannot.

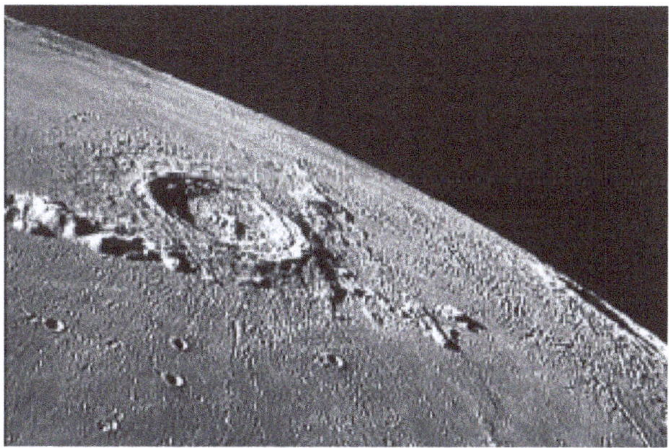

Fig.21.36: This crater is evidence for a small impact on our Moon. Curtsy - NASA

What can we learn from all this? Although evolutionists claim to base their model on science, the reality is quite different. The solar system as we see it today—in other words, the actual scientific evidence—contradicts the nebular hypothesis. To rescue their model from the facts, secular astronomers are forced to invent a long series of 'just so' stories.

There's also more than a hint of hypocrisy here. Creation scientists are often criticized for believing in Noah's Flood. Since the Flood was a one-time catastrophe, it is non-repeatable. Therefore, say many evolutionists, it is outside of science. But where is the outrage for the endless series of non-repeatable catastrophes in the nebular hypothesis?

DENYING HISTORICAL FACTS

When you deny the truth, you must believe a falsehood. As secular astronomers deny the Bible, they cannot base their model on the solar system's true history. Thus, their model cannot be correct. They're left with a series of just-so stories and self-contradictory assertions.

It's far better to acknowledge our Creator as we observe the magnificent solar system He has made. The heavens truly declare the glory of God (Psalm 19:1).

Do creationists deny that minor collisions have occurred in our solar system? Of course not. We see evidence for them in many places. Many heavenly bodies, including the earth itself, have features such as craters and impact basins.

Nevertheless, we know that these collisions occurred because they left *evidence*. Conversely, there is no evidence for most of the collisions that are necessary to rescue the nebular hypothesis from the facts.

Indeed, the opposite is often true. The evidence suggests that many of the alleged collisions could never have happened. As one example, a recent analysis of lunar soils revealed that the Moon cannot have come from an earth-shattering collision. As another example, the moons of Uranus cause great difficulties for believing in a collision that tilted the planet.

JUPITER - PLANETS' KING

Testament to our Creator

Jupiter is the largest planet in our solar system. This gargantuan object dwarfs the earth—indeed, over 1,300 Earth-sized objects could fit inside Jupiter. And its mass is 2.5 times that of all the other planets combined.

Where did this magnificent planet come from?

The Bible tells us that Jupiter, along with other heavenly objects, was created on Day 4 of Creation Week (*Genesis 1:14–19*).

However, evolutionary astronomers deny the biblical account. They claim that Jupiter formed by natural processes about 4.6 billion years ago. Unfortunately for them, Jupiter poses enormous problems for those wishing to deny the creation of our solar system.

Jupiter Giant Gas Planet

Unlike Earth, which is mostly rock, Jupiter is mostly gas. It might have a small rocky core deep inside—we don't know for sure.

Jupiter is about five times as far from the sun as Earth is. Because of this vast distance, Jupiter appears in our nighttime sky as a bright white star. However, thanks to its massive size, even a modest telescope will reveal that it isn't a star, but a planet—it looks like a disk, while stars look like points of light.

Jupiter's most famous feature is probably its Great Red Spot. This is an enormous, violent storm system—it's even larger than Earth! This storm has been raging continuously for at least 300 years. For all we know, it might have been in place since the planet's creation.

THE EXISTENCE OF JUPITER

Jupiter is more than just a beautiful object in the sky, Figure 21.37. Its very existence also poses an enormous challenge for those who want to believe in an evolutionary origin of our universe.

The standard evolutionary explanation for our solar system is that it formed from a swirling cloud of gas and dust. About 4.6 billion years ago, this cloud collapsed into a disk shape. The dust condensed into grains, the grains allegedly stuck together to become small rocks, and the small rocks stuck together to become larger rocks. But a major problem for this view is that fast moving rocks are more likely to bounce off each other rather than stick, Figure 21.37.

Fig.21.37: *Jupiter and Earth shown to scale. Curtsy - NASA*

According to the evolutionists, the rocky planets like Venus and Earth formed as these large rocks collected together. The gas giants like Jupiter and Saturn formed initially in the same way. Nonetheless, in contrast to the inner planets, the embryonic giant planets were far enough from the sun for ice to condense. Therefore, extra mass could accumulate—over 10 times as much material as the entire earth contains today. With the help of the ice, this accumulation had so much gravity that gas was pulled onto them, eventually forming the gas planets which we see today. Since the collection of rocky chunks became the cores of the gas planets, this idea is called the 'core accretion' model.

This story is still being told today on television, in books and magazines, in science videos, and so on. However, scientists have known for a long time that this model isn't true.

A FALSIFIED MODEL

The core accretion model has at least four fatal problems.

Problem 1.

It makes definite predictions about the chemical composition of Jupiter. However, back in 1995, the Galileo spacecraft dropped a probe into Jupiter's atmosphere. Evolutionists were shocked to discover that Jupiter's atmosphere contained high amounts of certain gases (argon, xenon, and krypton). Evolutionary models say that these elements can't be there in such high concentrations.

As one report by an evolutionary astronomer explained:

'Jupiter is the largest of all the planets. But results … now reveal the embarrassing fact that we know next to nothing about how—or where—it formed.'

Problem 2.

This model requires Jupiter to have a rocky large core—with 10 to 30 times as much mass as the entire Earth, Figure 21.38. However, the Galileo spacecraft discovered that Jupiter's core can't be this large—not even close. At most, it would be only as massive as six Earths. It might also not exist at all.

Fig.21.38: A storm on Jupiter has been raging continuously for at least 300 years. NASA

Problem 3.

The model requires at least 10 million years for enough rocks and gas to accumulate to form Jupiter. Some scientists say it would take even longer—several hundred million years at least. But scientists also acknowledge that a disk of dust and gas wouldn't have lasted around our sun for that long. Many scientists believe that such a disk would have dissipated in less than 5 million years—leaving no time for Jupiter to form.

Problem 4.

Even if the gas/dust disk lasted long enough, we still wouldn't get a Jupiter from it. Recent computer simulations have revealed a devastating problem with the core accretion model. As gas giants formed within the disk, they would have interacted gravitationally with the dust remaining in the disk. It turns out that these interactions would pull the developing planets inward, towards the sun.

Therefore, both Jupiter and Saturn would have swirled inwards until they slammed into the sun. And this would occur 'quickly' in evolutionary terms: only 300,000 years after they started to form.

Obviously, according to evolution, Jupiter shouldn't be there at all. It's no wonder that evolutionists make complaints like this one:

'Building Jupiter has long been a problem for theorists.'

Or this one:

'I don't think the existence of Jupiter would be predicted if it weren't observed.'

Jupiter is a wonderful illustration of an important principle in science.

REJECT THE TRUTH - ACCEPTING LIES

Evolutionists have already rejected the biblical account of Creation. Therefore, they have to accept the 'best' evolutionary alternative. Even though their model has been denied by multiple lines of evidence, it 'must' be true.

Over and over again, the evolutionary model has been exposed as a falsehood. As one evolutionist lamented recently, '*... most every prediction by theorists about planetary formation has been wrong.*'

Another complained, '*The thing we've learned in the last couple years is that the standard model cannot work.*'

Yet evolutionists still cling to this model, despite the overwhelming evidence against it. They're unwilling to accept the truth that Jupiter illustrates about its Creator.

Pity the self-deceived evolutionist who is so committed to a bankrupt theory that he cannot see his own Creator's handiwork in this majestic planet.

Truly, Jupiter's size, **Table 21.2**, magnificence, and grandeur are a wonderful testament to our Creator—the God of the Bible, who not only made the stars and planets (for our benefit—*Genesis 1:14–19*), but also us!

Table 21.2 : Jupiter facts*

Mean distance from sun	778.0 million km or 483.4 million miles (5.20 × Earth)
Eccentricity of orbit	0.049 (Earth's = 0.017)
Diameter	Equator 142,800 km (11.2 × Earth); polar 133,400 km
Mass	1.8986×10^{27} kg (317.8 × Earth, $^1/_{1,047}$ Sun)
Volume	1.43128×10^{15} km^3 (1321.33 × Earth, $^1/_{990}$ Sun)
Mean Density	1.33 g/cm^3 (24% Earth, 94% Sun)
Surface Gravity	24.79 N/kg (2.530 × Earth)
Escape Velocity	59.5 km/s (5.32 × Earth)
Sidereal orbital period(around sun, i.e. year)	11.86 Earth years
Orbital inclination	1°19′ (Earth's = 0 by definition)
Rotation period (day)	9 hr 55.5 min (41.5% × Earth)
Axial tilt	3°4′ (cf. Earth 23°27′)
Atmospheric composition	~90% H_2, ~10% He (Earth 78% N_2, 21% O_2, 0.9% Ar)

Magnetic field strength at equator	4.3 gauss (13.8 × Earth's surface field)
Number of moons	63 confirmed
* *'The Solar System, Jupiter',* The New Encyclopædia Britannica *27:561, 1992; Jupiter,* Wikipedia, *<en.wikipedia.org/wiki/Jupiter>, 7 February 2008; Jupiter Fact Sheet, NASA, 7 February 2008.*	

APPENDIX

i. The first nebular hypothesis is usually attributed to Pierre Laplace in 1796, although Immanuel Kant had proposed a similar idea 40 years earlier. 20 years before that, the mystic Emmanuel Swedenborg claimed he got a similar idea from a seance!

iii. Actually, the problem is even worse than it appears; we are not only lacking enough time to form Neptune, but the planetesimals, etc. from which to build it aren't around anymore. Notice that the models require that the planetesimals would have dissipated long ago (to explain the lack of them today), but simultaneously need them around for thousands of millions of years into the future, in order to eventually build Neptune.

iv. Recent attempts to solve the migration problem have only made matters worse. One suggested answer is that Jupiter is only the last of a series of gas giant planets, each of which did migrate inwards into the sun. Jupiter was merely the last one in the series, which happened to form right as the gas/dust disk was dissipating. Even many evolutionists are bothered by the obviously *ad hoc* nature of this fable—there's no evidence for it, and its only purpose is to rescue evolution from the facts. Plus, Jupiter couldn't have formed from a disk that was depleted by previous generations of gas giant formation. So this suggested solution solves nothing. Another suggestion is the new 'disk instability' model, which proposes that the gas/dust disk collapsed quickly into planets, before migration could occur. This model has a long list of fatal problems: Uranus and Neptune don't match its predictions, the comets and other trans-Neptunian objects don't match its predictions, and the model's proponents have yet to demonstrate that rapid disk collapse is even possible anyway.

v. According to the evolutionary model, these elements could only be present in such concentrations if Jupiter had formed further out in the solar system—at more than 10 times Jupiter's current distance from the sun. Some have suggested that maybe Jupiter formed at that distance, and then moved inwards later. However, this doesn't solve the problem. There wouldn't have been enough material at that distance for Jupiter to form. Also, Jupiter doesn't contain the heavy elements that it would have, if it had actually formed way out there.

vi. [At the time of writing] I am yet to be convinced that such *extrasolar planets* exist, because of many examples of 'crying wolf' and wishful thinking—see Kalas, P., Dusty disks and planet mania, *Science* **281**(5374):182–183, 10 July 1998. However, some creationists think that some are genuine, e.g. Spencer, W., The existence and origin of extrasolar planets, *TJ* **15**(1):17–25, 2001. All agree that they present new problems for evolutionary theories of the origin of the solar system. For example, the best established claims are of gas giants bigger than Jupiter but orbiting their star closer than Mercury does our Sun. At this distance, ice would evaporate, whereas it is thought that ice is essential so that a growing planetoid would become massive enough to attract gas. Also, gas would have been blown away when the early star goes through the T-Tauri phase. This is a problem even for the alleged evolutionary formation of the huge planets in our own system many times further from the Sun than Earth, so how much more would it be a problem for these alleged extrasolar gas giants? So speculative theories are proposed in which giant planets migrate millions of miles inward after formation. [I now think that genuine extrasolar planets exist, and are still problematic for evolutionary theories. See First light from extrasolar planets and Planets and migrating theories.]

vii. Jerome, who produced the Vulgate, the major Latin translation of the Bible, translated the Hebrew word *heylel* in Isaiah 14:12 as 'Lucifer', although it simply means 'morning star' or 'star of the morning'. Lucifer literally means 'light-bearer' and is one name of the angelic being who fell and became Satan (= 'adversary'). Some older English translations reflect the Vulgate influence in this verse.

viii. Creationist physicist Dr Keith Wanser pointed out that the rate of energy loss of a pulsar due to gravitational radiation is proportional to c, according to General Relativity. The 1993 Nobel Prize in Physics was awarded to Russell Hulse and Joseph Taylor for discovering a binary pulsar and showing that the observed energy loss matched the predictions of General Relativity to within 0.4%. But this indicates that c hasn't changed in the thousands of years since light left that pulsar.

x. The demonstrable usefulness of GR in the physics of time-keeping, for example, can be separated from certain 'philosophical baggage' that some have illegitimately attached to it, and to which some Christians have objected, thinking that such relativity in physics in some way supported relative morality. However, the fundamental postulate of relativity is the absoluteness of the speed of light; Einstein actually wanted to call it the 'Invariance Theory'.

xi. "None of the suggested mechanisms, including gas-drag, pull-down, and three-body capture, convincingly fit the group characteristics of the irregular satellites. The sources of the satellites also remain unidentified." Jewitt, D., and Haghighipour, N., Irregular satellites of the planets: products of capture in the early solar system, *Annual Review of Astronomy and Astrophysics* **45**:261–295, 2007. Abstract available at arjournals.annualreviews.org/loi/astro.

xii. Satellite scientist Dr Mark Harwood points out that time dilation is most relevant to GPS navigation, because the clocks in the satellites are faster by 38 microseconds per day than clocks at sea level. This doesn't sound like much, but would accumulate errors in position at a rate of 400 metres every hour. See creation.com/starlight2, 17 January 2009.

REFERENCES

1. The big bang hypothesis has many problems; see creation.com/bigbang.
2. This explains the mass-media excitement in early 2014 when cosmologists claimed proof for inflation in gravitational waves. See Williams, A., Big Bang blunder bursts the multiverse bubble; creation.com/multiverse-bubble-bursts, 12 June 2014.
3. Brooks, M., 13 things that do not make sense, New Scientist 2491:30–37, 19 March 2005.
4. Steinhardt, P., The inflation debate, Scientific American 304(4):36–43, April 2011.
5. Wieland, C., Speed of light slowing down after all? Journal of Creation 16(3):7–10, 2002; creation.com/cdk.
6. Lisle, J., Light-travel time: a problem for the big bang, Creation 25(4):48–49, 2003; creation.com/lighttravel.
7. Norman, T.G. and Setterfield, B., The atomic constants, light and time, privately published, 1990.
8. Magueijo, J., Faster Than The Speed of Light: The Story of a Scientific Speculation, Basic Books, 2003.
9. Many billions of stars exist, many just like our own sun, according to the analysis of the light coming from them. Such numbers of stars have to be distributed through a huge volume of space, otherwise we would all be fried.

10. Creationist physics professor, Dr John Hartnett builds the world's most precise clocks at present; see creation.com/hartnett-interview.
11. Gibbs, W.W., Profile: George F.R. Ellis—Thinking globally, acting universally, Scientific American 273(4):50–55, 1995.
12. Hartnett, J., Where are we in the universe? Journal of Creation 24(2):105–107, 2010; creation.com/location-in-universe.
13. Williams, A. and Hartnett, J., Dismantling the big bang; God's universe rediscovered, Master Books, US, 2005; creation.com/dtbb.
14. See papers listed under: What are some of the problems with the big bang hypothesis? creation.com/astronomy#bigbang.
15. Wieland, C., Secular scientists blast the big bang, Creation 27(2):23–25, 2005; creation. com/bigbangblast.
16. Eric Lerner and 33 other scientists from 10 different countries, Bucking the big bang, New Scientist 182(2448):20, 2004; cosmology.info/open-letter.
17. Humphreys, D.R., New time dilation helps creation cosmology, Journal of Creation 22(3):84–92, 2008 (technical); creation.com/dilation.
18. Humphreys, D.R., Flaw in creationist solution to the Pioneer anomaly? creation.com/ pioneer-anomaly-heat, 11 May 2013.
19. Carmeli, M., Cosmological Relativity: The Special and General Theories for the Structure of the Universe, World Scientific Publishing Company, 2006.
20. Hartnett, J., A 5D spherically symmetric expanding universe is young, Journal of Creation 21(1):69–74, 2007; creation.com/5d (technical).
21. See layman's summary: Wieland, C., Starlight and time—a further breakthrough, Creation 30(1):12–14, 2007; creation.com/starlight-time.
22. Cosner, L. and Bates, G., Did God create over billions of years? creation.com/billions, 6 October 2011.
23. Batten, D. and Sarfati, J., 15 Reasons to take Genesis as history, Creation Ministries International, Australia, 2006; creation.com/15r.
24. Zimmermann, A., The Christian foundations of the rule of law in the West: a legacy of liberty and resistance against tyranny, Journal of Creation 19(2):67–73, 2005; creation. com/christianlaw. Dr Augusto Zimmermann lectures in Law at Western Australia's Murdoch University and is a Vice-President of the Australian Society of Legal Philosophy.
25. Northrup, T. and Connerney, J., A micrometeorite erosion model and the age of Saturn's rings, Icarus 70:124–137, 1987; p. 124. 2. Pollack, J. and Cuzzi, J., Rings in the solar system, Scientific American 245(5):104–129, 1981; pp. 117, 125–126, 127, 129
26. Soderblom, L. and Johnson, T., The moons of Saturn, Scientific American 244(1):101–116, 1982; p. 101.
27. Pasachoff, J., Contemporary Astronomy Saunders, Philadelphia, p. 429, 1985.
28. Jeffreys, H., On certain possible distributions of meteoric bodies in the solar system, M.N.R.A.S. 77:84–92, 1916; p. 84.
29. Jeffreys, H., Transparency of Saturn's rings, J. British Astronomical Association 30:294–295, 1920; p. 295.
30. Alexander, A., The Planet Saturn, Faber and Faber, London, p. 320, 1962, reprinted, Dover, New York, 1980.
31. Kerr, R., Making better planetary rings, Science 229:1376–1377, 1985; p. 1377.
32. Burns, J., Hamilton, D. and Showalter, M., Bejeweled worlds, Scientific American 286(2):64–73, 2002; p. 73.

33. Hartmann, W., Astronomy, Wadsworth, Belmont, CA, p. 253, 1991.
34. Dikarev, V., Dynamics of particles in Saturn's E ring: effects of charge variations and the plasma drag force, Astronomy and Astrophysics 346:1011–1019, 1999; p. 1011.
35. Hartmann, ref. 10, pp. 252–253.
36. Brush, S., Everett, C. and Garber, E., Maxwell on Saturn's Rings, MIT, Cambridge, MA, p. 7, 1983.
37. Fix, J., Astronomy, WCB/McGraw-Hill, Boston, pp. 270, 275, 1999.
38. Pollack and Cuzzi, ref. 2, p. 129.
39. Fix, ref. 14, p. 274.
40. Eberhart, J., Saturn's 'ring rain', Science News 130(6):84, 1986.
41. Fix, ref. 14, pp. 289–290.
42. Esposito, L., The changing shape of planetary rings, Astronomy 15(9):6– 17, 1987; p. 15.
43. Snow, T., Essentials of the Dynamic Universe, West, St. Paul, p. 157, 1984.
44. Laplace, P., *Exposition du Système du Monde (Exposition of the System of the World)*, 1796. Return to text.
45. Jeans Mass (M_J) = $K\rho^{-1/2}T^{3/2}$, where K is a constant, ρ is the density, and T is the temperature. Alternatively, this can be expressed as $M_J \approx 45M_\odot\ n^{-1/2}T^{3/2}$, where $M_\odot$ is the solar mass, n is the density of atoms per cm^3, and T is the temperature in Kelvins. Return to text.
46. According to big bang theory, the temperature was about 3,000 and density about 6,000, therefore $M_J \approx 10^5\ M_\odot$. Return to text.
47. Wieland, C., and Sarfati, J., "He made the stars also"—interview with creationist astronomer Danny Faulkner, *Creation* 19(4):42–44, 1997. Return to text.
48. Quoted by Marcus Chown, Let there be light, *New Scientist* 157(2120):26–30, 7 February 1998. Return to text.
49. See also Stars could not have come from the big bang , *Creation* 20(3):42–43, 1998. Return to text.
50. Taylor, S., *Solar System Evolution: A New Perspective*, 2nd edition, Cambridge University Press, p. 64, 2001. Return to text.
51. Carroll, B., and Ostlie, D., *An Introduction to Modern Astrophysics*, pp. 890–891, Addison-Wesley, 1996. Return to text.
52. Worraker, W., The sun—the greater light to govern the day, *Origins* 37/38:11–15, 2004. Return to text.
53. Muir, H., Earth was a freak, *New Scientist* 177(2388):24, 29 March 2003. Return to text.
54. Psarris, S., Jupiter: king of the planets and testament to our Creator, *Creation* 30(3):38–40, 2008; creation.com/jupiter2. Return to text.
55. Psarris, S., Neptune: monument to creation: According to evolutionary ideas Neptune should not exist! What is its secret? *Creation* 25(1):22–24, 2002; creation.com/neptune. Return to text.
56. Naeye, R., Birth of Uranus and Neptune, *Astronomy* 28(4):30, 2000. Return to text.
57. Sarfati, J., Venus: cauldron of Fire, *Creation* 23(3):30–34, 2001; creation.com/venus. Cf. Spencer, W., The search for Earth-like planets, *J. Creation* 24(1):72–76, 2010. Return to text.
58. Sarfati, J., and Catchpoole, D., Backwards comet perplexes scientists, *Creation* 31(4): 38–39, 2009; creation.com/backwards-comet. Return to text.
59. Maugh, T., Distant planets rattle theories with their orbit, *Los Angeles Times*, 13 April 2010. Return to text.

60. Abell, G.O., Morrison, D. and Wolff, S.C., *Exploration of the Universe*, 6th edition, Saunders College Publishing, Philadelphia, USA, p. 186, 1993. Return to text.
61. Ref. 1, p. 185. Return to text.
62. Ref. 1. The name is from classical mythology and was suggested by Venetia Burney, an 11 year old school girl from Oxford, England. Return to text.
63. Using Newton's formulation of Kepler's third law. Return to text.
64. Ref. 1, pp. 288–289. Return to text.
65. See Spencer, W., Revelations in the Solar System, *Creation* 19(3):26–29, 1997.
66. Sarfati, J., The earth's magnetic field: Evidence that the earth is young, *Creation* 20(2):15-17, 1998.
67. Humphreys, D.R., The Creation of Planetary Magnetic Fields, *CRSQ* 21(3):140–149, 1984.
68. Cardno, S., The mystery of ancient man, *Creation* 20(2):10–13, 1998.
69. See also Sarfati, J., Solar system origin: Nebular hypothesis, *Creation* 32(3):34–35, 2010.
70. Alemi, A. and Stevenson, D., Why Venus has no moon, *Bulletin of the American Astronomical Society* 38:491, 2006. Abstract available at adsabs.harvard.edu.
71. "[T]he original size-frequency distribution of the irregular moons must have significantly evolved by collisions to produce their present populations." Nesvorný and two others, Capture of planetary satellites during planetary encounters, *The Astronomical Journal* 133(5):1962–1976, 2007; iopscience.iop.org/1538-3881/133/5/1962.
72. verity01.jpl.nasa.gov/sse/planets/profile.cfm?Object=Ura_Miranda, 24 May 2010.
73. Desch, S., and Porter, S., Amphitrite: A twist on Triton's capture, *LPI Contribution No. 1533*, 41st Lunar and Planetary Science Conference, held March 1–5, 2010 in The Woodlands, Texas, p. 2625. Available at www.lpi.usra.edu/meetings/lpsc2010/pdf/2625.pdf, 24 May 2010.
74. D., A biblically-based cratering theory, *Journal of Creation* 13(1):100–104, 1999; Spencer, W.R., Response to Faulkner's 'biblically-based cratering theory', *Journal of Creation* 14(1):46–49, 2000.
75. Water has been confirmed to exist in lunar soils. However, it would not be there if the moon had been formed in a giant collision. One of the scientists who discovered it said, "It's hard to imagine a scenario in which a giant impact melts, completely, the moon, and at the same time allows it to hold onto its water … That's a really, really difficult knot to untie." npr.org/templates/story/story.php?storyId=92383117&ft=1&f=1001, 24 May 2010.
76. See also Sarfati, J., Earth is 'too special'? *Creation* 28(3):42–44, 2006; <creation.com/earthspecial>.
77. Ball. P., Giant mistake, *Nature science update*, 18 November 1999, <www.nature.com/news/1999/991118/full/news991118-10.html>.
78. See comments from Alan Boss and Hal Levison in Mullen, L., *Birth of a giant: how did Jupiter get so big?*, 17 May 2001, <www.space.com/scienceastronomy/solarsystem/jupiter_origins_010517-3.html>.
79. Wetherill, G.W., How special is Jupiter? *Nature* 373(6514):470, 9 February 1995.
80. Wetherill, G.W., *The Formation and Evolution of Planetary Systems*, Cambridge University, p. 27, 1989, as quoted in Stuart Ross Taylor, *Solar System Evolution: A New Perspective*, Cambridge University Press, Cambridge, UK, p. 205, 2001.
81. Scott Tremaine, as quoted by Kerr, R.A., Jupiters like our own await planet hunters, *Science* 295(5555):605, 25 January 2002.
82. Harold Levison of the Southwest Research Institute as quoted in *Solar system makeover: wild new theory for building planets,* <www.space.com/scienceastronomy/solarsystem/planet_formation_020709-2.html>, 9 November 2007.

83. For more in depth on this topic, see *J. Creation* 14(1):3–4, 2000, and *J. Creation* 15(3):85–91, 2001.
84. Christiansen, E.H. and Hamblin, W.K., *Exploring the Planets*, 2nd Edition, Prentice-Hall Inc., New Jersey, p. 424, 1990.
85. See Humphreys, R., Beyond Neptune: Voyager II supports creation, icr.org, *Impact* 203, 1990, and Humphreys, R., The Creation of Planetary Magnetic Fields, *Creation Research Society Quarterly* 21(3):140–149, 1984. His published predictions in 1984 on the field strength were 100,000 times greater than the evolutionary ones, and his article said it would be a good test of his theory. The results were squarely in the middle of Humphreys' prediction.
86. R.N., Birth of Uranus and Neptune, *Astronomy* 28(4):30, 2000.
87. Taylor, S.R., *Destiny or Chance: our solar system and its place in the cosmos*, Cambridge University Press, Cambridge, p. 73, 1998.
88. Dormand, J.R. and Woolfson, M.M., *The Origin of the solar system: the capture theory*, Ellis Horwood Ltd, W. Sussex, p. 39, 1989.

EARTH'S MAGNETIC FIELD – EARTH AGE

FLIPPING EARTH'S MAGNETIC FIELD

A new science fiction movie is being released called *The Core*. In this movie, the Earth's core stops rotating and our planet's magnetic field collapses. Since the field shields us from dangerous charged particles from the sun, it must be restarted, so scientists are sent deep inside the Earth to jump-start the rotation.

Dr Larry Newitt, an expert on geomagnetism with the Canadian-Government–funded Geolab, says that *The Core*:

'… is composed of a few scientifically plausible ideas mixed with a large dosage of sheer nonsense. It should be fun.'

However, Dr David Whitehouse, *BBC News Online* science editor, said in an article Is the Earth preparing to flip?

'It is not just the plot for a far-fetched science-fiction disaster movie. Something unexplained really is happening to the Earth's magnetic field.'

Therefore, what are these unexplained phenomena?

EARTH'S MAGNETIC FIELD IS DECAYING

Dr Whitehouse says: *'But something else is happening to the Earth's magnetic field: it is getting weaker.'*

Also, Dr David Kerridge, of the British Geological Survey, said:

'There is strong evidence that the field is decreasing by about 5% per century.'

This is not so well known to the public, but it has been known to creation scientists since the 1970s, when electromagnetism expert (and physics professor) Dr Thomas Barnes pointed this out. By using standard physics, he showed that this was most likely due to a decaying electric current. And this decay could not have been happening for more than about 10,000 years, otherwise the current would have melted the Earth. This has long been powerful support for the Biblical timescale, Figure 22.1.

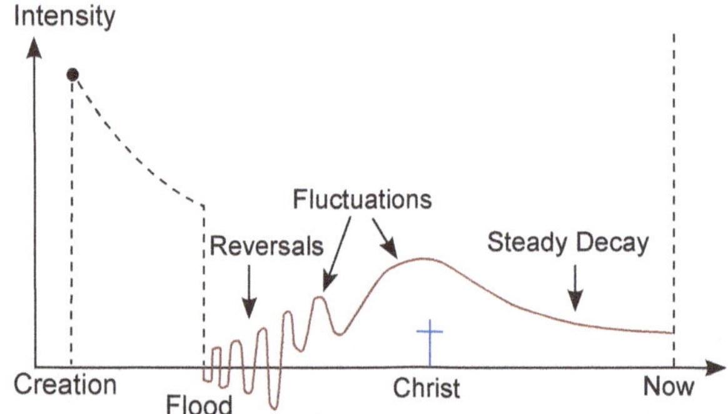

Fig.22.1: How the earth's magnetic field has changed since the earth's creation, its intensity decaying due to the energy loss of the freely decaying electric currents in the liquid outer core, but interrupted by rapid polarity reversals during the Flood and its aftermath due to the Flood's catastrophic plate tectonics changing the flow directions in the convection cells in the liquid outer core.

One way of long-agers avoiding this conclusion is to suppose that the magnetic field is generated by a self-sustaining dynamo (*electric generator*). Here, the magnetic field is somehow generated by circulating molten metal, and this of course relies on the Earth's rotation. *The Core* presupposes that this model is correct, because if the Earth's core stopped rotating, the field would be extinguished under this model, but not under Barnes' model. However, Barnes' model is based on sound physics and is rejected only because of the 'short' age implications; while there is no satisfactory dynamo theory, which is accepted mainly because it allows 'long' ages.

FIELD REVERSALS

Evolutionists have also claimed that magnetic field reversals are the answer to the problem Barnes raises, and claim that they occur over intervals of thousands of years, Figure 22.2. Whitehouse says, although not in the context of Barnes' theory:

'Looking back in the geological record it is clear that on average such events occur about every 250,000 years. However, it has been 750,000 years since the last reversal—so we are certainly overdue.'

However, evolutionists have no plausible mechanism. But creationist physicist Dr Russell Humphreys proposed that during the Genesis Flood, tectonic plates plunged down far enough to cool the outer core rapidly.

Fig.22.2: Reversals of Earth's Magnetic Field Explained by Small Core Fluctuations

The change in convection patterns would cause *rapid* field reversals, and cause the field energy to decay even faster. Humphreys predicted that such rapid reversals would be recorded in lava flows that cooled, from outside to inside, over only a few days. *This prediction was dramatically confirmed only three years later.*

In fact, Dr Whitehouse's comment is further support for this. As he says, under the uniformitarian theory, we are long overdue for a reversal. But if all reversals occurred during the Flood year, then there is a very good reason why we have not had any reversals since then.

DECLINING EARTH MAGNETIC FIELD - EXPLAIN EXPONENTIAL DECAY

'Some researchers suggest that it could be the start of a geomagnetic reversal, when the strength of the Earth's magnetic field decreases and then returns a few thousand years later with the north and south magnetic poles reversed.'

However, once again, the declining field strength is best explained by an exponential decay of the field due to a decaying electric current. While this decay will weaken the shielding effect, there is *no* experimental reason to believe it heralds another reversal.

It seems that Dr Newitt was right about *The Core*, but for some different reasons from what creationists would think. The movie should be worthwhile if it inadvertently alerts the public to the decay of the field, and its implications for the age of the Earth. It should also be a good opportunity to explain the best scientific explanation for the magnetic field and correct the errant notions of the dynamo theory promoted in the film.

THE EARTH'S MAGNETIC FIELD

Evidence that the Earth is Young

The Aurora Borealis is caused by charged particles from space striking the earth's atmosphere. These particles have been deflected towards the poles by the presence of the earth's magnetic field (which also diverts many such particles harmlessly into space), Figure 22.3.

Fig.22.3: The Aurora Borealis (Northern Lights). Curtsy Wikipedia.org

The earth has a magnetic field pointing almost north-south—only 11.5° off, Figure 22.4. This is an outstanding design feature of our planet: it enables navigation by compasses, and it also shields us from dangerous charged particles from the sun. It is also powerful evidence that the earth must be as young as the Bible teaches.

In the 1970s, the creationist physics professor Dr Thomas Barnes noted that measurements since 1835 have shown that the field is decaying at 5% per century (also, archaeological measurements show that the field was 40% stronger in AD 1000 than today), Figure 22.4. Barnes, the author of a well-regarded electromagnetism textbook, proposed that the earth's magnetic field was caused by a *decaying electric current* in the earth's metallic core. Barnes calculated that the current could not have been decaying for more than 10,000 years, or else its original strength would have been large enough to melt the earth. Therefore, the earth must be younger than that.

N

S

Fig.22.4: A 'force-field' Around the Earth.

EVOLUTIONIST ANSWERS

The decaying current model is obviously incompatible with the billions of years needed by evolutionists. Accordingly, their preferred model is a *self-sustaining dynamo* (electric generator). The earth's rotation and convection is supposed to circulate the molten nickel/iron of the outer core. Positive and negative charges in this liquid metal are supposed to circulate unevenly, producing an electric current, thus generating the magnetic field. But scientists have not produced a workable model despite half a century of research, and there are many problems.

However, the major criticism of Barnes' young-earth argument concerns evidence that the magnetic field has *reversed* many times—i.e., compasses would have pointed south instead of north. When grains of the common magnetic mineral *magnetite* in volcanic lava or ash flows cool below its *Curie point* of 570°C (1060°F), the magnetic domains partly align themselves in the direction of the earth's magnetic field *at that time*. Once the rock has fully cooled, the magnetite's alignment is fixed. Thus, we have a permanent record of the earth's field through time.

Although evolutionists have no good explanations for the reversals, they maintain that, because of them, the straightforward decay assumed by Dr Barnes is invalid. Also, their model requires at least thousands of years for a reversal. And with their dating assumptions, they believe that the reversals occur at intervals of millions of years, and point to an old earth.

CREATIONIST COUNTER-RESPONSE

The physicist Dr Russell Humphreys believed that Dr Barnes had the right idea, and he also accepted that the reversals were real. He modified Barnes' model to account for special effects of a liquid conductor, like the molten metal of the earth's outer core. If the liquid flowed upwards (due to convection—hot fluids rise, cold fluids sink) this could sometimes make the field reverse quickly. Now, the proposes that the plunging of tectonic plates was a cause of the Genesis Flood. Dr Humphreys says these plates would have sharply cooled the outer parts of the core, driving the convection. This means that most of the reversals occurred in the Flood year, every week or two. And after the Flood, there would be large fluctuations due to residual motion.

However, the reversals and fluctuations could not halt the overall decay pattern—rather, the total field energy would decay even faster, Figure 22.1.

Also, this model explains why the sun reverses its magnetic field every 11 years. The sun is a gigantic ball of hot, energetically moving, electrically conducting gas. Contrary to the dynamo model, the overall field energy of the sun is decreasing.

Dr Humphreys also proposed a test for his model: magnetic reversals should be found in rocks known to have cooled in days or weeks. For example, in a thin lava flow, the outside would cool first, and record earth's magnetic field in one direction; the inside would cool later, and record the field in another direction.

Three years after this prediction, leading researchers Robert Coe and Michel Prévot found a thin lava layer that must have cooled within 15 days, and had 90° of reversal recorded continuously in it. And it was no fluke—*eight years later*, they reported an even faster reversal. This was staggering news to them and the rest of the evolutionary community, but strong support for Humphreys' model.

The earth's magnetism is running down. This world-wide phenomenon could not have been going on for more than a few thousand years, despite swapping direction many times. Evolutionary theories are not able to explain properly how the magnetism could sustain itself for billions of years.

The earth's magnetic field is not only a good navigational aid and a shield from space particles, it is powerful evidence against evolution and billions of years. The clear decay pattern shows the earth could not be older than about 10,000 years.

Recently, geophysicist David Stevenson at the California Institute of Technology admitted the problems that the earth's magnetic field poses for long-age dogma:

"Right at this moment, there is a problem with our understanding of Earth's core and it's something that's emerged only over the last year or two. The problem is a serious one. We do not now understand how the Earth's magnetic field has lasted for billions of years. We know that the Earth has had a magnetic field for most of its history. We don't know how the Earth did that. We have less of an understanding now than we previously thought we had a decade ago of how the Earth's core has operated throughout history."

ORIGIN OF THE EARTH'S MAGNETIC FIELD

The Humphreys Proposal

Dr Humphreys proposed that God first created the earth out of water. He based this on several Scriptures, e.g., *2 Peter 3:5* which concludes that the earth was formed out of water and by water.

After this, God would have transformed much of the water into other substances like rock minerals. Now water contains hydrogen atoms, and the nucleus of a hydrogen atom is a tiny magnet. Normally these magnets cancel out so water as a whole is almost non-magnetic. But Humphreys proposed that God created the water with the nuclear magnets aligned. Immediately after creation, they would form a more random arrangement, which would cause the earth's magnetic field to decay. This would generate current in the core, which would then decay according to Barnes' model, apart from many reversals in the Flood year as Humphreys' model states.

Supporting Observation

Dr Humphreys also calculated the fields of other planets (and the sun) based on this model. The important factors are the mass of the object, the size of the core and how well it conducts electricity, plus the assumption that their original material was water.

His model explains features which are deep puzzles to dynamo theorists. For example, evolutionists refer to 'the enigma of lunar magnetism'—the moon once had a strong magnetic field, although it rotates only once a month. Also, according to evolutionary models of its origin, it never had a *molten* core, necessary for a

Fig.22.5: The planet Neptune as photographed by the Voyager probe

dynamo to work. Also, Mercury has a far stronger magnetic field than dynamo theory expects from a planet rotating 59 times slower than Earth.

Even more importantly, in 1984, Dr Humphreys predicted that the field strength of Uranus was about 100,000 times the evolutionary predictions from their 'dynamo' theory. The two rival models were tested when the Voyager 2 spacecraft flew past these planets in 1986 and 1989. The field for Uranus was just as Humphreys had predicted. Dr Humphreys' prediction for the field strength of Neptune aligned with the evolutionary prediction, because this planet has a high heat outflow, Figure 22.5. But Neptune's field was still different in other important ways from the evolutionary dynamo prediction: the tilt and offset. Yet many anti-creationists call creation 'unscientific' because it supposedly makes no predictions!

Humphreys' model also explains why the moons of Jupiter that have cores have magnetic fields, while Callisto, which lacks a core, also lacks a field.

The Cause of the Earth's Magnetic Field

Materials like iron are composed of tiny *magnetic domains*, which each behave like tiny magnets. The domains themselves are composed of even tinier atoms, which are themselves microscopic magnets, lined up within the domain. Normally the domains cancel each other out. However, in magnets, like a compass needle, more of the domains are lined up in the same direction, and so the material has an overall magnetic field.

Earth's core is mainly iron and nickel!

Could its magnetic field be caused by the same way as a compass needle's? **No, it could not, because** —above a temperature called the *Curie point*, the magnetic domains are disrupted. The earth's core at its coolest region is about 3400–4700°C (6100–8500°F), much hotter than the Curie points of all known substances.

However, in 1820, the Danish physicist H.C. Ørsted discovered that an electric current produces a magnetic field. Without this, there could be no electric motors. Therefore, could an electric current be responsible for the earth's magnetic field? Electric motors have a power source, but electric currents normally decay almost instantly once the power source is switched off (except in superconductors). So, how could there be an electric current inside the earth, without a source?

Fig.22.6: Michael Faraday (1791-1867)

The great creationist physicist Michael Faraday answered this question in 1831 with his discovery that a changing magnetic field *induces* an electric voltage, the basis of electrical generators, Figure 22.6.

Imagine the earth soon after creation with a large electrical current in its core. This would produce a strong magnetic field. Without a power source, this current would decay. Thus, the magnetic field would decay too. As decay is change, it would induce a current, lower but in the same direction as the original one.

Therefore, we have a decaying current producing a decaying field which generates a decaying current ... If the circuit dimensions are large enough, the current would take a while to die out. The decay rate can be accurately calculated, and is always exponential. The electrical energy doesn't disappear—it is turned into heat, a process discovered by the creationist physicist James Joule in 1840. This is the basis of Dr Barnes' model.

'In the most accurately recorded period, from 1970 to 2000, the total (dipole plus non-dipole) energy in the earth's magnetic field has steadily decreased by 1.41±0.16%. At that rate, the field would lose at least half its energy every 1500 years, give or take a century or so. This supports the creationist model that the field has always been losing energy—even during magnetic polarity reversals during the Genesis flood—ever since God created it about 6000 years ago.

'The evolutionists, on the other hand, have no workable, mathematically-analyzable theory of reversals.

MAGNETIC NAVIGATION

Earth's Magnetic Field for Navigation

Many dog lovers, appreciate a heart-warming story of a dog that has been lost and has found its way home, Figure 22.7. Some dogs have travelled remarkable distances, through even obstacles such as a fast-flowing river. After posting flyers and contacting animal shelters, the dog's family is overjoyed when Fido shows up on their doorstep.

Some researchers suggest that lost dogs, with their sensitive noses, estimated to be thousands of times more sensitive than human noses, can use familiar scents to find their way home. Apparently, they can sense scents from farther than 15 kilometers (9.4 miles) away.

Dogs may use earth's magnetic field for navigation

Also, there appears to be evidence that dogs can use the earth's magnetic field to help them align their journey in the right direction.

CZECH STUDY

Czechs' Scientists equipped 27 hunting dogs with GPS collars and action cams, let them freely roam in forested areas, and analyzed components of homing in over 600 trials.

The inbound track during scouting started mostly with a short (about 20 meters) run along the north-south geomagnetic axis, irrespective of the actual direction homewards. Performing such a 'compass run' significantly increased homing efficiency.

More controlled experimental work may be required to eliminate the possibility that the dogs were using visual cues (e.g., the location of the sun). And the researchers plan some control experiments that place magnets on the dogs to see if the navigation is thrown off. Regardless, the initial work seems to indicate that dogs can sense the earth's magnetic field.

Fig.22.7: Dog's use Earth's magnetic field like a compass. Curtsy - Miami Herald

OTHER ANIMALS – EARTH MAGNETIC FIELD

Other Animals use earth's magnetic field for navigation. A search on 'ScienceDaily.com' using the words 'magnetic' and 'navigate' produced dozens of summary reports. These studies document animals' use of the earth's magnetic field to align their migrations or other activities. Among the animals and insects which are reported to use the magnetic field are birds, sea turtles, bats, ants, salmon, Figure 22.8, and sharks.

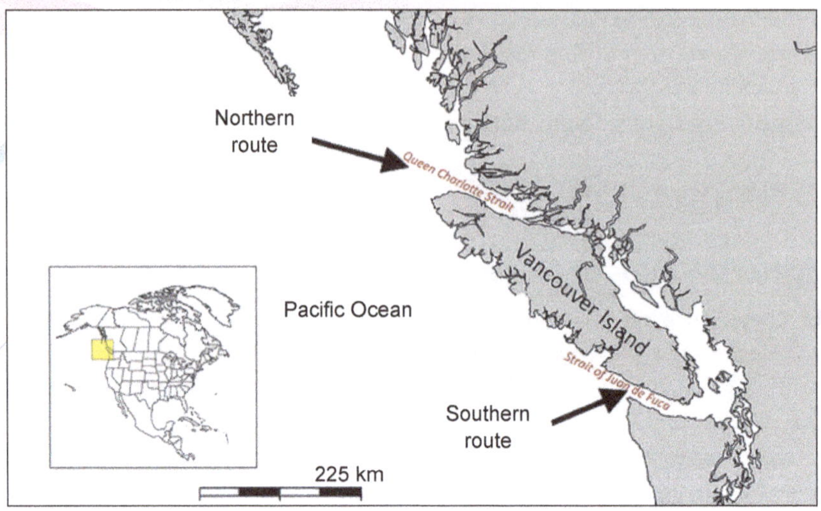

Fig.22.8: Magnetic Memories Guide Salmon Home

Some bacteria have also been found to maneuver by aligning with the earth's magnetic field. Plants may also react to changes in the earth's magnetic field.

DECAYING EARTH'S MAGNETIC FIELD

It has been known for centuries that the earth is a giant magnet. Early magnetic compasses, using lodestones, which aligned to the earth's magnetic field appear in the historical record around the 12th century. However, it wasn't until 1820 that Hans Christian Ørsted, a Danish physicist, identified the probable source generating the magnetic field as an electric current running around the earth.

Earth's Magnetic Shield

Earth's magnetic field "shields us from dangerous charged particles from the sun [solar wind]." In turn, the solar wind shields us from dangerous cosmic ray particles, Figure 22.9. Also, the field provided mariners and continental explorers with compass directions before the invention of GPS, and helps animals and insects navigate.

Fig.22.9: Radiation - Earth's magnetic field shields Us – Curtsy Physics Stack Exchange

Creation Ministry International CMI, has been reporting for decades on the fact that the earth's magnetic field is decaying exponentially and far too rapidly to support the long-ages claims of evolutionary cosmologists and biologists. Scientists who hold to the long-ages view admit that there is a measurable decay of the magnetic field, but they speculate that it regenerates on a cyclical basis. However, they are unable to provide empirical evidence for this supposed regeneration.

The decay of the earth's magnetic field and a similar observed decay process associated with other planetary objects in our solar system is evidence that these objects are not billions of years old. Rather, they were created during the creation week documented in Genesis 1.

The decay of the earth's magnetic field, which at minimum serves as a protective barrier for life on earth and provides guidance information for animals, presents a serious challenge for proponents of deep time.

GOD'S PROMISE

However, we have God's promise:

"I will never again curse the ground because of man, for the intention of man's heart is evil from his youth. Neither will I ever again strike down every living creature as I have done. While the earth remains, seedtime and harvest, cold and heat, summer and winter, day and night, ***shall not cease***." (*Genesis 8:21–22*)

This promise provides us with the confidence that we can trust that God will provide a solution, possibly with the renovation of the created order; *2 Peter 3:13*, before humans and animals will have difficulty surviving on this earth.

Turtles Read Magnetic Maps

Magnetic compasses have been vital to navigation, as they enable people to use the earth's magnetic field to tell direction, Figure 22.10. Now, recent experiments have demonstrated that some organisms also navigate with their own 'compasses.'

Fig.22.10: *Loggerhead Sea Turtle - Photo Mike Gonzalez, Wikipedia.org*

At different places on the earth, the strength of the earth's magnetic field and its inclination, which is the angle at which it intersects the earth's surface are different. Accordingly, if organisms could sense these changes, they would have something like longitude and latitude readings on a map.

This is important to young Loggerhead Turtles (*Caretta caretta*) which must stay within the North Atlantic Gyre, the circular ocean current system that surrounds the Sargasso Sea. Researchers Kenneth and Catherine Lohmann of the University of North Carolina have shown that the turtles use magnetic measurements to stay in the gyre.

They placed turtles in water tanks surrounded by computerized electric coils generating an artificial magnetic field. When the field's inclination was the same as that of the northern boundary of the gyre, the turtles would swim south, as if back into the gyre. Conversely, when the inclination was the same as the southern boundary's, the turtles swam north-northeast, again as if away from the danger boundary.

In other experiments, they also kept the inclination constant but varied the magnetic field strength in the tank. When the strength was the same as that of the western boundary of the gyre, the turtles swam east, again as if into the gyre and away from the danger boundary. And they swam west when the strength was the same as the eastern boundary.

The famous British evolutionist (and communist) J.B.S. Haldane claimed in 1949 that evolution could never produce 'various mechanisms, such as the wheel and magnet, which would be useless till fairly perfect.' Therefore, such machines in organisms would, in his opinion, prove evolution false. These turtles which use magnetic sensors have indeed fulfilled one of Haldane's criteria. Also, the 'simple' bacterium propels itself by a filament called a flagellum, which is propelled by a rotary motor — a type of wheel, thus fulfilling Haldane's other criterion. I wonder whether Haldane would have had a change of heart if he had been alive to see these discoveries …

Stars Guiding Moths

Exceptional, as it seems, navigation by the stars is likely to be the case for some moths and other creatures, Figure 22.11.

In fact, the whole subject of animal navigation is enthralling evidence that it is not only human beings who are '*fearfully and wonderfully made*.'

Many creatures perform annual return migrations involving many thousands of miles—organisms as different as the green sea turtle, the salmon, and the garden warbler bird.

Many of these accomplishments of navigation are truly amazing, and appear to involve inbuilt, genetically programmed skills. What do these creatures use to orientate themselves, and how do they know when to go? There are still many unsolved mysteries about the mechanisms involved, but research shows that in varying degrees and differing combinations among different species, the following are various factors which may be involved.

Fig.22.11: An adult male pine processionary moth. Curtsy - Alvesgaspar wikimedia commons

Timing Migration

- An *internal calendar* which helps the particular creature to 'know' when the time is approaching and helps it to deposit fat as a store of fuel for the journey, for instance. This calendar can be fine-tuned, Figure 22.12.
- A sensitivity to changes in the number of hours of daylight, especially among birds
- Certain whale migrations appear to be triggered by a seasonal *drop in water temperature.*

Fig.22.12: Geese Migrating

Hints to Direction

- Sensing the position of *the earth's magnetic field.* Some species not only use this as a simple compass, but can detect local variations in the earth's magnetic field (of less than 0.2%!) and use them as familiar landmarks.
- *The direction of air and water currents* can be used as points of orientation.
- *Visual landmarks*
- *Odors* (minute particles) in the air.
- *The position of the sun in the sky.* Because the sun moves through the day, as seen from earth, this works only if the creature 'knows' what time of day it is. This is done with another internal clock.

- *The pattern of polarized light.* The proportion of light which is scattered (and thus partially polarized, as compared to direct unpolarized sunlight) varies with the position of the sun. This may be particularly important when the sun is hidden by cloud.
- *Star patterns in the night sky.* Some young birds, for instance, are capable of 'learning' the pattern of rotation of constellations of stars and using it to tell which way is north. (It is not being suggested that they logically deduce this, but many innate migratory instincts appear to be genetically programmed to flexibly respond to learned cues.) The discovery that not only birds, but even insects (some moth species) are capable of such stellar navigation is astonishing.

Migration Instinct

Fig.22.13: *Male Blackcap Eating from an Olive Tree. Curtsy Cayambe wikimedia commons*

The genetic program can vary within a species. Blackcap birds, for instance, migrate in several different directions. If you cross two of these which migrate in separate directions, the hybrid offspring migrate in a separate direction again from either parent. In each instance, they are able to cope by being able to develop a 'map sense' using certain landmarks, learning staging areas along the way, and so on, Figure 2.13.

Of course, where there is variation within genetic information, both breeding and natural selection can act on a population. Research has shown that some blackcaps have undergone a change in their migration patterns in only 30 years. This built-in ability for a population to respond and adapt to changing circumstances involves utilizing the built-in programmed information already present in its kind, not 'evolving' any extra information. It is crucial in conserving populations of wild animals.

In the case of migration patterns, such flexibility makes great sense. In a devastated, post-Flood world, climate and vegetation patterns would have been changing rapidly for centuries, making it necessary for migratory habits to change. If these populations had been created only with an inflexible 'map' and a rigid migration program, with no genetic variation and no built-in potential for 'learning', they would have rapidly become extinct. In any case, such a created 'map' would have been superseded after the Flood had changed the face of the earth.

Fig.22.14: *Bees know how to make a special food capable of miraculously turning an ordinary bee into a super bee—the queen. Curtsy Photo by stockxpert*

INTRIGUING INSTINCTS

Many creatures perform the most amazing accomplishments. For example, accurate migration routes of butterflies, turtles, birds and fish; navigation of ants; hitch-hiking by insects; and in nest designs of birds, web designs of spiders, and dam and lodge building of beavers. How do they manage this? Often, it is 'explained' by the catch-all term 'instinct', but this hides the ingenuity behind these behaviors.

HONEY BEES

Bees make use of some dazzling technology in the workings of the hive. Some of the larvae in the hive are fed a special food that the bees prepare (royal jelly) which performs the incredible achievement of turning what would have been an ordinary bee into a larger and very different bee—a queen, Figure 22.14.

How do the bees know they need queens and where did the nursery bees get the hi-tech recipe for royal jelly? Because bees only live a matter of weeks, it all has to work. Some bees in the hive are ventilation technicians—they hold on tight and buzz their wings until their wings eventually wear out. It is their life's work to create a draught of fresh air through the hive. A bee also has a fascinating dance that tells her fellow workers where to find nectar, and an amazing navigation system that robotics engineers' envy.

PARENTAL INSTINCTS

The Young Wolf

Nowhere do we see this life-and-death critical instinctive knowledge more dramatically displayed than in the lives of parents and their offspring.

'Knowing' that the time has arrived to give birth to her first litter of pups, the young wolf retreats to her prepared den where, as each cub is born, she 'knows' to chew the umbilical cord in half, wash the cub by licking, Figure 22.15, which stimulates her milk production and guide the new born to the teat where each cub knows to suck. How do mother and baby know what to do?

Fig.22.15: *Mother Wolf Nurtured her Babies*

Beavers

Beavers, when their lodge is built, know to make a vent hole in the top to provide vital oxygen Figure 22.16.

Fig.22.16: *Beaver's Instinct - Photo by stock.xchng*

Mother Kangaroo

When a mother kangaroo gives birth to her joey, he is only about the size of the tip of a human little finger. Immediately he begins a monumental climbing journey from the birth canal through her fur to his mother's pouch, Figure 22.17. Climbing down inside the pouch, he homes in on her nipple where he 'locks on' for the bumpy weeks ahead. In fact, he seals on so firmly that to pull him off would damage his mouth and his mother. The joey does not randomly crawl all over his mother's body until he accidentally finds her pouch. Instead, he follows a plan of what to do to achieve a lock on, or he would quickly die—he is extremely vulnerable at this point (if he didn't know what to do, then there would be no more 'roos).

Fig.22.17: Mother Kangaroo

The Mallee Fowl

Mr. Mallee Fowl uses a compost heap of leaves to incubate eggs. He displays vital working knowledge of organic heating—produced by the decay in a compost oven. For example, the fowl uses *solar heating*—uncovering layers to heat in the sun; *insulation*—covering the heated layers to retain the heat and *ventilation*—if overheating occurs, Figure 22.18. He regularly probes the 'oven' to test the temperature and keep the eggs within the critical range to hatch. How did he learn all this? If he gets it wrong, he may have to explain to his mate, "he cooked the kids.'

Fig.22.18: Mr. Mallee Fowl Uses a Compost Heap of Leaves to Incubate Eggs

The Magnificent Migrating Monarch

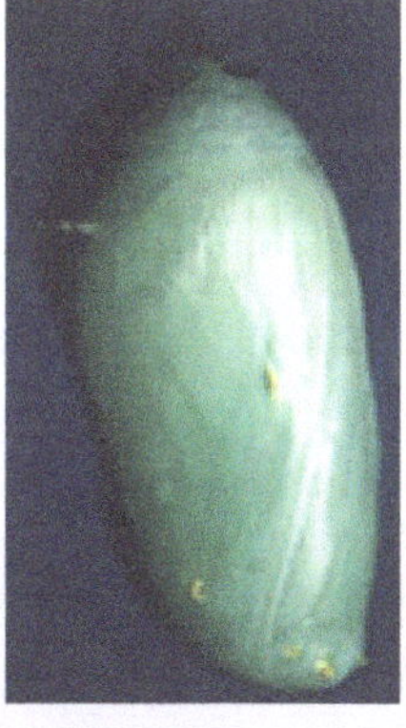

Fig.22.19: The chrysalis from which the adult monarch emerges after its spectacular transformation from a caterpillar

THE STUNNING MONARCH BUTTERFLY NAVIGATION

Scientists designed parts for some very sophisticated electronic gadgetry, including navigation equipment for various space and defense projects. The level of technology in the circuits that guided men to the moon is phenomenal, Figure 22.19. However, the navigation equipment packed into the brain of the monarch butterfly shows, through the incredible achievements of migration performed by that creature, that there is a far greater level of technology involved. And it is all packed into a brain no bigger than a pinhead!

This tiny, yet beautiful, insect can perform a migration flight of thousands of miles, navigating unerringly to reach a place it has never seen. For instance, some monarchs fly from Nova Scotia, Canada to the mountains west of Mexico City, some 3,000 miles in all. Not just to the very same place to which their forefathers migrated, but each one often to the very same tree!

Monarch butterflies can fly in still air at a speed of around 30 miles per hour, and considerably faster with a tail wind. They usually fly close to the ground, but have been found as high as 12,000 feet. They have been known to fly more than 375 miles over water non-stop in 16 hours. Their 3,000 miles migration takes them eight to ten weeks, travelling only in daylight.

Monarchs can be taken hundreds of miles off course and still find their way to their destination. How do they perform this amazing accomplishment?

To this day, no scientist knows for certain. It looks as if they have magnetic material in their head and thorax region, so they may use the earth's magnetic field to help them. However, most scientists believe they only use the earth's magnetism to give them the general direction. They use the sun's position for most of their navigation, to locate where they are on the earth's surface.

It is instructive to look at the two basic systems by which humans can fix their position using the sun. Monarchs wouldn't use exactly the same sorts of systems as humans do, but they would have to solve essentially the same basic problems.

- **Method 1:** To work out the latitude of your position on the earth (i.e., how far you are north or south of the Equator), you note the time the sun just rises above the eastern horizon in the morning on a given date, and look up the relevant tables for that day in the *Navigator's Almanac*. To work out your longitude, you compare the time where you are with the time in Greenwich, England. Dividing the difference (in minutes) by four gives the degrees of longitude.
- **Method 2:** This requires that at least two sightings of the sun's position relative to the horizon are taken at two different times. From the first measurement (and by using an almanac which tells you what the sun's position is on a certain date and knowing the Greenwich time) a line can be drawn and the navigator can say 'I'm somewhere along that line.' The second measurement gives another line, and where they intersect gives the navigator's approximate position.

The monarch butterfly, about 1.5 inches long and weighing half a gram. Each one of its four wings has about 1.4 million scales. Every scale is filled with air to provide buoyancy and make it easier to fly. Design engineer Jules Poirier, who has worked on various space projects, said: 'only a great Creator could have designed such a wonderful flying machine.'

It is likely that the monarch butterfly uses something like the second method to determine its position on the earth. God must have programmed the monarch with a simplified almanac of the sun's position relative to a date and time. This means that this butterfly also has to have an accurate internal clock. Monarchs can detect different polarizations of light, so even on a cloudy day, they can measure the angle to the sun from the horizon. Humans had to discover mathematics and a host of other things to be able to navigate much more crudely than this insect, which has it all programmed into it from birth.

The monarch's life cycle is an amazing story in itself: the adult lays an egg, and from the egg, a caterpillar hatches. After it has grown enough, it makes a chrysalis and enters the pupa stage Figure 22.19. Inside the chrysalis, the caterpillar's tissues disintegrate and re-form into the adult butterfly. This process is powerful evidence of a designing hand, because until it was fully operational, the creature could not have reproduced.

Imagine a hypothetical evolutionary intermediate stage: a caterpillar evolves the materials and instinct to make a chrysalis, then another enzyme to dissolve all its tissues. What happens then? A container full of a soup of cells that can do nothing—it can't eat, drink, or reproduce. So, these amazing changes will not be passed on to any offspring. No, the creature needs both the ability to make a chrysalis and dissolve its tissues, and the ability to reform into the intricate flying insect, all at once. There can be no series of small changes which could accomplish this, one step at a time.

What makes the migration all the more sensational is that due to the short life cycle of the monarch butterfly, many have never before been to the place to which they are headed despite stopping many times on the way to drink nectar, and being often blown off course, they unerringly make the necessary corrections to get back to the place from which their parent (or sometimes even grandparent) commenced the journey.

THE INCREDIBLE JOURNEYING OF MONARCH BUTTERFLIES

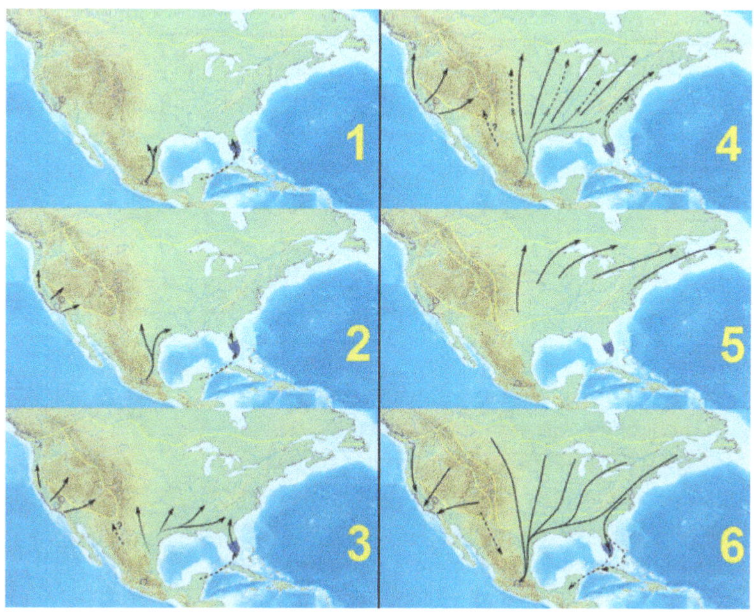

Fig.22.20: Monarch Butterfly Migration

The monarch butterfly is the only insect known to migrate annually over major continental distances, Figure 22.20. There are two basic migrating groups on the North American continent. The Eastern population is based east of the Rockies; some 300 million of these butterflies migrate from as far north as northern Nova Scotia to about 13 sites covering 25 hectares (40 acres) in the Neo-volcanic Mountains in Mexico [150–225 miles] west of Mexico City). Individual trees can harbor as many as 100,000 at a time, while sites can contain as many as 50 million!

Some are blown eastward to Bermuda. From here they fly nonstop to the Bahama Islands, and thence to Cuba, Jamaica, Hispaniola and Puerto Rico, and on to winter sites in Guatemala. The Western population live in valleys west of the Rocky Mountains, and winter in a number of sites stretching from Bodega Bay in northern California to Baja, Mexico. Little is known about the South American monarch migrations. There are monarchs in other parts of the world which do not migrate.

It is not likely that monarchs would have had programmed into them, from the beginning of creation, the exact route of their journey, but rather the capacity to establish such migration patterns and carry them out. The world has changed significantly since the great global Flood, and Figure 22.20. monarchs have clearly made adjustments to their patterns over time. This requires an even greater level of design sophistication. Not only does the monarch have to have a built-in clock, almanac, and navigational computer, it has to have the programmed capacity to make and remake its own internal maps.

In addition, somehow that learned information (for example which tree it came from) has to be passed to the next generation, who have never flown over that route before. In the light of all that is known today about the processes of inheritance, how that could possibly be is a major mystery.

Designing navigation equipment to take men beyond the confines of this planet and safely back again took an enormous amount of intelligent effort. The fact that the monarch can do these unbelievable accomplishments with such an amazingly miniaturized 'control center' reveals a level of design engineering which demands an overwhelmingly great intelligence. The Bible indicates clearly who that intelligent Creator was—Jesus Christ, the eternal Word, (John 1:1–3) God the Son, 'in whom are hid all the treasures of wisdom and knowledge' (Colossians 2:3).

The same Bible makes it clear that at the return of Christ, those who now arrogantly scoff at and reject the Creator of the universe will be left with no excuse—no arguments, since to those who are not willfully blind, the evidence of this almighty power is everywhere. At that time 'every knee shall bow' to Him (Isaiah 45:23). Those who bend the knee now in repentance will receive His loving forgiveness—those who refuse to do so now will do it later, receiving His righteous judgment instead.

HIBERNATION, MIGRATION AND THE ARK

A report of a year-long hibernation in a tiny marsupial raises a subject worth revisiting.

A recent news item caused a flurry of interest among creation scientists. It was based on an article in the German journal *Naturwissenschaften* (Natural Sciences), about a marsupial able to hibernate for more than a year, Figure 22.21. It is evidence that 'animals could have hibernated during the year of the Flood'.

It's worth exploring just how this does or does not add to the apologetic arguments about the feasibility of the Flood account. First, some more detail on the report.

The Discovery

The animal concerned was the pygmy possum, *Cercartetus nanus*, a marsupial. This is an 'opportunistic non-seasonal hibernator.' In the right circumstances, it is able to put on substantial fat reserves which enable it to go into prolonged torpor. The research in this instance was directed to seeing whether the pygmy possum, given the right conditions, would be able to prolong its hibernation, existing only on its own body fat, well beyond winter.

Fig.22.21: *Cercartetus Nanus - Curtsy - Image Wikipedia*

The outcome was impressive—the prolonged hibernation lasted 310 days on average within various of those creatures, with one reaching 367 days.

Relating to the Flood Account

On the surface, it seems somewhat obvious. Faced with the problem of caring for thousands of animals for a year, it would make things far simpler if the animals went into some sort of prolonged shutdown for most of the journey.

For one thing, it would dramatically reduce the amount of food and water required. For example, in the case of the pygmy possum mentioned above, the average energy expenditure during hibernation was reduced to about 2.5% of normal. This in turn drives the waste load way down, too.

For another, one can imagine the fear and restlessness among all those animals, confined in dark quarters for months while the ship was subject to all manner of 'boat-rocking' external forces, despite the demonstrated high stability of its proportions. Not to mention the sound of driving rain for days on end—and presumably wind, waves, and thunder, too.

It's not surprising, then, that as far back as the classic *The Genesis Flood*, creationist authors have been suggesting that the animals may have entered a state of prolonged hibernation. This suggestion is perfectly reasonable, and the *possibility* is not being questioned here. But to point to some present-day hibernation accomplishments as evidence supporting the Flood account is not as simple and straightforward as a quick glance might suggest.

Observe that even without hibernation of any sort, the Ark journey still 'works,' despite being a much more problematic for its inhabitants. This is reinforced by the detailed figures and rigorous arguments in Woodmorappe's classic work, *Noah's Ark: a feasibility study*. In short, raising the issue of hibernation for apologetic purposes is not so much a crucial necessity as it is a helpful nicety.

The Need for Miracles

In pointing to such things as present-day hibernation, in fact in all such 'Ark feasibility' studies, one is actually attempting to minimize the need for supernaturalism, to try to explain it naturally without the need for a miracle. This is an understandable goal, to want to avoid multiplying the number of miracles required in some arbitrary fashion.

That is not to be confused with bowing to naturalism and liberal theology in denying the miraculous in Scripture. The description of the Flood/Ark in Genesis reinforces what Henry Morris has called the 'economy of miracle' seen in the Bible in general. God could, for instance, easily have suspended all the animals and Noah's family above the clouds during the year of the Flood. But He chose to use natural laws such as the principles of buoyancy involved in a floating ship. Even then, He could have materialized a readymade Ark of safety, but instead chose to give detailed instructions for its presumably long and laborious construction.

The absence of a flurry of capricious miracles in the Bible (apocryphal gospels have an abundance of these) is actually one hallmark of its authenticity. It makes the rare, special-purpose miracle, like raising Lazarus from the dead, or feeding the five thousand, stand out all the more. This is in part why we feel more comfortable when we have a 'naturalistic' explanation for an Ark-feasibility problem, as per Woodmorappe's book. Having to postulate miracle after miracle, especially ones the Bible does not mention, would seem awkward and would in practice make the Bible account less believable to skeptics.

Clearly, having the animals go to sleep makes the journey far less problematic. What is being discussed here is the appeal to existing 'natural instincts,' as part of this drive to minimize the supernatural.

Migration—a Parallel to Hibernation

Creation apologists, dealing with the issue of the animals travelling to the Ark, have similarly sought naturalistic explanations where possible. They often point to 'the migration instincts in various animals', and/or their instinct to travel to safety if there is impending danger. But both here and in the case of hibernation, the appeal to existing instincts is problematic. As we will see, it cannot avoid the need for the miraculous, pure and simple—and in substantial doses, in fact.

- First, present-day migration instincts are nowhere near universal among animals. So even if God may have used the existing instinct somehow in some species, that still leaves the overwhelming majority of those that needed to be on board, which show little trace of a migration instinct. Therefore, if supernatural action is needed for that majority, why not the lot? How much, then, has the 'instinct' argument really helped the 'explanation'?
- Second, existing instincts do not direct animals towards a man-made boat.
- Third, *even if* all animals had a migratory instinct, and *even if* all were programmed to migrate towards large man-made vessels, why did only those particular ones from each type make the journey?

Clearly, a mighty miracle was involved. We correctly talk of Noah *taking* the various creatures on board, but it should not be overlooked that these were ones which God *sent* (*Genesis 6:20*). Noah did not have to roam the world with lassoes and animal traps. In fact, the sight of pairs of animals migrating to the Ark would likely have been an awesome testimony to onlookers that the hand of the miracle-working creator God was here to be seen—notwithstanding the fact that hearts remained hard.

Often the argument is worded such that God could have 'modified existing migratory instincts' in certain creatures. OK, He *could* have done that, and we're not told either way. But it's clear from the earlier discussion that in any case, many animals would have needed to have such instincts specially created at the time. And those that already had them needed them extensively reprogrammed—and then only in those chosen for this journey. The degree of supernatural specificity is so extensive that bringing up the argument in this way seems, on analysis, to be of little help. Why not accept that God directly and supernaturally commanded the animals He wanted to travel to (and board) the Ark to do so? In short, pointing to some migratory instincts to attempt to make the account more feasible is not exactly an apologetic 'coup.' It does not avoid or in any tangible way mitigate the need for supernaturalism, despite perhaps giving such an impression.

HIBERNATION PROCESS

We similarly see creation scientists claim that God 'could have used or modified existing hibernation instincts.' Nonetheless, here, too, we cannot escape from the raw fact that to put all those animals to 'sleep' for the year of the Flood would have involved a substantial dose of supernaturalism. Many mammals do hibernate each year, often for about six months at a time. (Even hibernating for half the journey would help, of course.) But many animals do not hibernate at all. So, why should those on the Ark hibernate, and why at that particular time? Here, too, it is almost redundant to talk of 'modifying existing instincts,' since it might have been just as much trouble for God to put the animals directly into a torpid state.

Then of course there is the issue of whether, even when put to sleep by God, most of the animals could have had sufficient fat reserves to last them for a year without further supernatural help. This is possibly why the observation concerning the pygmy possum stirred some interest. If some animals can be induced to hibernate for up to a year or more without running out of 'body fuel,' then maybe this could be true for all? But the pygmy possum is already programmed to hibernate, and more importantly in this context, is designed with the capacity to 'fatten up' adequately in anticipation of extended periods without food or drink.

Presumably, those creatures which do not hibernate would once again require special intervention to 'fatten up' to the extent needed in anticipation of a year-long journey.

So here, again, one might ask, 'Why not just go straight to the obvious? God did it supernaturally.' We are not talking about 'any old god' here, but about the God of Genesis, who created a complex universe (including all of its contained migration and hibernation instincts) in six earth-rotation days in the first place. Though His normal operation in the world today is via what modern science *describes* as the physical laws, He is certainly not constrained by them. He has described instances in the Bible in which He has overridden (or perhaps better, added to) them, for special purposes at special times in history. And the Flood certainly was an incredibly special time.

The Flood, almost by definition, would have required a mixture of natural and supernatural activity—both for its causes and particularly for the survival of the Ark and its crew. In the understandable tendency to seek 'natural' explanations wherever possible, one can easily overlook the fact that most animals would not normally and naturally either head off to and then board the Ark, or go to 'sleep' once there, let alone have already stored up enough fat for all or most of the journey.

To talk of God 'modifying' certain instincts (whether migration or hibernation) overlooks/sidesteps the fact that not all animals have those instincts. And for those which do, the degree of specificity and complexity involved in the necessary 'reprogramming' would seem to make the existence of any previous instincts almost superfluous.

The appeal to existing instincts therefore falls somewhat short of a robust explanation. If one makes that appeal in a way that gives the impression that these mechanisms can be conveniently conscripted in a 'naturalistic' fashion, one risks using it largely as rhetorical window-dressing.

One needs to remember that the instincts in today's creatures are in any case there as a result of supernatural programming during Creation Week. So, notwithstanding all the caveats in this article, it would still seem reasonable to point to these present-day instincts in a discussion on Flood issues. Provided, that is, that one does so as mere analogy to what was required during the Flood, rather than giving the impression that a naturalistic 'blanket solution' has been provided.

MIGRATION AFTER THE FLOOD

How did plants and animals spread around the world so quickly?

Fig.22.22: The Bible provides a better explanation for the observed patterns of biogeography than evolution - Curtsy - Wikipedia.org

The main issue in migration is biogeography—the study of where on the earth we find the different kinds of plants and animals, Figure 22.22. It is also about two competing views of Earth history:

1. The secular, evolutionary ancient Earth view
2. The biblical creationist young Earth view.

1. The Secular, Evolutionary, Ancient Earth View

According to this, the earth is billions of years old, and natural processes have been slowly changing the earth's continents and slowly changing life on Earth over many millions of years. If we go back a couple of hundred million years, so we're told, the earth looked like this. All the continents were together in one great continent they call Pangaea. And the world we see today, we're told, formed as this single land mass split up, as the continents we know today slowly drifted apart.

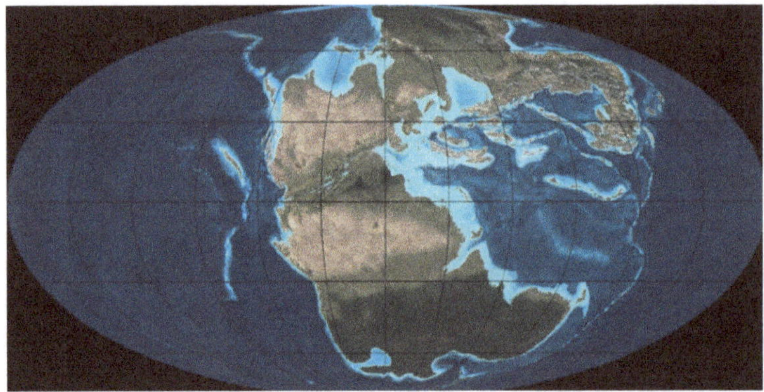

Fig.22.23: *The Separation of the Continents - Credit: Ron Blakey, NAU Geology*

As this was happening, we're told—as the continents were slowly drifting apart over millions of years—dinosaurs went extinct, reptiles evolved into birds and mammals, non-flowering plants evolved into flowering plants and, Figure 22.23, of course, apes evolved into people. Therefore, the evolutionists tell us, when we study biogeography, we discover too much evidence supporting this view. Indeed, there's a mountain of evidence they say, from biogeography, showing that evolution is true; and a mountain of evidence confirming that the continents split apart millions of years ago, separating the various plants and animals that lived on the earth around that time.

When we go to Africa, we find leopards, rhinoceroses, giraffes and gorillas. In America, we don't find any of these. Instead, we find raccoons, jaguars, armadillos and opossums. When we go to Australia, we find marsupials like kangaroos. Evolutionists claim that the reason we find these different animals on these different continents is that they evolved in the different parts of the world. Thus, they say, gorillas are found in Africa and not America, because they evolved in Africa and not in America. Armadillos are found in America, because they evolved there and not anywhere else.

Evolutionists also claim that strong evidence for evolution can be found from studying the biogeography of islands. For example, evolutionists make much of the different species of finches found on the Galapagos Islands.

One of the prominent features of the Galapagos finches is their different beak shapes, and of particular significance is that the finches have beaks best suited to the kind of food found on the islands where they live. Some have strong stubby beaks for crushing hard seeds; others have thinner beaks for probing flowers or fishing insects from the crevices in trees. And so, the story goes—and there's much to be said for it—one original species flew to these islands from the mainland and, over time, diversified into all the different species. This is often termed 'speciation'—where a number of different species come from one original species.

Actually, much more impressive examples of speciation can be found on the Hawaiian Islands, right out in the middle of the Pacific. Around 500 unique species of fruit fly can be found here. Also, Hawaii is home to more than 1,000 different species of snails and slugs, species that, again, are found nowhere else in the world. Evolutionists claim that what we see on the Galapagos and Hawaiian Islands provides absolutely 'irrefutable' evidence supporting their theory of evolution.

Within mammals, there are two main groups: the placentals and the marsupials. Placental mammals, such as humans, complete their embryonic development in the womb, joined to the mother by a placenta. Marsupial mammals, such as kangaroos, have a very different reproductive system, where the mother carries and suckles her young in a pouch at the front of her body. We can see a few more placentals on the left and a few more marsupials on the right.

There are around 140 species of marsupial in Australia, most of which are not found anywhere else; according to evolutionists, there's a very straightforward explanation for this. Evolutionists tell us—these marsupials evolved in Australia!

According to Professor Richard Dawkins, "The pattern of geographical distribution [of plants and animals] is just what you would expect if evolution had happened" and Jerry Coyne, who is Professor of Biology at the University of Chicago, had this to say: "The biogeographic evidence for evolution is now so powerful that I have never seen a creationist book, article or lecture that has tried to refute it. Creationists simply pretend that the evidence doesn't exist."

2. The Biblical Creationist Young Earth View

Many creation scientists believe that there was, originally, one great continent—when God first created the world. In Genesis 1:9, God said, "Let the water under the sky be gathered to one place, and let dry ground appear."

The Bible also tells us that around 1600 years after creation, God judged the world with a global flood—and this would have been much more than just a deluge. There would have been much geological activity, volcanism and ground movements as well. Probably a majority of creation scientists—not all, but probably a majority—would agree that the continents we see today did split apart from the one original land mass. And they would say that this happened at the time of the Genesis Flood, when all this geological activity was going on. Of course, they wouldn't say this happened slowly, over millions of years, but rapidly—not by 'continental drift,' but 'continental sprint'. And it's important to note that, according to this view, the movements of the continents would have occurred beneath the flood waters. This, we wouldn't have had living populations of plants and animals being split by this continental separation.

During the Flood, the whole of the pre-flood world was destroyed. In Genesis 6:17, God said, "I am going to bring floodwaters on the earth to destroy *all* life under the heavens, *every* creature that has the breath of life in it. *Everything* on Earth will perish" (my emphasis). Therefore. the world we see today grew up after this global catastrophe, and over the last 4,500 years or so. Plants left floating on the surface of the waters would have recolonized the areas where they finally settled, after the flood waters receded, and the animals that disembarked the Ark would have migrated to the places they now inhabit.

Which view best fits the data?

The question we might ask is this: "Which view is best supported by the scientific evidence, the data? The secular evolutionary ancient Earth view, or the biblical creationist young Earth view?"

Firstly, the finches on the Galapagos and the many fruit flies and snails on the Hawaiian Islands are not a problem for creationists. We believe that God designed animals with the capacity to vary within their kinds, and with the capacity to change and adapt to different environments; and when we study this carefully, we actually find a problem for the evolutionists. This is because there is growing evidence that this kind of adaptation occurs quickly—it doesn't require hundreds of thousands or millions of years.

There's too much evidence that plants and animals can change. Finches can become other species of finch, fruit flies can become other species of fruit fly, snails can become other species of snail and so on. But this is hardly scientific evidence that an amphibian can become a reptile or that a reptile can become a bird. Nor is it scientific evidence that an ape can evolve into a man.

Have the continents really been separated for millions of years?

Now the evolutionary view is that we have different animals on different continents because they evolved in different parts of the world.

According to this view, jaguars and lions descended from a common evolutionary ancestor that lived around three million years ago. And after three million years of evolution, they say, we got jaguars in South America and lions in Africa. But it's possible to mate a jaguar and a lion and get a hybrid, a jaglion. If these two species, jaguars and lions, were really separated by three million years of evolution, it is most unlikely that their mutated DNA would allow them to hybridize.

Evolutionists face an even bigger problem trying to explain hybrids between jaguars and leopards. This is because the female of this kind of hybrid is fertile. Think about it. Three million years of separate evolution—half the time it allegedly took for ape-like creatures to become people—and the hybrid is still fertile. This seems very unlikely. But the ability of these big cats to hybridize fits the biblical account of history very well.

If jaguars, lions, leopards and tigers all descended from a pair of cats that disembarked from the Ark around 4,500 years ago, it is not surprising that they can mate and produce offspring.

Arguably, evolutionists face an even greater problem with the iguanas of the Galapagos Islands. The land and marine iguanas supposedly separated from their common evolutionary ancestor around 10 million years ago. But, as it was pointed out in Darwin documentary, land iguanas can mate with marine iguanas and produce offspring. This amazed evolutionists when they saw it.

Were the Continents Actually Joined for Millions of Years?

In the evolutionary view, South America and Africa were joined for millions of years. Why then are more seed plants common to South America and Asia than to South America and Africa? Of around 200 seed plant families native to Eastern South America, only 156 are common to Eastern South America and Western Africa, but 174 are common to Eastern South America and Eastern Asia.

If South America and Africa had actually been joined for millions of years, we would expect to see exactly the opposite. We would expect there to be more plants common to S. America and Africa than to S. America and Asia.

According to Dr Simon Mayo, of the Royal Botanic Gardens, Kew - near London, "The overall similarity of the floras [plants] of the two continents is surprisingly low given such a clear geophysical background." Dr Mayo expresses surprise that there are so few plants found in both South America and Africa. Why is he so surprised? Well, he believes these two continents were joined for millions of years. The actual locations of plants, however, doesn't support this view.

Some biogeographers have found this kind of data so puzzling that they have argued against the idea of a super continent, Pangaea, where South America and Africa were joined, and proposed a different arrangement—they have suggested instead that there was a super continent they call Pacifica, where South America and Asia were joined.

If the ancient earth view were correct, we would expect the data and the models of ancient earth geologists to fit with the data and models of ancient earth biogeographers. We would expect them to be harmonious, but they're not; often they're inconsistent and contradictory.

NEW WORLD MONKEYS

We find monkeys in South America, Africa and Asia. For example, we find the Spider Monkey in South America, the Olive Baboon in Africa, the Langur in India and Macaques in Japan.

Now evolutionists tell us that monkeys evolved in Africa. Well, it's not difficult to see how they might have migrated to Asia. But how did they get to South America? According to evolutionary theory, South America split off from Africa millions of years before monkeys had evolved. Evolutionists have the same problems explaining why rodents and some flowering plants are found on these two continents, because, again, they are said to have evolved after South America split from Africa.

BERINGIAN FOSSILS

Currently, the western tip of Alaska is very close to the eastern tip of Asia. They're separated only by the narrow Bering Strait. In fact, many people believe that in the recent past, the two were connected. But, according to evolutionists, and their theory of slow continental drift, Asia and Alaska have only been close to one another for ten million years or so; not for very long in their thinking.

Fig.22.24: *Beringian Fossils*

Prior to this, in the supercontinent Pangaea, Alaska and Asia are understood to have been separated by thousands of miles of ocean—by all the water on the far side of the globe. However, we find plant fossils of the same species in rocks either side of the Bering Strait; and these rocks were laid down in what evolutionists would call the Jurassic period. However, the Jurassic period, we're told, ended around 150 million years ago. So according to this thinking, these identical plant fossils were buried at least 150 million years ago.

Do you see the problem here for the ancient earthers? In their thinking if we go back over 150 million years, Eastern Asia and Alaska weren't close to one another as we see in Figure 22.24.

They were separated by thousands of miles of ocean, Figure 22.25.

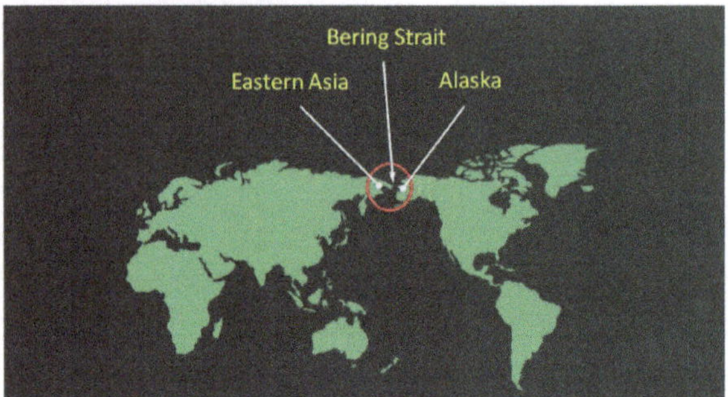

Fig.22.25: Eastern Asia and Alaska weren't Close to one another

Why, then, do we find plant fossils of the same species buried in Jurassic rocks in these two very distant regions? This is yet another example of the kinds of conflicts that arise between ancient earth geology and ancient earth biogeography.

Millions of Years of Evolution?

There are many similar plants and animals found in eastern Asia and eastern North America, but not in the regions between them, Figure 22.26.

Now evolutionists try and explain this by saying that millions of years ago the northern regions were warmer and these two areas of Asia and America were part of one continuous plant and animal distribution—Figure 22.27.

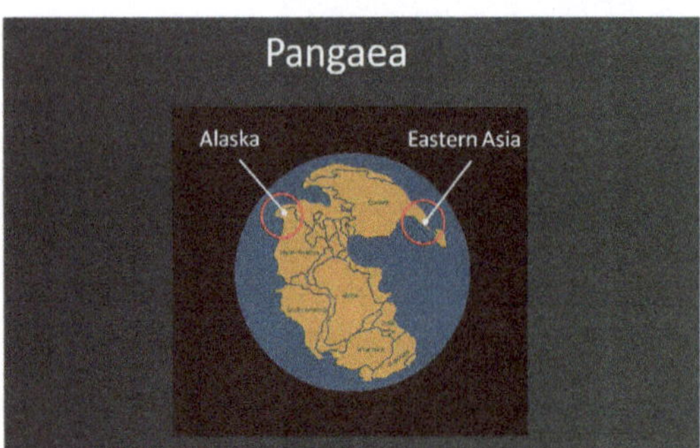

Fig.22.26: Alaska and Eastern Asia Separated by Thousands of Miles

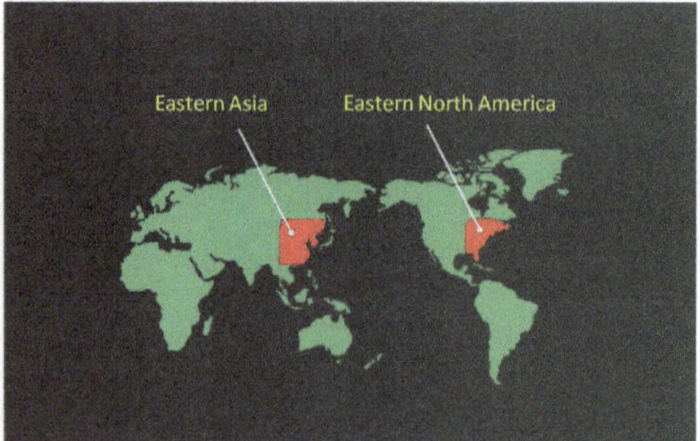

Fig.22.27: Asia and America were part of one continuous plant and animal distribution

And a few million years ago, they say, the climate then cooled and the plant and animal life was separated into the two zones, Figure 22 .28.

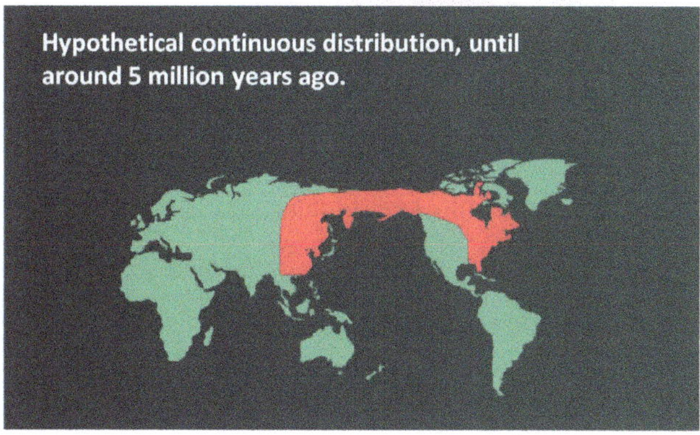

Fig.22.28: Plant and Animal Life was Separated into the Two Zones

But, again, their millions of years' scenario hits a problem. And it's a considerable problem, because many of the plants in these two regions are regarded as being the same or virtually the same species. How, then, could they have been separated for millions of years? If all these plants had really been separated for millions of years, they would not be expected to have retained their similarities to the point that they would still be considered to be the same or virtually the same species, Figure 22.29.

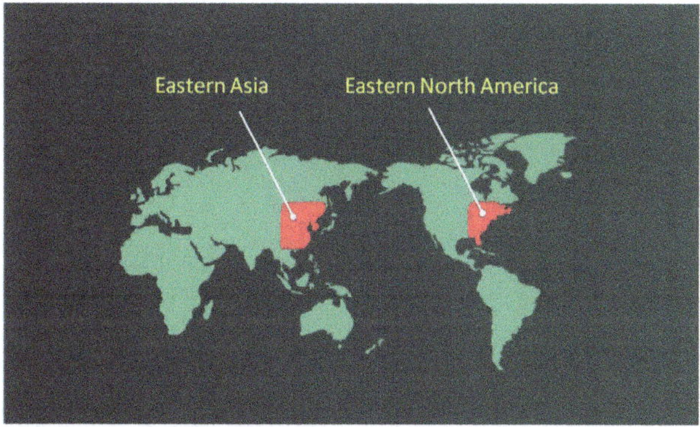

Fig.22.29: Plants in these Two Regions are Regarded as being the Same or Virtually the Same Species

Creationists would expect them to change because plants and animals appear to be designed by God to vary within their kind; and evolutionists would expect them to change because of genetic mutations and their understanding of the evolutionary process

In evolutionary thinking it took only six million years for ape-like creatures to evolve into people. Make no mistake, there are considerable differences between apes and people. But to the evolutionist this is easily explained: the evolutionary process, they say, is so powerful it can bring about these kinds of remarkable changes in just a few million years. Why then did all these plants and animals in Asia and America not evolve too and change significantly over the same period?

From a creationist point of view, though, it would seem perfectly reasonable to understand that there was a continuous plant and animal distribution linking these two parts of the world—but not of course millions of years ago, but in fairly recent history.

MORE BIOGEOGRAPHIC PUZZLES FOR EVOLUTIONISTS

Many plants and animals are found only in the northern regions and southern regions and not in between. Crowberries are one example, Figure 22.30.

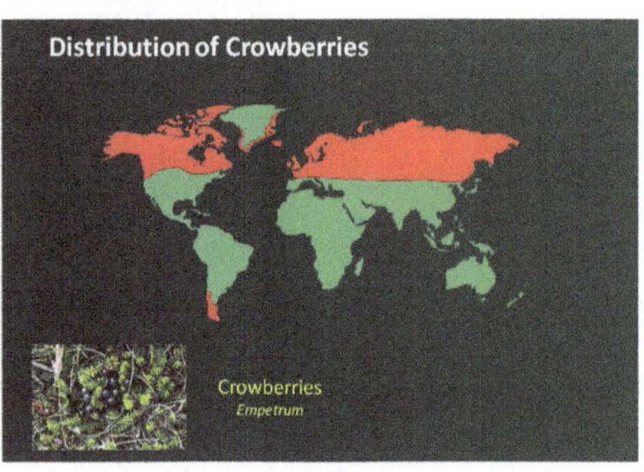

Fig.22.30: Plants and animals are found only in the northern regions and southern regions and not in between.

It certainly can't be said of these that they are found where they are because that's where they evolved. There are so many plants and animals found only in the northern and southern regions that some biogeographers have made the astonishing suggestion that these parts of the world were once in contact with each other. They have seriously suggested, based on biogeographic data, that the arrangement of the continents in the past was such that the northern regions were adjacent to the southern regions. Hence, again, we see the thinking and views of ancient earth biogeographers conflicting with the views of ancient earth geologists. Few of the latter would have much time for the idea that the northern regions were once adjacent to the southern regions.

So, how might we explain biogeography within the framework of biblical Earth history?

One process by which plants and animals could have spread around the world after the Genesis Flood is rafting—that is, on log mats driven by ocean currents. Actually, a growing number of evolutionists are proposing rafting as an explanation for how some plants and animals dispersed from one island to another, and even from one continent to another.

When Mt St Helens erupted in 1980, a tsunami was generated in the nearby Spirit Lake, and this caused around a million trees to be uprooted from the surrounding hillside. These eventually settled on the lake as an enormous log mat, Figure 22.31.

Fig.22.31: Rafting—Log Mats Driven by Ocean Currents

Following the great earthquake off the coast of Japan in 2011 and the resulting tsunami, a trail of debris formed in the Pacific Ocean, around 70 miles (100 km) long and covering an area of over 2 million square feet (186,000 square meters).

Now the effects of the Mt St Helens and Japanese tsunamis were nothing as compared with the destruction that would have been wrought by a global flood. The flood we read of in the book of Genesis would have resulted in billions of trees floating on the surface of the oceans. These log mats would have been like enormous floating islands and, regularly watered by rainfall, they could have easily transported plants and small animals' great distances. Some creationists believe that the pre-Flood world included great floating forests, a bit like the quaking bogs we know today. Perhaps these were broken up during the Flood and became rafts too.

The ability of ocean currents to distribute floating objects around the world was seen recently, when thousands of bathtub rubber ducks were lost off a container ship in the North Pacific. Within just a few months, these had floated to Indonesia, Australia and South America, and subsequently into the Arctic and Atlantic oceans. Figure 22.32.

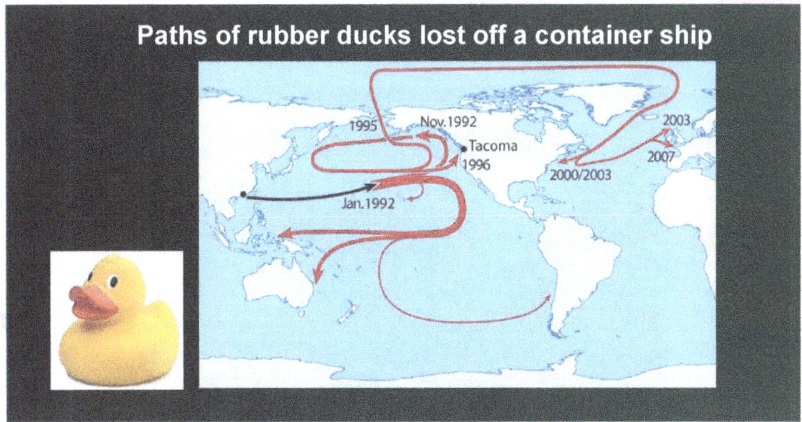

Fig.22.32: Ocean Currents to Distribute Floating Objects around the World

Often, we find plants distributed along coastlines and islands. The distribution of the Sago palm can be. It's found in East Africa, Madagascar, the tip of Indian and parts of Indonesia and Australasia, Figure 22.33.

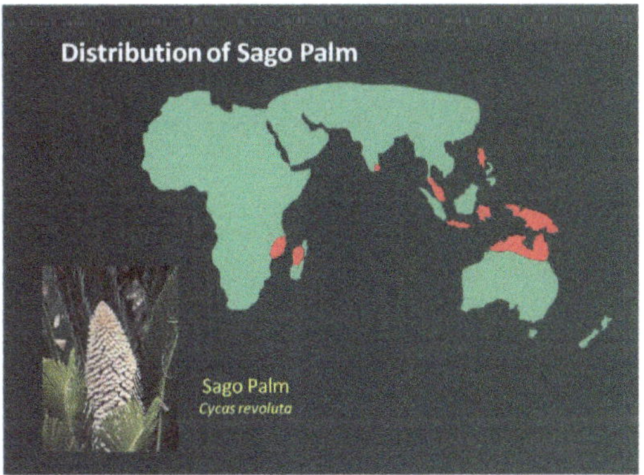

Fig.22.33: The Distribution of the Sago Palm

Pelargonium is another example. It's found right out in the Atlantic, in South Africa, Figure 22.34.

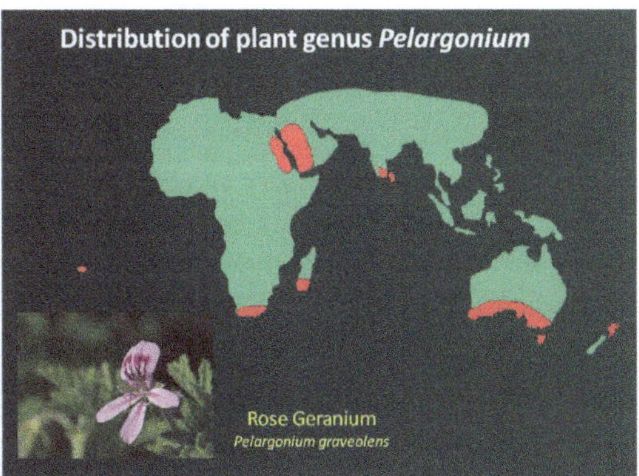

Fig.22.34: Pelargonium Found in the Atlantic, in South Africa

Madagascar, east Africa, India, Sri Lanka, southern Australia and New Zealand. Another example is a type of fern plant, Strangeriaceae, Figure 22.35.

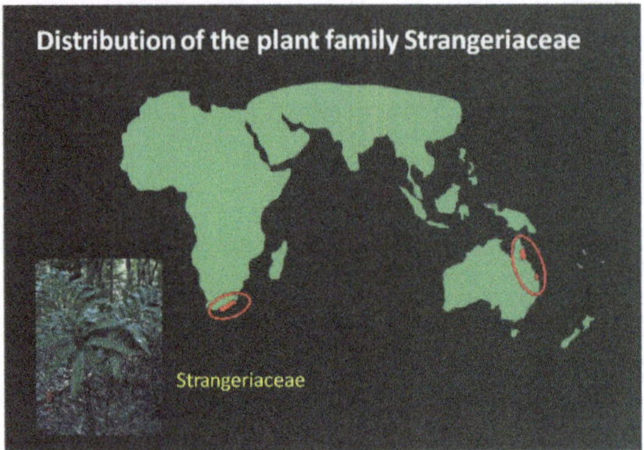

Fig.22.35: Fern Plant in Strangeriaceae

It's found in South Africa and along the eastern coast of Australia. This shows the distribution of a plant called Hook and Arn, a member of the carrot family. Figure 22.36.

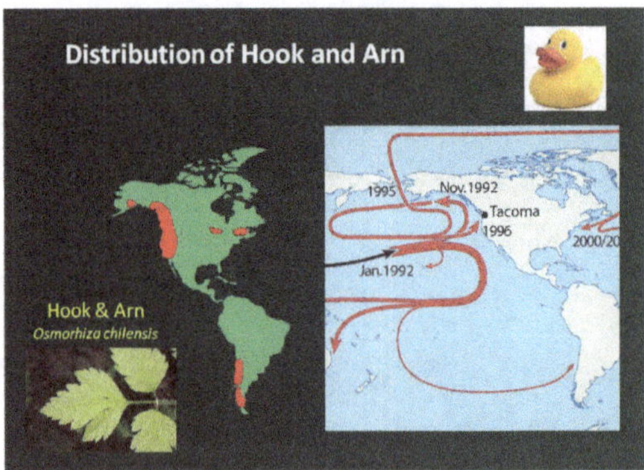

Fig.22.36: Plant Called Hook and Arn, a member of the Carrot Family

Based on the routes taken by the rubber ducks, it seems very reasonable to believe that rafting explains this. We can see how the rubber ducks floated to North and South America.

TRACKS OF DISPERSAL

A prominent feature of biogeography is 'tracks of dispersal.' This is an example. It shows how *Oreobolus* plants dispersed throughout Indonesia and Australasia and across the Pacific Ocean. And what's so significant about this is that many other plants (and small animals) have followed a similar route, Figure 22.37.

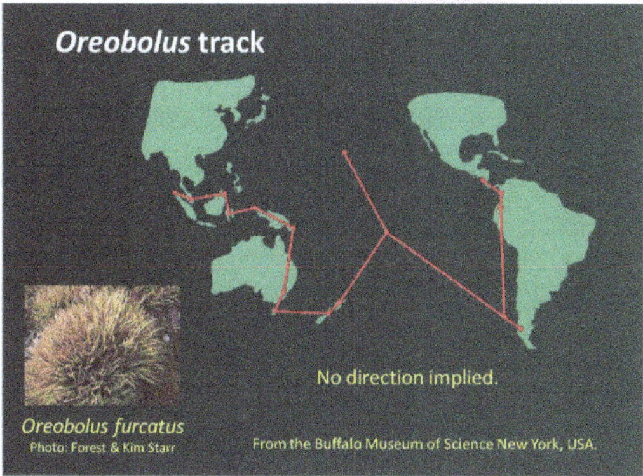

Fig.22.37: *Oreobolus Plants Dispersed Throughout Indonesia and Australasia and Across the Pacific*

Observe, too, that all the plants and animals following this track are found either side of the Pacific Ocean. Now, when the habitats of particular plants or animals are broken or split by land or water, it's called a disjunction. So, we might say that all these plants and animals are disjunct (or split) across the Pacific Ocean.

AREAS OF ENDEMISM

Text books typically show the main biogeographic regions. This is the map for animals, showing six main faunal (or animal) regions—regions where we tend to find the same sort of animals. But these kinds of diagrams are really over simplifications. A more realistic picture is where we find many different plants and animals concentrated in small regions known as 'Areas of Endemism,' Figure 22.38.

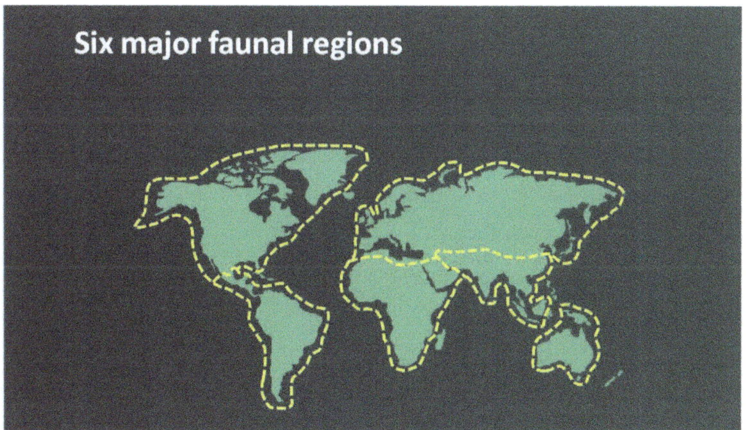

Fig.22.38: *The Map for Animals, Showing Six Main Faunal (or animal) Regions*

Now, endemic means native or restricted to a particular area, and an area of endemism is one where there are a high number of endemic species—where many different species are found in the same small distinct region. And here, each color represents one of these regions.

Interestingly, areas of high plant endemism often coincide with areas of high animal endemism. So, these areas where we find lots of different plant species concentrated together tend to be the same as the areas where we find lots of different animal species concentrated together. Many areas of endemism are coastal regions and islands.

For example, the tropical Andes, enclosed in red on the left, is the richest and most diverse plant region on Earth, Figure 22.39.

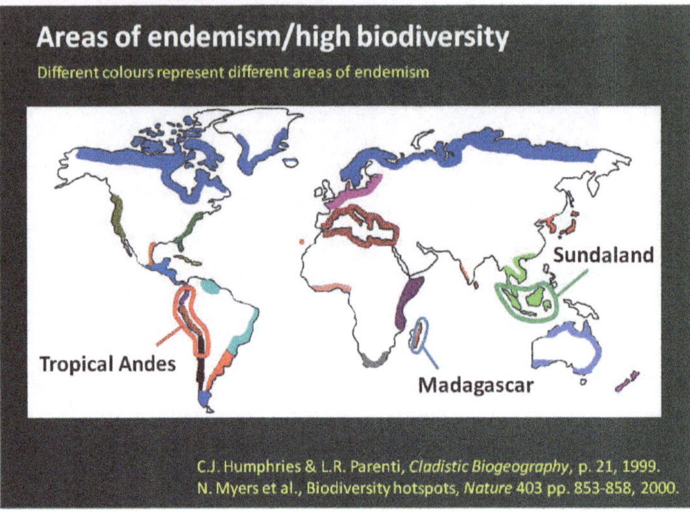

Fig.22.39: The Tropical Andes, Enclosed in Red on the Left

It contains around 15% of the all the world's plant life in less than 1% of the world's land area. Around 20,000 of its 40,000 plant species are endemic. The area of Sunderland, enclosed in green to the right, contains 15,000 endemic plant species and the island of Madagascar, enclosed in blue, has over 9,000 endemic plant species.

We also find many similarities between these regions—where the same plants and animals are distributed around or either side of an ocean. There are numerous patterns of disjunction like this, where, again and again, we find the same plants and animals in the same widely separated areas. In the last century, a man called Leon Croizat plotted many tracks like these showing dispersal routes across the world, Figure 22.40.

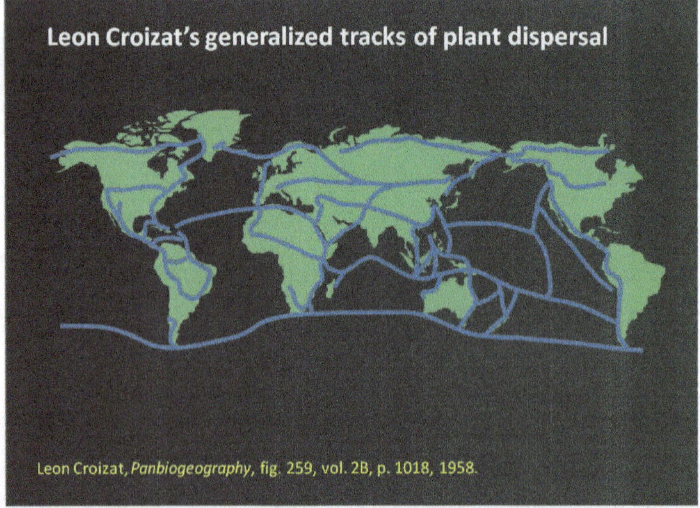

Fig.22.40: Leon Croizat Plotted Many Tracks Showing Dispersal Routes across the World

Each of these lines is known as a 'generalized track', where many different plants follow the same route. Croizat's work made clear that many tracks of dispersal either cross oceans or follow coastlines. A good case can be made for the biogeographic regions being oceans rather than continents! Christopher Humphries of the Natural History Museum in London and Lynne Parenti of the Smithsonian Institution wrote,

"Characteristically, many disjunct patterns span ocean bottoms, to the point that the oceans have been characterized as the natural biogeographic regions and the continents represent the land areas around the periphery."

To me, this is very strong evidence supporting the rafting hypothesis—that transport across oceans explains much of the biogeography of the world. I believe that many plants and small animals were transported on great log mats left over from the Genesis Flood and that the areas of endemism we see today correspond to the landing places of these rafts. Interestingly, researchers from Bryan College, Tennessee, showed that the intersections of ocean currents with the continents tend to coincide with these areas of endemism, Figure 22.41.

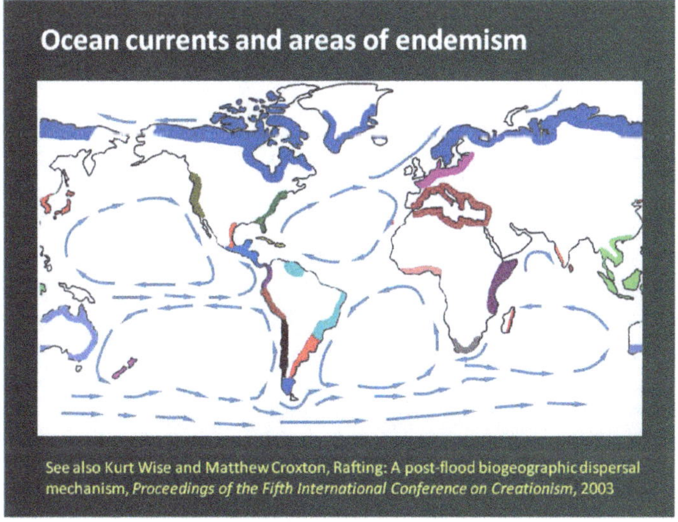

Fig.22.41: *Post-Flood Rafting Biogeographical Disposal Mechanism*

MIGRATION ACROSS LAND BRIDGES

Another way animals could have spread around the world is through migration across land bridges that are now below sea level, Figure 22.42.

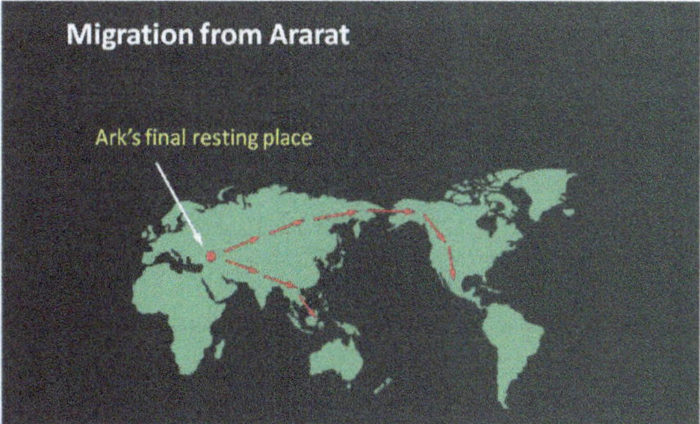

Fig.22.42: *Animal Migration across Land Bridges*

We believe that soon after the Genesis Flood, there was an Ice Age. We don't believe in many ice ages, but just one. Conditions would have been ideal for an Ice Age at the end the Flood. The oceans would have been warm, due to hot underground water being added to them at the beginning of the Flood, and this warming of the oceans would have caused much water to evaporate into the atmosphere. At the same time, volcanoes erupting during and after the Flood would have thrown lots of dust into the air and this would have blocked some of the sun's heat, keeping the continents cool. So, the large amounts of water vapor in the atmosphere would then have fallen as snow, building up significant amounts of ice on the land. Sea levels would then have fallen as the oceans' water evaporated and was trapped as ice sheets on the continents. And land bridges would have appeared as the sea level dropped.

There was likely a land bridge across what is now the Bering Strait and a number of land bridges linking parts of Indonesia and perhaps even Australasia. Of course, we don't see these land bridges today, because much of the ice has now melted and sea levels have risen again. Also, due to continued geological activity after the Flood, ground movements could have caused other land bridges to fall below sea level. This, it's not difficult to imagine how animals could have migrated from Ararat to many places throughout the world. Also, some plants and animals, especially those useful for farming, may have been transported by man, particularly during the dispersal from Babel.

I think, too, there was probably some rafting of animals to Australia and South America, Figure 22.43. In fact, North and South America may not have been connected until sometime after the end of the Flood. You can see on this diagram that I've shown South America detached from North America.

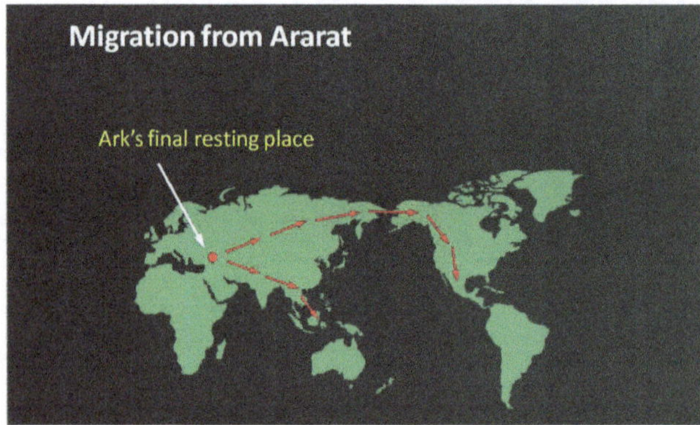

Fig.22.43: *Final Resting Place - Ararat*

MARSUPIAL DISTRIBUTIONS

Living marsupials are found in Australia, New Guinea and parts of America—Figure 22.44.

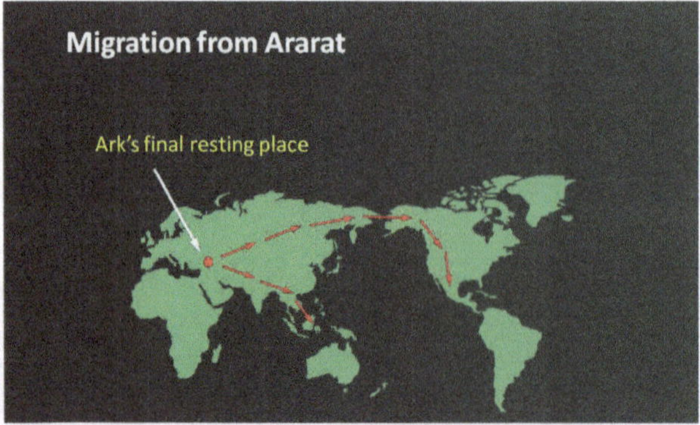

Fig.22.44: *Migration From Ararat*

Most marsupials, however, are found in Australia and New Guinea and many of these are found nowhere else. So how can we explain this within the framework of biblical Earth history? Well, I don't have any firm answers, but I can outline one possibility that I think makes sense.

Now, you remember we're dealing here with two types of mammals—placentals and marsupials. And there's some evidence to suggest that placentals tend to outcompete marsupials when they share the same habitats. From a biblical point of view, this does seem plausible. Marsupials stepped off the Ark alongside the placentals and must have lived around the Middle East and the surrounding continents. Yet only placentals live in these places today. It seems significant, too, that North America has only one marsupial—the Virginia opossum. Marsupials have become well established in Australia but largely in the absence of placentals.

Perhaps competition from placentals drove marsupials to migrate away from the Ark ahead of placentals. Marsupials then gained an early foothold in Australia and South America and, without competition from placentals, they thrived in those places. And perhaps, as the log rafts broke up and sea levels rose and covered the land bridges, Australia and South America became almost completely isolated before very many placentals had made their way to those continents, Figure 22.45.

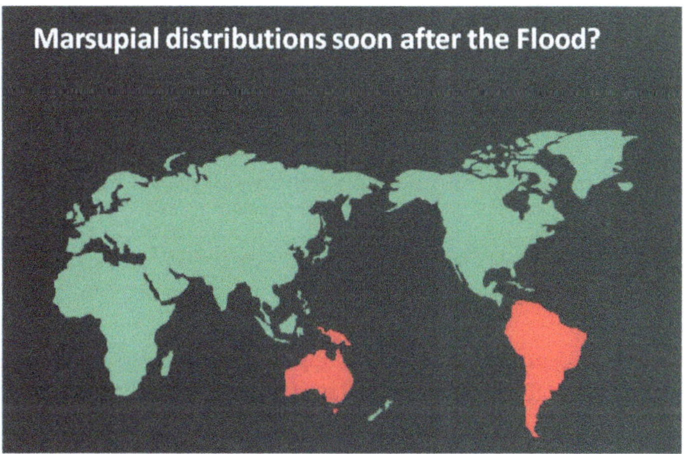

Fig.22.45: *Marsupial Distribution after the Flood*

Accordingly, driven by competition from placentals, marsupials could have migrated to Australia and South America and then been protected from placental competition as these continents were cut off from the rest of the world. Fossils of marsupials are found on every continent. So, in evolutionary thinking, marsupials died out in all the continents except the ones where we see them today. Why can't creationists simply argue the same?

There are some interesting twists in the evolutionary story about marsupials. If, in evolutionary thinking, we go back to the late Cretaceous rocks, supposedly around 65 to 80 million years ago, we don't find any marsupial fossils in Australia and South America. They are found only in Europe, Asia and North America. An article in *Science* journal said this:

"Living marsupials are restricted to Australia and South America ... In contrast, metatherian [marsupial] fossils from the Late Cretaceous are exclusively from Eurasia and North America ... This geographical switch remains unexplained."

Hence, again, according to evolutionists, 65 million years ago marsupials lived in Europe, Asia and North America, but they then died out in those areas and now live in Australia and South America. If evolutionists can have marsupials dying out in Europe, Asia and North America, why can't creationists?

Another problem for evolutionists is the 'Little Mountain Monkey' of South America, Figure 22.46. DNA comparisons suggests that this little South American marsupial is more similar to the Australian marsupials than to other American ones. How did it end up in America? It seems very difficult to argue that it evolved there.

Fig.22.46: 'Little Mountain Monkey' of South America

In Conclusion:

- The main evidence for evolution from biogeography is speciation—a fact of biology that is better explained by the creation model.
- The distribution of plants does not support the belief the continents were joined for millions of years.
- Fertile hybrids between animals on different continents indicate that they have not been separated by millions of years.
- Plants and animals are concentrated in small regions of high biodiversity along coastlines and islands. While these correspond, to some degree, with areas of high rainfall, this does not explain why there are many patterns of disjunction where the same plants and animals are found in the same widely separated areas of endemism.
- Such biogeographic observations, however, appear to be well explained by transport across oceans.
- Log rafts left over from the Genesis Flood could have provided the means of rafting.
- Migration across land bridges may explain other biogeographic distributions.
- Following the confusion of languages at Babel (*Genesis 11:1-9*), when humanity spread out around the world, people probably took with them plants and animals for farming and other purposes.

As with other branches of science, the data appear to fit the biblical account of Earth history very well.

Fig.22.47: Genesis 1 – 'In the Beginning God Created'

Interpreting Genesis 1

Like any discipline, Bible interpretation methods (hermeneutics) can suffer from being used to solve problems which lie outside its sphere. Much of the modern discussion of interpretation methods is inconclusive because it involves an interchange between men who differ not at the level of method but at the more fundamental level of religious beliefs. Hence to put this writing in its proper context it must be stated that the investigation begins by assuming that the view of Scripture associated with evangelical Christianity is valid, Figure 22.47.

INTERPRETING SCRIPTURE FROM OUTSIDE

In considering the interpretation of the early chapters of Genesis it is important that our own historical situation be clearly in view. We are not the first Christians to be troubled by the teaching of Genesis. Simply because the Bible has a different view of origins to those put forth in human philosophy there is a period of conflict whenever the church comes under the influence of a human philosophical system. Thus, any defender of the beliefs of Plato in Augustine's day or of Aristotle in the late Middle Ages found himself in trouble with Genesis.

It is a gross oversimplification to act as though we alone face a problem here. Nevertheless, the problem for most Christians today is generated by a specific challenge, namely that of biological evolution and related theories. It is believed that there are deeper problems than merely the problem of Genesis.

If we take the theory of evolution as established and modify our interpretation of Genesis accordingly, then we introduce a problem for the doctrine of Scripture. It is nonsense to speak of the unique and total authority of Scripture at the same time as we change our interpretation of Scripture to accord with theories drawn from outside Scripture.

Hence evangelicals have tended to seek for principles within Scripture itself which will allow them to interpret Genesis in a way that is compatible with evolution. If Scripture itself forces us to such an interpretation, then we are not subjecting Scripture to evolutionary theory. It is with these attempts to find such principles within Scripture that this writing is mainly concerned.

RELIGION AND SCIENCE

However, there is need to establish first that the basic problem can really be reduced to methods of interpretation. Particularly this must be demonstrated when there has been a tendency to solve the problem by regarding the Biblical and the evolutionary descriptions as complementary rather than conflicting.

This may be expressed in many different ways but the basic idea is a distinction between religious, theological and/or naive explanations as distinct from scientific, technical ones. It is argued that there is no conflict because the two approaches are in separate spheres or on separate levels. It must be emphasized that this in itself does not solve the basic problem. It merely shifts the point to be proven.

If we interpret Genesis in terms of this religious/scientific distinction we may be just as guilty of imposing an alien authority upon the Scriptures. We must first establish that such a distinction is warranted by Scripture. The distinction itself looks suspiciously like Kant' distinction between the realm of ideas (noumena) and the realm of sensory experience (phenomena). It makes little difference in principle if the foreign authority is that of Kant rather than Darwin.

Kant' Categories:

"Kant" ultimately distinguishes twelve pure concepts of the understanding, divided into four classes of three:

- Quantity. Unity. Plurality.
- Quality. Reality. Negation.

- Relation. Inherence and Subsistence (substance and accident) Causality and Dependence (cause and effect)
- Modality. Possibility. Existence.

In saying that the distinction must be demanded by Scripture itself before it can validly be employed one misconception must be avoided. If someone approaches the Scripture already accustomed to seeing things in terms of the Kant' categories, then the basic question has already been decided. Is Scripture a book of religious truths or a textbook of geology? We naturally tend to say it is the former. Yet this question may pose a false dilemma. There is always the possibility that it is a book of religious truths which lays down basic principles which are relevant, even mandatory, for geology.

If the question is posed so as to exclude this last alternative, and Kantian philosophy so poses the question, then the basic problem has been solved not by appeal to the explicit teachings of Scripture but by a philosophical presupposition drawn from outside the Scriptures.

GENERAL REVELATION

A second way in which an attempt is made to solve the problem, without having to resort to the difficult task of establishing internal guidelines for the interpretation of Genesis, is by appeal to general revelation. It is claimed that since the creation is itself revelatory of God, we do not impose an outside authority when we interpret Scripture in terms of science. However, once again, the basic problem is not solved but merely camouflaged. Is our concept and use of general revelation a valid one or is '*general revelation*' merely a label which allows us to ignore or destroy Biblical teaching?

The question can only be decided by establishing a correct view of *general revelation* on the basis of Scripture. One may say categorically that a Biblical view of *general revelation* gives no support to the common use of science to determine our interpretation of Genesis.

- First there is no indication in the Bible that *general revelation* tells us about the means God used in creating the earth and life upon it. The passages which theologians appeal to in establishing a doctrine of *general revelation*, such as Psalm 19; Romans 1, etc., tell us that creation reveals the nature of God. We may argue that the creation reveals the glory and power of its creator. We have no basis for saying that it 'reveals' scientific theories.
- Secondly Romans 1 is adamant that sinful man suppresses and distorts the revelation of the creation. Any view of the creation that commands a consensus among non-Christians must be suspect. The appeal to certain scientific theories as though they are to be treated as revelation is completely invalidated by the Biblical teaching on general revelation.
- Finally, even if one were to grant that the creation does clearly reveal the manner in which God created the heavens and the earth, we would have to maintain the distinction between what the creation reveals and what people say it reveals. This is equivalent to the distinction between infallible Scripture and fallible later theologies.

Thus, we would have to decide whether evolution etc. was actually what was revealed by creation. Discussion of this question lies beyond the realm of this writing but a few remarks may be made. In order to conclude that a scientific theory is a correct interpretation of *general revelation* one must be certain that the method by which it was established was not in any way contrary to Biblical teaching. We certainly cannot say this for a science which systematically excludes any supernatural factors. There is no logical alternative to evolution once the intervention of God has been excluded. Furthermore, even among those who metaphysically accept evolution there is no certainty that it has been proven.

'THE THOUGHT-FORMS OF THE DAY'

Another of the attempts to solve the problem is that which claims that God expressed himself in the thought-forms of the day. It would therefore be wrong to attempt to make these categories authoritative for our scientifically sophisticated age. The same reservation is valid here as previously. This assertion about the way in which God revealed the history of creation must itself be justified by Scripture.

Parenthetically, it should be noted that this argument is formally identical with that used by Bultmann in his appeal for the demythologization of the resurrection narratives. He similarly argues that the resurrection narratives are expressed in terms of concepts held in that day which cannot be taken literally today. Here evangelicals typically maintain a great inconsistency, being ready to accept a form-critical method when it applies to the Old Testament but not to the New Testament.

To return to the main point, the argument being considered has a number of serious weaknesses. In order to apply it consistently one must first make some sort of a distinction between the cosmology implied in the terms used and the theological truth conveyed by the use of those terms. That is to say, unless one wants to remove the whole of *Genesis 1:11* from the Bible, one argues that theological truths can be separated from the views of the physical universe implied. Such a distinction is just a variant on the Kantian noumena/ phenomena distinction discussed above.

It would greatly help the discussion if this supposed use of concepts common to the era was more carefully specified and defined. One would like more than the bare assertion that the Bible employed the common concepts of the day. For the argument to be valid this would have to be carefully established.

Once again this lies outside the main subject of this writing but a few remarks are necessary. One must first reckon with the fact that certain ideas or stories may be shared by the Bible and surrounding cultures because they are both based on a historical event. For example, it would be rather ridiculous to argue that God chose to convey certain theological truths in terms of the flood concepts already possessed by the Mesopotamians. Obviously both Bible and Sumerian traditions mention a flood because there was a flood.

As in the case of evolutionary theory there is a problem created by the fact that much work in the ancient Near Eastern field specifically excludes God's activity. Hence the ideology and concepts of Israel must be considered as derived from its neighbors. As long as this view is prevalent the uniqueness of Biblical thought is depreciated and denied.

A more mundane problem is the fact that when the ancient Near East History was a younger study it was natural to use the known to illuminate the unknown. Problems were solved by the use of Biblical analogies and the impression thus created of a greater degree of common ground than was warranted. More investigation has a tendency to remove this false overlap.

If supernatural intervention in the history of Israel is rejected, the most plausible explanation for the religion of Israel derives it by a process of ideological evolution from Israel's neighbors. It follows then that the concepts of Israelite thought must be those common at the time. However, if we do not make this assumption, and Scripture will not allow us to make it, then we must carefully investigate the thought of the ancient Near East in order to see if the same concepts are used as in the Biblical text. Even this search is fraught with problems of personal bias. Some version or other of the flood story was known in Mesopotamia. There was also a memory of the fact that at one time man had a common language though to my knowledge the confusion of tongues was not connected with the tower of Babel.

One resemblance which is often referred to is that between the creation of the heaven and the earth in Genesis and the splitting of Tiamat to form the heaven and the earth in the Mesopotamian *Enuma Elish* legend. The tree often depicted on cylinder seals has been connected with the tree of life.

These last two examples raise another set of problems. When it is said that God employed symbols common in that day is it meant that both the symbol and what is symbolized were already known or that only the symbol was known with a completely different connotation? The distinction is an important one. For this argument to be convincing the former must be the case.

Otherwise, one is saying that God gave the symbol a completely new meaning. And if he did that, we are no longer dealing with symbols common at the time, but with new symbols. Then the necessity of interpreting them against the Near Eastern cultural background is removed. Whether there is any ultimate relationship between Biblical and Babylonian accounts as we now have them, they belong to different ideological worlds.

The symbols are not the same because the ideology is different. The goddess Tiamat defeated in a war by the god Marduk, if she may be called a 'symbol', must be seen as a symbol within the context of Babylonian polytheism whereas the creation of heaven and earth belongs within the context of Biblical thought. It is meaningless to say that God used the same symbol but changed its meaning. It is then no longer the same symbol.

Furthermore, there are important elements in the early chapters of Genesis with no real counterpart in contemporary thought. Of course, it is quite possible that such a counterpart existed and has been lost. However, the onus of the proof lies on those who confidently affirm that Genesis employs the common symbols of the day. There is no real counterpart to the fall into sin in literature of that time.

'NAIVE COSMOLOGY'

Sometimes, it seems that those who claim that the Bible used the symbols of its day are merely trying to say that it used a naive as opposed to a scientific cosmology, or, to put it more popularly, it did not bother to correct the prevalent three-story cosmology. If we assume for the sake of the argument that this is the case, then it should be clearly recognized that all we have established is that scientific dogma should not be made out of Biblical cosmology. The argument has no relevance to other parts of the account like the creation of animals, man, etc.

Unfortunately, this argument is generally used without this careful delimitation. Generally, it is argued that the fact that one element shows the use of nonscientific concepts proves that the whole uses naive ideas whose details may not be pressed.

Yet once more the validity of the basic premise must be questioned. Was there ever a pure 'three-story universe' idea in antiquity? For the pagan contemporaries of the Bible writers, cosmology was theology. The heavens expressed and were controlled by the various divinities. The sort of abstract spacial/mechanical interest involved in the idea of a three-story universe is a product of the demythologization of Greek rationalism and Euclidian spacial concepts.

One should not try to project a late idea back into Biblical times in order to explain the Bible. In its rejection of polytheism Biblical cosmology is of necessity radically different to its surroundings. It is not popular cosmology.

Secondly, what is so wrong about a 'naive cosmology'? It is probably as close to the ultimate truth as modern cosmology. If we had not defied modern science, we would not be embarrassed by those points in which Biblical thinking diverges from prevailing modern ideas. Certainly, Biblical cosmology fits into a different structure of thought from modern cosmology, but it is the validity of that very structure of thought that is at issue. We tend to assume that the assumptions underlying modern physics are unquestionable. If we assume the validity of the structure of physics from any period with its philosophical presuppositions and concomitants we run the risk of accepting a structure which, because of its ultimate origin in a total humanistic philosophy, must clash with a Biblical worldview.

What has generally happened is that the structure and method of modern science has been accepted as truth. When the conflict between this and a Biblical view has been appreciated, an attempt has been made to give the Biblical view a validity in some sort of restricted religious sphere. The basic question is whether our interpretation of the Bible is to be determined by the Bible itself or by some other authority.

Once science has been set up as an autonomous authority it inevitably tends to determine the way in which we interpret the Bible. From the point of view of this discussion the outside authority may be Newton or Hoyle just as well as Darwin or Kant. The issue involved is still the same.

Somewhere in this sort of discussion poor Galileo is always dragged in. Yet, if we want to learn from history we should at least begin with good history. There is nothing particularly Christian about Aristotelian cosmology. In fact, there are points at which it cannot be reconciled with the Bible. How did the church find itself in the position of defending Aristotelian cosmology against the new Copernican cosmology? It found itself in that position because it accepted the argument of Aquinas that the Biblical texts which contradicted Aristotle should not be pressed as the Bible was not written in technical philosophical language.

Moses spoke the language of his day. This is not to say that the church should have accepted readily the new astronomy. In its neo-Pythagorean mysticism, it was no more Biblical than Aristotle was. Those who want to say that the Bible is written in the popular language of its day and should not be pressed where it differs from modern philosophical-scientific structures cannot claim to have learnt from the Galileo affair. They are merely repeating the arguments that helped to put the church in that situation.

METHODS OF INTERPRETATING—GENESIS 1–11

One of the fundamental beliefs of evangelical Christianity is that "the Bible is its own interpreter." This belief comes under attack when Christians try to interpret *Genesis 1–11* so as to fit into evolutionary biology and uniformitarian geology. Various justifications are given for this, e.g., "the Bible is teaching religion and not science," or "we must interpret Genesis by *general revelation* (=science)", or "the Bible was written in terms of the naive unscientific beliefs of its day." When each of these arguments is examined, it is found that it involves interpreting the Bible according to ideas drawn from outside the Bible. Thus, the Bible is no longer its own interpreter.'

Is there any explicit teaching within the Bible itself that suggests its details are not to be pressed in matters of the physical creation?

We know of no such teaching. When reference is made to the original creation, the creation narrative is treated as fact without any reservation. Consider Peter's argument in 2 Peter 3:5–7, 'But they deliberately forgot that long ago by God's word the heavens existed and the earth was formed out of water and with water. By water also the world of that time was deluged and destroyed. By the same word the present heavens and earth are reserved for fire, being kept for the day of judgment and destruction of ungodly men.' Observe that the apostle did not shrink from reliance upon some of the details of the Genesis narrative. Other examples of Biblical references back to Genesis (e.g., Exodus 20:11; Matthew 19:4; Romans 5:12-19; 1 Tim. 2:13-14), to be considered in more detail below, show a similar reference to specific details such as creation in seven days (Exodus 20:11) and creation of woman from the man (1 Timothy 2:13–14).

This should in itself be enough to dismiss the frequent statement that we may not press the details of the account. Yet, most people would want to interpret scripture upon a basis of Kantian philosophy. But this philosophy itself is not sanctioned by Scripture. No clear distinction is ever made in the Bible between statements concerning the physical creation and theological statements.

One influences and determines the other. Note that in the Biblical references given above, the form which the original creation took is made the basis of theological and/or ethical teaching. The separation between physical creation and theology is one that has to be imposed upon the text by us. It is not naturally there in the Bible.

THE LITERARY CHARACTER OF GENESIS 1

It seems a more serious attempt at exegesis when appeal is made to the literary nature of Genesis 1. Even here care is needed that an outside standard be not imposed. One cannot simply define Genesis 1 as poetry by using a standard of poetry drawn from outside the Scripture, without assuming the very point at issue.

Even if Genesis 1 were poetry, we would still be entitled to inquire what truth it conveys. Our answer to that question would have to be framed in terms of the rest of Scripture. If we take the passages referred to above, we obtain enough to place us in conflict with modern evolutionary approaches. Thus, the claim that *Genesis 1* is poetic does not resolve the problem.

Furthermore, by what criteria do we call *Genesis 1* poetic? The parallelism of days 1–3 to 4–6 is often cited. This however, is merely parallelism that makes up Hebrew poetry. Hebrew poetry consists of a series of couplets or triplets exhibiting complementary, climatic or antithetic parallelism e.g., in Psalm 5:1, *'Give ear to my words, O Lord'* is complemented and paralleled by *'Consider my meditation'*. This is clearly different from the fact that on days 1–3 God creates the environment and on days 4–6 the creatures who are to live and rule in the respective environments. One is a parallel of ideas in successive *stichoi*, the other a parallel of ideas which may be several verses apart.

Nevertheless, it may be argued that the very fact that Genesis 1 exhibits such a structure proves that it is not to be taken literally. Surely, to state this argument is to refute it. Short of some sort of metaphysical presupposition that regards history as totally random and all order in historiography as being a result of arbitrary human imposition, I cannot see how one would ever prove such a proposition. The attempt to make a case by analogy from the book of Revelation is quite beside the point. If we took elements of Revelation as symbolical without explicit Biblical warrant then we would be guilty of imposing an outside standard upon the Scripture. Revelation itself tells us that we are meant to see symbolism in its pictures: *'the great city, which is allegorically called Sodom and Egypt, where their Lord was crucified' (11:8); 'And a great portent appeared in heaven' (12:1); 'and on her forehead was written a name of mystery, 'Babylon the Great ... I will tell you the mystery of the woman ... This calls for a mind with wisdom. The seven heads are seven mountains ... and there are also seven kings ... The waters that you saw, where the harlot is seated, are peoples and multitudes ... And the woman that you saw is the great city which has dominion over the kings of the earth'* (17:5–18). It is the lack of a similar interpretation of the 'symbolism' of Genesis which so sharply distinguishes Genesis and Revelation.

STRUCTURED HISTORY

Even though there is no logical reason why the presence of a structure should prove that a passage is not to be taken literally, this idea seems, to have great emotive appeal. The whole question of structured history needs to be examined more closely. The title of this writing limits discussion to Genesis 1–11. This is because among evangelicals anyway there is a willingness to accept the historicity of the patriarchal narratives. However, the patriarchal narratives are structured history in the same way as the earlier chapters of Genesis. They fit within a framework created by the heading *'These are the generations of...'* (2:4; 5:1; 6:9; 10:1; 11:10; 11:27; 25:12, 19 etc.). There are clear instances of parallel structure. Thus, the experiences of Isaac parallel those of Abraham. Both have barren wives (15:2; 16:1; 25:21). Both lie concerning their wives (20:2; 26:7). Both face famine in the promised land (12:10; 26:1). Both make a covenant with the Philistines (21:22–34; 26:26–33). If parallelism of structure proves that a passage is not historical then the patriarchal narratives are not historical.

This of course is the conclusion of many liberal exegetes, but evangelicals once more maintain an inconsistency, being willing to apply a higher-critical principle in one area of Scripture but not in another.

If one looks carefully at these structured histories, one sees that the structure is theological. Abraham and Isaac both face barrenness and famine because they both experience the trial of faith in being forced to believe the promise of God contrary to the physical situation (Romans 4:17–18; Hebrews 11:8–12). The structure that underlies the parallelism of Genesis 1 is that of covenant vassal and suzerain. On days 1–3 the environment or vassal was created and on days 4–6 the appropriate creature or suzerain to live and rule in that environment. This notion of covenant head and vassal underlies also the story of the Fall in that on the fall of the suzerain the vassal is placed in rebellion against its Lord (3:17–19). Further the idea of covenant structures the whole of history into old and new covenant each under their respective heads (Romans 5:12–21; 1 Corinthians 15:45–49). For the historian who proceeds on antitheistic assumptions such a theological history must be rejected. He must assign all such histories to the category of theological subjectivism.

A theologically structured history presupposes a God who actively shapes history so that it conforms to his plan. A liberal exegete who denies the existence of such a God must dismiss as true history all Biblical accounts which see theological patterns in history.

The evangelical has no basis for such an a priori dismissal of structured history. The fact that Genesis 1 displays a structure in no way prejudices its claim to historicity.

SCRIPTURAL INTERPRETATIONS OF THE GENESIS ACCOUNT

So far, the views discussed have consisted of statements about Scripture which were not themselves based on Scripture. An a priori statement about the Bible cannot claim Biblical authority. Discussion of this area has been obscured by the number of these statements and there is a need to return to interpreting Scripture by Scripture and not by hypothesis. There are a number of passages which reflect upon the original creation. Some have been referred to in other connections above.

Exodus 20:8–11 is significant in that it gives us a clear answer to the debated question about whether the 'days' of Genesis are to be taken literally. The commandment loses completely its cogency if they are not taken literally.

This passage is also important in giving a proper direction of our thought. It is often said that the creation is described in seven days because this is the pattern of labor to which the Hebrews were accustomed. The text however says the very reverse. The Hebrews are to become accustomed to a seven-day week because that is the pattern that has been set by God. The point is an important one as it is crucial to the distinction between true and false religion. The oft-repeated claim that human thought and custom has created the categories through which, of necessity, all God's activity must be viewed is a denial of the spirit of Biblical religion. It gives to man the priority which rightly belongs to God.

Psalm 104 deserves more consideration in this question than it usually receives. The Psalm follows in a general fashion the order of the creation days. The one point that is of particular interest is that the psalmist has integrated the account of *Genesis 1* with that of the creation of springs in Genesis 2:4–6. The reference to springs falls where one would logically expect it between the account of the creation of dry land (*Psalm 104:6–9*) and that of vegetation (Psalm 104:14–17). The problems of relating the accounts of *Genesis 1* and 2 is outside the scope of this writing but any attempt must begin with *Psalm 104*. Unfortunately, some evangelicals have accepted too readily the assertion of the documentary hypothesis that they are independent accounts of creation. The psalmist knew better. A number of passages which refer to the original creation of man and woman and their relationship may be considered together (Matthew 19:4; 1 Corinthians 11:8–9; 1 Timothy 2:13–14).

Observe that the account is taken literally and made the basis of teaching on the relation of man and woman. Even if in only this point we take issue with evolutionary theory we find ourselves in complete antithesis to naturalistic evolution. If on the authority of Scripture, we hold to the Biblical account of the creation of man and woman then we can give up all hope of a harmony between the Bible and 'science'. The proper subject of this writing is the interpretation problem and these passages are adduced to show that the rest of Scripture sees the early chapters of Genesis as literal history. It may be objected as a last resort that only those details of the account mentioned as literal by the rest of Scripture may be taken literally. Even if this point be granted there is still enough contained in just these few verses to reopen the battle with evolutionary theory.

However, the argument that only those passages in Genesis 1–11 referred to elsewhere as literal accounts are to be taken as such may be summarily dismissed. The early chapters of the Bible are clearly a unity and whatever interpretation method is valid for part is valid for all. This fact has been realized by those who have sought by various arguments to find evidence of 'poetry' in one part and to extend it to all. Yet all these attempts in so far as they were not attempts to see how the rest of Scripture treated the chapters in question must be condemned as methodologically faulty. Scripture is its own interpreter.

Against this one might argue that even though the NT treats Genesis 1–11 as literal, this should not be taken as proving that it is a literal description. One may argue that the NT writers were accommodating themselves to the beliefs of the time or that these passages are referred to only as illustrations and that their literalness is not implied by the NT usage. The first alternative must be rejected as involving a denigration of Christ and his apostles. The accommodation argument when used as a way of avoiding the implications of Christ's use of the OT for the doctrine of Scripture has been rightly rejected by evangelicals.

It is inconsistent to attempt to revive it to avoid the implications of NT teaching on another subject. Furthermore. the fundamental objection against a rule of exegesis drawn from outside Scripture applies here also. If the accommodation idea is to be allowed in the discussion, then it must

- first be demonstrated that it is itself taught by Scripture.
- The second alternative will not bear examination.

Clearly in 1 Corinthians 11:8-9 and 1 Timothy 2:13-14 the argument of Paul would collapse if the details of the account to which he refers did not happen as recorded. It is foolish to suggest that his point would still be valid even if woman was not created after and from the man and even if Eve was not beguiled into sin. Similarly, Peter's point is without cogency if the world was not destroyed by the Flood (2 Peter 3:5-6).

The thrust of this writing has been to direct discussion away from theoretical pre-exegetical arguments over the interpretation of Genesis and to concentrate on the way the rest of Scripture interprets it. We meet simple literalism in the scriptural exegesis of Genesis. Certainly not every detail of the bible chapters in question is referred to elsewhere but when they are literalism prevails.

If this be the case why has so much discussion been concentrated on arguments which are not only inconclusive but also diminish the right of Scripture to be its own interpreter?

We suspect that the real debate is not interpretation at all. If it were, then it would have been decided long ago by a comparison of Scripture with Scripture.

The real problem is that we as Christians have in a double sense lost our historical perspective. We have forgotten that the church has always been under pressure to allegorize Genesis so that it may conform with Plotinus or Aristotle or some other human philosophy. We have treated the problem as though it were a modern one, as though we alone have had to face the onerous task of holding to a view of cosmic and human origins which is out of sympathy with the philosophical premises of our culture.

The second sense in which we have lost our historical perspective is that we have forgotten that until our Lord returns, we face strife and conflict in this world. We have sought to avoid that conflict in the intellectual realms. We have accepted the claim of humanistic thought that its scholarship is religiously neutral when the Bible teaches us that no man is religiously neutral. Man, either seeks to suppress the truth in unrighteousness or to live all his life to the glory of God. In that total warfare scholarship is no mutually declared truce.

APPENDIX

a. Thompson, J.A., *Genesis l-3 Science? History? Theology?* Theology Review 3/3 p. 16.

b. The attempt to explain these parallel incidents in terms of the documentary hypothesis is shown to be ridiculous if an attempt is made to assign each parallel to a different source in every case in which a parallel exists. The cases of both Abraham and Isaac lying concerning their wives is often used as proof of the documentary hypothesis. However, inconsistently, the theory attributes both barrenness accounts and both famine accounts to J. The inconsistencies become more evident if the parallels in the life of Jacob are also considered. Basically, the documentary hypothesis is able to make a plausible case by ignoring most of the incidents of 'duplicate' narratives. When all are taken into account then it is clear that the 'duplicate' narratives and the other 'criteria' for dividing documents come into conflict.

c. John Murray (in *Principles of Conduct* [London: IVP; Grand Rapids: Eerdmans, 1957], p. 30) claims that Genesis 2:2 refers to 'the seventh day in the sphere of God's action, not the seventh day in **our** weekly cycle' (emphasis his). Consideration of this question would involve a lengthy treatment of the meaning of God's seventh-day rest. The frequent affirmation that the seventh day of Genesis 2:2 is still continuing needs to be proven. Murray unfortunately omits such proof. Briefly it may be argued that the text gives no indication of such a sphere distinction. The text is not concerned with God as He is in Himself but with God's activity in a temporally conditioned creation. Even the seventh day refers not to God in Himself but to God in relation to His creation. At this point I can agree with Murray (ibid., p.31): God's rest is the rest of delight in the work of creation accomplished. '*And God saw all that which He made, and behold, it was very good*' (Genesis 1:31). This is expressly alluded to in Exodus 31:17 in connection with God's Sabbath rest, '*On the seventh day He rested and refreshed Himself*' and means surely the rest of satisfaction and delight in the completed work of creation.

d. Packer, J. I., *'Fundamentalism' and the Word of God* (London: IVP, 1958), pp. 59–61.

e. Scattered light is polarized; the direction of polarization enables the butterfly to determine the direction of the light source even if obscured by clouds.

f. This is far from being a new situation. Many techniques of literary and form criticism were used first in the Old Testament field and later created much greater opposition when consistently applied in the New Testament. Gunkel himself was moved to the Old Testament field from New Testament when it was realized that his methodology could be applied there and incur less opposition.

g. Presumably not many would postulate that all animals around the world were sent to Noah for his selection; the most reasonable inference from the Genesis account by far is that only those intended for the journey were compelled to make the trip to the boat.

h. It is possible that the Mesopotamian parallels are the results of distortions of the original creation narrative to fit a polytheistic system. If that is the case they would then belong to the same category as the flood account. The argument that the Mesopotamian accounts must be the originals because our extant documents of the Mesopotamian versions are older than the extant Biblical texts (Speiser, E.A., *Genesis* [Garden City: Doubleday, 1964] p. 10) is utter nonsense.

REFERENCES

1. Benediktová, K. *et al.* Magnetic alignment enhances homing efficiency of hunting dogs, ncbi.nlm.nih.gov, 16 Jun 2020.
2. American Physical Society, Cryptochrome protein helps birds navigate via magnetic field, sciencedaily.com, 27 Feb 2015.
3. Lund University, even non-migratory birds use a magnetic compass, sciencedaily.com, 18 May 2017.
4. Lund University, how birds can detect Earth's magnetic field, sciencedaily.com, 6 Apr 2018.
5. Bangor University, Birds can 'read' the Earth's magnetic signature well enough to get back on course, sciencedaily.com, 12 Feb 2021.
6. Sarfati, J., Migratory birds use magnetic GPS, *Creation* 44(2):16–17, 2022.
7. University of North Carolina at Chapel Hill, Baby sea turtles use earth's magnetic field to navigate across Atlantic Ocean and back, sciencedaily.com, 16 Oct 2001.
8. University of North Carolina at Chapel Hill, Genetic evidence that magnetic navigation guides loggerhead sea turtles, April 12, 2018, sciencedaily.com.
9. Public Library of Science, Bats use magnetic substance as internal compass to help them navigate, sciencedaily.com, 27 Feb 2008.
10. University of Würzburg, Navigating with the sixth sense: Desert ants sense Earth's magnetic field, sciencedaily.com, 26 Apr 2018.
11. Oregon State University, Magnetic pulses alter salmon's orientation, suggesting navigation via magnetite in tissue, sciencedaily.com, 4 May 2020. Return to text.
12. Cell Press, Sharks use Earth's magnetic fields to guide them like a map, sciencedaily.com, 6 May 2021.
13. Daniel Pfeiffer *et al.* A bacterial cytolinker couples positioning of magnetic organelles to cell shape control, pubmed.ncbi.nlm.nih.gov, 30 Nov 2020.
14. O'Hanlon, L., Do plants feel Earth's magnetic field?, Discovery News, 24 Jan 2014; abc.net.au.
15. Sarfati, J., The earth's magnetic field: evidence that the earth is young, *Creation* 20(2):15–17, Mar 1998; updated Aug 2014. Return to text.
16. Sarfati, J., Solar wind protects us from cosmic rays: startling discoveries from the Voyager 2 space probe, creation.com/heliopause, 3 Mar 2020.
17. Snelling, A., The Earth's magnetic field and the age of the Earth, *Creation* **13**(4):44–48, 1991.
18. Magnetic fields and the science of biblical creation, creation.com, 6 Jul 2013., Return to text.
19. K.L. McDonald and R.H. Gunst, 'An analysis of the earth's magnetic field from 1835 to 1965,' *ESSA Technical Report, IER 46-IES 1*, U.S. Govt. Printing Office, Washington, 1967.
20. R.T. Merrill and M.W. McElhinney, *The Earth's Magnetic Field*, Academic Press, London, pp. 101–106, 1983.
21. T.G. Barnes, *Foundations of Electricity and Magnetism*, 3rd ed., El Paso, Texas, 1977.
22. Measurements of electrical currents in the sea floor pose difficulties for the most popular class of dynamo models—L.J. Lanzerotti *et al.*, Measurements of the large-scale direct-current earth potential and possible implications for the geomagnetic dynamo, *Science* 229:47–49, 5 July 1986. Also, the measured rate of field decay is sufficient to generate the current needed to produce today's field strength, meaning that there is no dynamo operating today, if it ever did.

23. D.R. Humphreys, Reversals of the earth's magnetic field during the Genesis Flood, *Proceedings of the First International Conference on Creationism*, Creation Science Fellowship, Pittsburgh, **2:**113–126, 1986. The moving conductive liquid would carry magnetic flux lines with it, and this would generate new currents, producing new flux in the opposite direction. See also the interview of Humphreys in *Creation* 15(3):20–23, 1993.
24. Humphreys, D.R., Physical mechanism for reversals of the earth's magnetic field during the flood, *Proceedings of the Second International Conference on Creationism*, Creation Science Fellowship, Pittsburgh, 2:129–142, 1990. Dr Barnes, who had opposed field reversals because no mechanism could be demonstrated, responded (p. 141): 'Dr Humphreys has come up with a novel and physically sound approach to reversals of the magnetic field.'
25. D.R. Humphreys, Discussion of J. Baumgardner, Numerical simulation of the large-scale tectonic changes accompanying the Flood, *Proceedings of the First International Conference on Creationism*, Creation Science Fellowship, Pittsburgh, 2:29, 1986.
26. R.S. Coe and M. Prévot, Evidence suggesting extremely rapid field variation during a geomagnetic reversal, *Earth and Planetary Science* 92(3/4):292–298, April 1989. See also the reports by Dr Andrew Snelling, Fossil magnetism reveals rapid reversals of the earth's magnetic field, *Creation* 13(3):46–50, 1991 The Earth's magnetic field and the age of the Earth, *Creation* 13(4):44–48, 1991.
27. R.S. Coe, M. Prévot and P. Camps, New evidence for extraordinarily rapid change of the geomagnetic field during a reversal, *Nature* 374(6564):687–692, 1995; see also A. Snelling, The principle of 'least astonishment', *Journal of Creation* 9(2):138–139, 1995.
28. Cited in: Folger, T., Journeys to the Center of the Earth: Our planet's core powers a magnetic field that shields us from a hostile cosmos. But how does it really work? *Discover*, July/August 2014.
29. D. Russell Humphreys, The creation of planetary magnetic fields, *Creation Research Society Quarterly* 21(3):140–149, 1984.
30. The Voyager measurements were 3.0 and 1.5 x 10^{24} J/T for Uranus and Neptune respectively. N.F. Ness *et al.*, Magnetic fields at Uranus, *Science* 233:85–89, 1986; A.J. Dessler, Does Uranus have a magnetic field? *Nature* 319:174–175, 1986; R.A. Kerr, The Neptune system in Voyager's afterglow, *Science* 245:1450–51.
31. Dr Humphreys had predicted field strengths of the order of 10^{24} J/T—*Creation Research Society Quarterly* 27(1):15–17, 1990. The fields of Uranus and Neptune are hugely off-centred (0.3 and 0.4 of the planets' radii) and at a large angle from the planets' spin axis (60° and 50°). A big puzzle for dynamo theorists, but explainable by a catastrophe which seems to have affected the whole solar system (see Revelations in the solar system).
32. Humphreys, D.R., Physical mechanism for reversals of the earth's magnetic field during the flood, *Proceedings of the Second International Conference on Creationism*, Creation Science Fellowship, Pittsburgh, 2:129–142, 1990.
33. For the classic statement of the viewpoint that underlies this paper see Warfield, B.B., *The Inspiration and Authority of the Bible* (Philadelphia: Presbyterian and Reformed Publishing Co., 1964).
34. *E.g.* Jeeves, M.A., 'Towards the Recovery of Harmony Between Science and Christian Faith', *Theolog Review* **3**(2):15-23, McKay, D.M. (Ed.), *Christianity in a Mechanistic Universe* (IVP, 1965).
35. Lest this strike the reader as fundamentalist rhetoric I would draw attention to the very important symposium, *Mathematical Challenges to the Neo-Darwinian Interpretation of Evolution*, (Eds.) Moorhead,

P.S., and Kaplan, M.M. (Philadelphia: Wistar Institute Press, 1967). On page 79 C.H.Waddington answers M.P. Schutzenberger's argument that evolution according to Neo-Darwinian principles is statistically impossible by arguing that it must be possible because the only alternative would be special creation.

36. E.g. Thompson, J.A., '*Genesis 1-3*, Science? History? Theology?' *Theolog Review* **3**(3):16.
37. To use a trivial example, Philadelphia University Museum used to caption the well-known offering-stand from Early-Dynastic Ur which shows a billy-goat standing with its forelegs on the branches of a tree. (Frankfort, H., *The Art and Architecture of the Ancient Orient* [Harmondsworth: Pelican, 1954] p. 31 and pl. 28) as the 'ram caught in a thicket'. Saner minds seem to have prevailed and this caption has been removed.
38. Similarly the tendency of research is often to emphasize the discontinuity rather than the relatedness of animal groups (Kerkut, *op. cit.*, p. 149).
39. See Pritchard, J.B., (Ed.), *Ancient Near Eastern Texts* (Princeton: Princeton University Press, 1955), p. 67 for translation of this text.
40. For discussion see Frankfort, H., *Cylinder Seals* (London: Macmillan, 1939), pp. 205ff. He argues that on Assyrian seals it is a symbol of the god Assur. It is hard to see any connection between this symbol and the trees of Eden.
41. It is significant that Speiser who is convinced that the Biblical story was derived from Mesopotamian prototypes *(ibid.*, p. 1v) cannot find a better parallel than the 'Civilization' of Enkidu by a prostitute *(ibid.*, pp. 26f. For translation of this supposed parallel see Pritchard. *op. cit.*, p. 75).
42. For discussion of the philosophical presuppositions of physics, old and new, see Capek, M., *The Philosophical Impact of Contemporary Physics* (Princeton: Van Nostrand, 1961).
43. Fritz Geiser, 'Yearlong hibernation in a marsupial mammal', *Naturwissenschaften* 94(11):941–944 November 2007.
44. Dawkins, R., Global Atheist Convention, Melbourne, Australia, 2010.
45. Coyne, J., *Why Evolution is True*, Penguin, p. 263, 2009.
46. Catchpoole, D. and Wieland, C., Speedy species surprise, *Creation* 23(2):13–15, March 2001; creation.com/speedy–species-surprise.
47. Thorne, R.F., Floristic relationships between tropical Africa and tropical America, in *Tropical Forest Ecosystems in Africa and South America: A Comparative Review*, Smithsonian Press, 1973.
48. Mayo, S.J., Aspects of aroid geography, in George, W. and Lavocat, R., eds., *The Africa-South America Connection*, Clarendon Press, Oxford, p. 44, 1993.
49. Humphries, C.J. and Parenti, L.R., *Cladistic Biogeography*, Oxford University Press, UK, 2nd ed., 1999, pp. 131–135.
50. Smiley, C.J., Pre-tertiary phytogeography and continental drift—some apparent discrepancies, in Gray, J. and Boucot, A., eds., *Historical Biogeography, Plate Tectonics and the Changing Environment*, Oregon State University Press, 1976, pp. 311–319.
51. Qian, H., Floristic relationships between eastern Asia and North America: Test of Gray's hypothesis, *The American Naturalist* 160(3) 2002, pp. 317–332.
52. Yih, D., *Land Bridge Travellers of the tertiary: the eastern Asian–eastern North America floristic disjunction*; http://arnoldia.arboretum.harvard.edu/pdf/articles/2012-69-3-land-bridge-travelers-of-the-tertiary-the-eastern-asian-eastern-north-american-floristic-disjunction.pdf.

53. Some species of fungi are similarly disjunct. See Hongo, T. and Yokoyama, K., Mycofloristic ties of Japan to the continents, *Memoirs of the Faculty of Education of Shiga University* 28:75–80, 1978; http://libdspace.biwako.shiga-u.ac.jp/dspace/bitstream/10441/3581/2/SJ07_0028_076A.pdf.
54. Queriroz, A., The resurrection of oceanic dispersal in historical biogeography, *Trends in Ecology and Evolution* 20(2):68–73 February 2005.
55. Wieland, C., Forests that grew on water, *Creation* 18(1):20–24, December 1995; creation.com/forests-that-grew-on-water.
56. Ford, P., Drifting rubber duckies chart oceans of plastic, *Christian Science Monitor*, 31 July 2003.
57. Clerkin, B., Thousands of rubber ducks to land on British shores after 15 year journey, *Daily Mail*, 27 June 2007.
58. Myers, N., *et al.*, Biodiversity hotspots, *Nature* 403: 853–858, 2000.
59. Nelson, G. and Platnick, N., *Systematics and Biogeography: Cladistics and Vicariance*, Columbia University Press, 1981, pp. 368, 524.
60. Cox, C.B., The biogeographic regions reconsidered, *J. of Biogeography* 28(4):511–523, 2001.
61. Croizat, L., *Panbiogeography*, fig. 256, p. 1018, 1958.
62. Beu, A.G., Gradual Miocene to Pleistocene uplift of the Central American Isthmus: evidence from topical American Tonnoidean Gastropods, *Journal of Paleontology* 75(3):706 720, 2001; http://si-pddr.si.edu/jspui/bitstream/10088/1386/1/Beu.pdf.
63. Gish, D., *Evolution: the fossils still say NO!*, ICR, California, 1995, pp. 178–183.
64. Cifelli, R.L. and Davis, B.M., Marsupial origins, *Science* 302:1899–1902, 2003.
65. *New Scientist*, 'Inside Science' supplement #56, 1838, September 12, 1992.
66. Sutherland, L.W., Genes map the migratory route, *Nature* 360(6405):625–626, 668–670, December 1992 | doi:10.1038/360625a0.
67. has been recently confirmed by experiment. See A Sun compass in monarch butterflies, *Nature* 387:29, 1 May 1997. Return to text.

CAN SCIENCE PROVE THE AGE OF THE EARTH

THE UNIVERSE

The Earth

There is no scientific method can *prove* the age of the universe or the earth. All calculated ages involve making assumptions about the past: the starting time of the 'clock,' the speed of the clock and that the clock was never disturbed.

There is no independent natural clock against which we can test the assumptions. For example, the amount of cratering on the moon, based on currently observed cratering rates, suggests that the moon is quite old. However, to draw this conclusion we have to assume that the rate of cratering has always been the same as it is now. There is now good reason to think that cratering might have been quite intense in the past, so the craters do not indicate an old age at all.

Age calculations assume the rates of change of processes in the past were the same as we observe today—called the principle of uniformitarianism. If the age calculated disagrees with what the investigator thinks the age should be, he/she concludes that the assumptions did not apply in this case, and adjusts them accordingly. If the calculated result gives an acceptable age, the investigator accepts it.

Examples of *young* ages listed here also rely upon the same principle of uniformitarianism. Long-age proponents will dismiss any evidence for a young earth by arguing that the assumptions about the past do not apply in these cases. In other words, age is not really a matter of scientific observation but rather an argument over our assumptions about the unobserved past.

We cannot *prove* the assumptions behind the evidences presented here. However, such a wide range of different phenomena, all suggesting much younger ages than are generally assumed, makes a strong case for questioning those ages (about 14 billion years for the universe and 4.5 billion years for the solar system).

A number of the evidences don't give an estimate of age but challenge the assumption of slow-and-gradual uniformitarianism, upon which all deep-time dating methods depend. They thus bring into question the vast ages claimed.

Creation scientists discovered many of the young age indicators when researching things that were supposed to 'prove' long ages. There is a lesson here: when skeptics throw up some challenge to the Bible's timeline, don't fret over it. Eventually that supposed 'proof' will likely be overturned and turn out to be evidence for a younger creation. On the other hand, with further research some of the evidences listed here might turn out to be ill-founded. Such is the nature of historical science, because we cannot do experiments on past events.

Science entails observation, and the only reliable means of telling the age of anything is by the testimony of a reliable witness who ***observed*** the events. The Bible claims to be the communication of the only One who witnessed the events of Creation: The Creator Himself. As such, the Bible is the only reliable means of knowing the age of the creation.

In the end, the Bible will stand vindicated and those who deny its testimony will be confounded. That same Bible also tells us of God's judgment on those who reject His right to rule over them. But it also tells us of His willingness to forgive us for our rebellious behavior. The coming of Jesus Christ, who was intimately involved in the creation process at the beginning (*John 1:1–3*) into the world, has made this possible.

Here are 18 evidences from various fields of science.

1. Lazarus bacteria—bacteria revived from salt inclusions supposedly 250 million years old, suggest the salt is *much* younger.
2. The decay in the human genome due to multiple slightly harmful mutations added each generation is consistent with an origin several thousand years ago.
3. Dinosaur blood cells, blood vessels and proteins are not consistent with their supposed age, but make more sense if the fossils are young.
4. Thick, tightly bent rock strata with no signs of melting or fracturing. These wipe out hundreds of millions of years of time and are consistent with extremely rapid formation during the biblical Flood.
5. Polystrate fossils—for example, broken vertical tree trunks in northern and southern hemisphere coal that traverse many strata indicate rapid burial and accumulation of the organic material that became coal, eliminating many millions of years.
6. Flat gaps—where one rock layer sits on another rock layer but with supposedly millions of years of time missing, yet the contact plane lacks significant erosion. E.g., Redwall Limestone / Tapeats Sandstone in the Grand Canyon (more than a 100-million-year gap).
7. The amount of salt in the world's oldest lake contradicts its supposed age and suggests an age consistent with its formation after Noah's Flood.
8. Erosion at Niagara Falls and similar places is consistent with a few thousand years since the Flood.
9. Measured rates of stalactite and stalagmite growth in limestone caves are consistent with an age of several thousand years.

10. Carbon-14 in all coal suggests that the coal is only thousands of years old.
11. The amount of helium, a product of decay of radioactive elements, retained in zircons in granite is consistent with an age of 6,000±2000 years, not the supposed billions of years.
12. The amount of lead in zircons from deep drill cores vs. shallow ones is similar. But there should be less in the deep ones due to the higher heat causing higher diffusion rates over the long ages supposed. If the ages are only thousands of years, this would explain the similarity.
13. Evidence of recent volcanic activity on Earth's moon contradicts the supposed vast age—it should have long since cooled if it were billions of years old.
14. Presence of magnetic fields on Uranus and Neptune, which should be "dead" according to evolutionary long-age beliefs. Assuming a solar system age of thousands of years, physicist Russell Humphreys accurately predicted the strengths of the magnetic fields of Uranus and Neptune.
15. Methane on Titan, Saturn's largest moon—it should all be gone in just 10,000 years because of UV-induced breakdown to ethane. And the large quantities of ethane are not there either.
16. Speedy stars are consistent with a young age for the universe. For example, many stars in the dwarf galaxies in the Local Group are moving away from each other at speeds of 10–12 km/s. At these speeds, the stars should have dispersed in 100 million years, which, compared with the supposed 14-billion-year age of the universe, is a short time.
17. Spiral structure in galaxies should be lost in much less than 200 million years. This is inconsistent with their claimed age of many billions of years. The discovery of 'young' spiral galaxies highlights the problem of the assumed evolutionary ages.
18. The existence of short-period comets (orbits of less than 200 years), is consistent with an age of the solar system of less than 10,000 years.

THE AGE OF THE EARTH

Biblical - Approach

"*Why are you so dogmatic about the age of the earth? Just deal with evolution and leave the age of the earth out of it.*"

We have often been told comments like this from well-meaning acquaintances, friends, and even members of our families. Sadly, ambivalent Christians when dealing with the origins issue. It's often said that raising this issue will also be a stumbling block to evangelism. Nothing could be further from the truth. A 4.5 billion-years-old Earth is actually an icon of evolutionism, a cause of unbelief and a wholesale undermining of the authority of God's Word, Figure 23.1.

Fig.23.1: *Biblical Approach to the Age of the Earth*

Our Authority

Many theologians and lay Christians try to add 'deep time' to the Bible because they were educated in the public realm—where we've all been taught an ancient Earth, billions of years in age. It's virtually impossible to escape the **millions of years** (*MOY*) mantra. It's everywhere—in nature documentaries, TV news, movies, books, magazines, and even in children's cartoons.

We are *indoctrinated* in it. The idea of MOY's is so strong that even when we present strong biblical and scientific arguments against it, it seems we are talking past each other.

On one occasion I was highlighting the theological problems of 'gap theory' with a gentleman. When he was shown the correct grammar and context of Genesis, plus what the New Testament authors believed, he had no answers except to say, *"I just believe the Bible!"* The problem was *he really didn't!* One does not get the idea of deep time from Scripture. There is nothing in the text to indicate a gap between Genesis 1:1 and 1:2, or a Lucifer's Flood, and nowhere will you see the slightest hint of MOYs *in Scripture.*

THE INITIATION OF 'MOY' - (MILLIONS OF YEARS)

Most of us have heard theological teaching that adds deep time to the Scriptures such as gap theory, day-age, framework hypothesis, or local flood theory. It might even have come from a respected 'theologian.' It can then be difficult to think it might be wrong when one has accepted a certain interpretation for most of one's Christian walk. And the massive discord of MOYs compared to a few thousand years adds to the difficulty.

Accordingly, I usually start by asking, "Do you know where the idea of MOYs comes from?"

EVEN BIBLE SCHOLARS—DON'T KNOW

If an answer to defend deep time is offered, it is usually radiometric dating, because it is assumed that scientists can do tests to establish the age of things. This is simply not the case, as there is no scientific test that can prove the age of anything that existed in the past!

The Millions of Years Belief!

There are rock strata seen all over the earth, which often-contain millions of fine sedimentary layers. Before Darwin, 'Hutton and Lyell' had decreed that we must explain the past by what's happening today. And because a flood like Noah's is not happening today, they decreed (not proved!) that it was inadmissible as an explanation. They asserted that these rock layers must have taken millions of years to slowly accumulate. Darwin avidly absorbed this view because he thought it gave him enough time (MOYs) for his slow and gradual concept of biological evolution to have taken place.

Same Facts, Different story

Creationists and evolutionists have different concepts of history that influence the way we interpret the facts that exist in the present. Yes, there are sedimentary layers that contain fossils all over the earth. But this oft-cited supposed evidence for deep time is one of the easiest for us to explain *if we truly believe the Bible as an historical record.* Eyewitnesses lived through the globe-reshaping Flood of Genesis 6–8. And Jesus and the New Testament authors affirmed it was a real historical event. 2 Peter 3: 1–6 declares that "scoffers" will ignore the evidence for a global Flood, and this is exactly what is happening today—because the implications are devastating for evolution theory.

The Genesis Flood Washes Away the Millions of Years

The Bible says the Flood lasted 12 months. So, if most of the geologic layers were laid down in one year, then simply there are no MOYs! Thus:

- There is no time for the alleged trillions of 'experiments' that supposedly organized random chemicals into the first living cell.

- There is no time for the amazing diversity of life to have developed from that supposed first primordial cell.
- There is no time for slow genetic mutations to turn apes into humans.
- There is no 'age of dinosaurs' 243 to 66 MOY ago.

No Time for Evolution to Happen

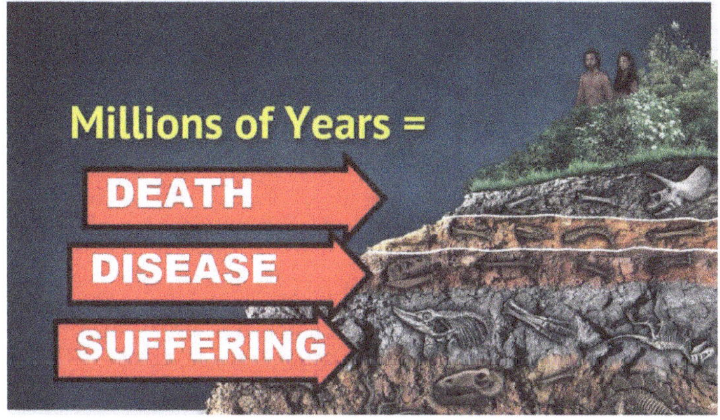

Fig.23.2: Putting millions of years into the Bible puts death before Fall

The above list could expand substantially. Nonetheless, the main issue is; the Flood of Genesis-6 not only washes away the idea of MOYs but everything that evolution has to offer gets washed away with it, Figure 23.2.

Adding MOYs to the Bible actually imposes a secular ('non-Christian') idea upon the biblical texts. And even if one does not believe in evolution, but tries to add MOYs to the Bible, it creates a massive theological problem:

- The MOYs come from the rock layers; there are fossils in those rock layers (dead matters).
- Therefore, adding MOYs into Genesis 1 necessarily puts death and disease before Adam's sin (contra Gen. 3).
- This undermines the very Gospel itself.

Accordingly, the age of the earth is an important issue *for all* Christians, and it's a lot easier to simply accept the Bible's plain teaching on the issue, rather than attempting all sorts of theological chicanery to insert MOYs into Scripture.

The Biblical Age of the Earth

Fig.23.3: Time Span Corresponds to Date of Crucifixion – Leading to the day of Creation

Purpose -Time Span

Careful consultation with the biblical record reveals a series of timespans linking creation to the Crucifixion, Figure 3.3. As a thought experiment, we combined these timespans to estimate the minimum and maximum allowable date of creation.

Our goal was not to contradict the existing body of literature on the subject, but to put constraints on what is and is not biblically allowable. Implied precision and potential cultural differences (e.g., calendar systems, birthday conventions, and rounding conventions) mean we cannot pinpoint the age of the earth to a single year, yet the accumulated imprecision from those sources is limited to a maximum range of 308 years.

TEXTUAL VARIATIONS AND DEBATES

Even including textual variants and debates over interpretation does not allow for dates approaching 10,000, let alone billions of years of Earth history. Accounting for all presently known relevant details and assuming the Babylonian Captivity began in 587 or 586 BC, we can say with confidence that the Bible places limits on the year of creation between 5665 and 3822 BC.

The uncertainty within this range is mainly driven by textual considerations. The Masoretic/LXX debate creates a 1,326-year dichotomy, the Long vs Short Sojourn positions differ by 215 years, and various interpretations of the lists of the kings of Judah and Israel equates to around 54 years of additional uncertainty.

Christians should avoid dogmatic claims of dating precision greater than intended by the Bible that could be refuted with new evidence, causing some to mistakenly believe the Bible itself has been refuted. Yet the combined tally of all the available data gives us fairly tight constraints on the age of the earth.

THE DATE OF CREATION

Using the Bible to estimate the date of creation has a long and rich history. The early Church Fathers put numbers on it, as did scientific greats like ***Sir Isaac Newton (about 4000 BC)*** and ***Johannes Kepler (3992 BC)***. The great academic and ***Archbishop James Ussher's date of 'Oct 23, 4004 BC'*** is perhaps the most famous estimated date in history, although he has been much maligned by scoffers in recent years.

Mining the Bible for Chronological Details

Most scholars veered away from biblical fidelity in the 19th and 20th centuries and very few seemed interested in mining the Bible for chronological details. In more recent years, however, the pages of this journal have been filled with many contributions on the subject.

1984, Osgood

Starting with the first issue in 1984, Osgood began publishing a series of papers that stretched out over the next several issues, eliciting responses and counter responses from various people. Over the years, more than a dozen different authors have published papers on chronology in this journal. They disagree on some of the details and there have been several sharp disputes, but two things unite them: a belief in the perspicuity of Scripture and a desire to systematically derive a consistent biblical dating scheme.

THE RANGE OF ACCEPTABLE DATES

We set out not to put a specific date on creation, but to put limits on the range of acceptable dates. And, while we certainly have strong opinions on how to resolve several of the historic debates, we wanted to know the 'worst case' scenario rather than to assume those opinions are correct.

We acknowledge the great amount of work that has already been done and we are indebted to the prior body of publication. However, there are several factors that have not yet been systematically outlined and these have a small but important effect on all date calculations. This paper was foreshadowed by one of the earlier contributors, Pete Williams, whose paper "Some remarks preliminary to a biblical chronology" appeared in these pages in 1998.

NUMERICAL LOCKS

There are some specific dates given in the Bible that are not up for debate. When a biblical author says a person was X years old when something happened, if we do not take that as a historical statement we quickly get to the point where words have no meaning. Many such numbers can be found throughout the Bible. For instance, we know that Caleb was 40 years old when he was sent with the other spies to Canaan (*Josh. 14:7*), and we know that he was 85 when he approached Joshua after the invasion of Canaan was completed to request Hebron for his inheritance (*Josh. 14:10*). We also know that the spying was done in the fall because it occurred during the grape (and pomegranate) harvest (*Num. 13:20, 23*). Statements like these are a very important source of data for biblical chronology.

Span of Time Events

There are other statements which give us a span of time between events. For example, in the time of the Judges, the Ammonites attempted to lay claim to the Reubenite territory just south of Ammon and east of the Dead Sea. Jephthah taunted the Ammonite king, saying, "While Israel lived in Heshbon and its villages, and in Aroer and its villages, and in all the cities that are on the banks of the Arnon, 300 years, why did you not deliver them within that time?" (*Judg. 11:26, ESV*). Thus, it is clear that the Israelites had occupied that area for 300 years. This probably does not mean 'exactly' 300, but it proscribes any attempt to reduce the period of the Judges to much smaller values.

There are other numbers in the Bible, however, that are more ambiguous, and when we string together multiple dates and date ranges, each with a certain degree of built-in ambiguity, we must acknowledge certain limits to precision.

Factors, Limiting Dating Precision

To generate a potential range of dates for creation, there are several sources of imprecision for which we must account. Some of these sources are inherent in the way humans report numbers. Others come from ambiguous statements in the biblical text (such as Terah's age at Abram's birth, **Table 23.1**. Still others come from the fact that we do not know which time-keeping conventions the ancients may have used.

Cumulative Imprecision

Williams used the phrase 'cumulative imprecision' to describe the problem. We will copy his terminology, but by 'imprecision' we do not mean 'error' or that the biblical authors were sloppy with their reporting. On the other hand, we should not read biblical time statements as though the intent of the authors was to build a minute-by-minute timeline of Earth history. Most of the time statements are simple reports of major happenings, and they tied those to a general series of datable events (like a man's age at the birth of a son). Sometimes, but not always, a series of dates can be bridged by a spanning statement that reduces the cumulative imprecision. And considering that most dates are given in 'years', we should not consider these to be an exact day count. This is what we mean by 'imprecision'.

Accounting for each source of imprecision widens the potential range of dates for creation, and there are many factors to consider, yet each source of imprecision has a limited effect. Therefore, the extent of the accumulated imprecision is also limited. We will consider each source of imprecision in turn.

IMPLIED PRECISION

When humans report measurements, the context or style of the report often implies the precision of that measurement. If someone were to claim a structure was 100 m long, but it turned out to be 101 m long, it would be false to claim the person said it was *exactly* 100 m long. One cannot arbitrarily change significant figures when reporting numbers. Another source of ambiguity deals with rounding of numbers, and we should not assume ancient writers used modern rounding conventions (e.g., anything ≥ .5 gets rounded up to the next integer). For all we know, they may have always rounded down.

An example can be found in *1 Kings 7:23* concerning the Bronze Sea Solomon commissioned to be made for the Temple: "it was round, ten cubits from brim to brim, and five cubits high, and a line of thirty cubits measured its circumference" (ESV). Modern scoffers often claim the Bible wrongly teaches the value of π (the circumference of a circle divided by its diameter) to be '3.0' rather than the correct '3.14 … '. They are claiming a greater precision than was specified. Ignoring whether the Bronze Sea was a perfect circle, and whether the diameter measurement was for inside or outside, it could be anywhere from 9.5 to 9.7 cubits in diameter (e.g., '10') to give it a circumference of between 29.8 and 30.5 cubits (e.g., '30') using the correct value of π. When our interpretation includes a correct understanding of implied precision, we find that the value of π derived from operational science agrees with the record of *1 Kings 7:23*.

CALENDAR SYSTEMS

In addition to the uncertainties generated by implied precision, one must also consider the time-keeping convention used by the people reporting those dates. Many ancient societies used lunisolar calendar systems, where months are tied to the lunar cycle, but an occasional 13th intercalary month is added to keep months aligned with seasons, since 12 lunar months are 11 days short of a solar year. Some societies also standardized the process with the addition of 7 deliberately placed intercalary months within 19-year cycles. This was more predictable than the 'as needed' method but still required an additional intercalary month every 80 years to keep months aligned with the seasons. However, standardization would often take centuries and different localities have often used conflicting systems. While we do not know the exact antediluvian method used, we do know that the Jews have used a lunisolar calendar since the Exodus, when Moses was directed by God to institute a new system (*Exo. 23:16, Lev. 23:39*, etc.).

There are abundant examples of cultures changing time conventions. Before Islam, the Arabs used a lunisolar system, but Muhammad arbitrarily abolished the use of intercalary months in the Qur'an (9:36–37).

Muslim Hijri Calendar

Modern Muslim countries such as Saudi Arabia use a 12-lunar-month Hijri calendar, where a month in summer one year will be in winter 17 years later. Their year numbering starts with Mohammed's move from Mecca to Medina in 622 AD. On the New Year's Day 2014 AD (1,392 solar years later), the Hijri year was 1435.

Wrestling with Time Measurement

There are many other examples of societies wrestling with time measurement. For instance, the Romans arbitrarily changed the date of the New Year to January 1 in the second century BC. The 'years of confusion' which followed were resolved by the Julian calendar, which re-aligned the months to the seasons by having one year with 445 days. Many ancient cultures began their year at the vernal equinox, while others began at the autumnal equinox. Various European localities up to the Middle Ages used a diversity of days to begin the year after the Council of Tours outlawed New Year's celebrations in 567. Even the Gregorian calendar system, with January 1 as New Year's Day, was not adopted uniformly across Europe, with the British Empire holding out until 1752, and some jurisdictions even longer than that.

Ancient peoples living in temperate latitudes presumably measured tropical years instead of sidereal years. However, ancient peoples living near the equator (or in places with no pronounced seasonal differences, e.g., the way many people imagine the antediluvian world) might be expected to default to a sidereal year, when the sun/stars/Earth return to the same alignment. Since a sidereal year is only about 20 minutes longer (1 sidereal year = 1.00003878 tropical years), this would have no significant impact on any age calculations, to the nearest year, adding at most one hour every three years, or just under 14 days in 1,000 years. However, this would have affected ***Ussher's 'Oct 23, 4004 BC'*** date, so the reader is cautioned.

The Mayans used multiple simultaneous calendars, including a 260-day divine calendar (the most important), a long-count calendar similar to the Julian Calendar (with which they dated past and future events), a civil calendar similar to the Gregorian calendar, and a 584-day calendar based on the position of Venus (where five Venusian years are about eight solar years, or 99 lunar months).

The point of this brief survey is to illustrate the fact that we do not know which convention was used in the ancient past, and we do not know if all biblical data are reported with the same convention, Figure 23.4.

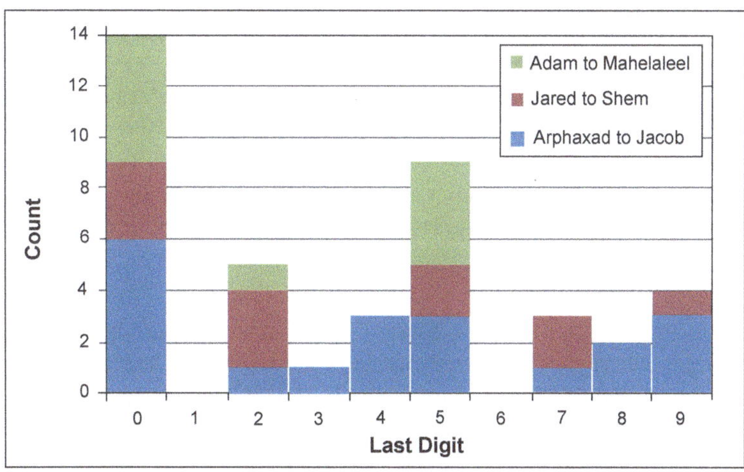

Fig.23.4: Histogram of last digits of Patriarch age data reported in Genesis. Not included are any ages back-calculated from the text (e.g., age of Noah, Terah, and Jacob when Shem, Abram, and Joseph were born). Note that the only '9' comes from Methuselah's age of death, which may have been back-calculated in the original, for the year of his death was quite obvious and significant. This is clearly not a random distribution, but the final digits appear more random as time progresses. After the Flood, most digits appear and the distribution appears more or less random, with the exception of more zeros than expected.

Years may have been reported in systems other than ones that align with solar years, and multiple possible shifts of six months or more may have occurred when societies switched or reformed their calendar systems.

CULTURAL DIFFERENCES IN BIRTHDAY CONVENTIONS AND COUNTING AGE

This far, we have considered imprecision in number reporting and a diversity of changing calendar systems, but we must also consider how ages are reported. In some East Asian cultures, newborns are traditionally said to be 1 year old (better translated 'in his first year') and ages are advanced at the lunar New Year, instead of on the birthday. It is possible that a child in these cultures could be '2 years old', while native English speakers would say '1 month'. In addition, some cultures count age from conception rather than birth.

People sometimes keep track of multiple time conventions simultaneously and can flip from one to the other at will, meaning it is often difficult for an outsider to keep up, and context is of utmost importance. Therefore, we must allow for two fewer years than the reported biblical 'ages' in order to account for unknown birthday conventions.

ROUNDING IMPRECISION

There are detailed genealogical lists in the Bible (e.g., Matt. 1, Luke 3). Some, however, come with specific dates and ages (e.g., Gen. 5, Gen. 11). The latter are more properly called 'chrono-genealogies' and they are of utmost importance, for they allow us to build a straightforward history of the time period they cover. Yet, there are certain facts about these numbers for which we must account. The chrono-genealogies of Genesis are not based on a calendar system. The years are given as the age of the father, not the age of the earth (*anno mundi* or AM). If, as in modern English-speaking cultures, they used zero-based ages incrementing on birthdates, since a child can be born anywhere within that one-year span, each generation should add an average of six months to the calculations. It is unlikely that a series of children would all be born on each successive father's birthday or on the day before those birthdays. But, accounting for the possibility of both extremes allows us to better estimate the range of dates for creation. Assuming random birthdates and that the ages were zero-based, 10 generations would carry about 5 extra years. But if ages were one-based (babies are in their '1st year'; Gen. 7:6 and 11 hint this was their convention), we should subtract about 5 years for every 10 generations instead. Many scholars of the past, including Ussher, have failed to recognize what we call 'date slippage.'

Table 23.1. Minimum (Min), maximum (Max), and simple additive (Add) dates for Adam to Noah from Genesis 5, accounting for potential differences in birthday and rounding convention. Additive dates were generated by simply adding up the given numbers in the text. Minimum dates take into account potential rounding and the possibility of a 1-based birthday convention. Maximum dates take into account the possibility of a ratcheting scheme with a 0-based birthday convention. See text for an explanation of the adjustment values at each generation.

Table 23.1 : Age at Birth of Son

Birth Year (*anno mundi*)						
Person	**Min**	**Add**	**Max**	**Min.**	**Add**	**Max.**
Adam	0	0	0	126	130	135
Seth	126	130	135	101	105	110
Enosh	227	235	245	86	90	95
Cainan	313	325	340	66	70	75

Birth Year (*anno mundi*)						
Person	**Min**	**Add**	**Max**	**Min.**	**Add**	**Max.**
Mahalalel	379	395	415	61	65	70
Jared	440	460	485	160	162	165
Enoch	600	622	650	63	65	67
Methusaleh	663	687	717	185	187	190
Lamech	848	874	907	180	182	185
Noah	1028	1056	1092			

To test the effects of date slippage over the number of reported generations between Adam and Noah, we created a simple Excel spreadsheet and populated it with pseudo-random numbers representing the month of birth of consecutive children over 10 generations. After 1,000 trials, fully 92% of the replicates (nearly 2 standard deviations) had a total slippage of 4–6 years and only 1.5% had a slippage of as few as 2 or as many as 8 years. This works for both positive (0-based) and negative (1-based) date slippage. Clearly, this is a factor that needs to be taken into account when attempting to date creation, but it primarily applies to the pre Exodus chrono-genealogies.

ROUNDING OF AGES

Consider the first five biblical Patriarchs listed in Genesis 5. Their ages at the birth of the next in line and at death are listed, but nine of the ten ages end in a 0 or a 5, Figure 23,4. This suggests the numbers may have been rounded to the nearest five. Or they may have used a 5-year ratcheting scale, where the age was only incremented every five years, meaning Adam could have been nearly 135 and still truthfully report his age as '130'. The lone '2' is Seth's age at death. From Jared to Shem, we see two additional digits, giving the appearance that they rounded down and to the nearest '2'. The lone '9' is Methuselah's age at death. Interestingly, in both cases the distribution of the reported numbers is evenly balanced (i.e., about the same number of zeros and fives from Adam to Mahalaleel and about the same number of zeros, twos, fives, and sevens from Jared to Shem).

We are not trying to prove these dates are rounded or ratcheted, but since the numbers are so odd (i.e., not what one would expect from a random sampling, as even the post Flood patriarchs have three times more zeros than expected), we must allow for the possibility. In order to account for potential changes in rounding conventions, we will allow for a 5-year rounding convention from Adam to Mahalaleel, a 2–3-year rounding convention from Jared to Shem (Table 23.1), and 1-year rounding (i.e., the modern convention) after that (Table 23.2).

Table 23.2: Birth Year (*anno mundi*)

Age at birth of son						
Person	**Min**	**Add**	**Max**	**Min**	**Add**	**Max**
Arphaxad	1628	1659	1697	33	35	36
Salah	1661	1694	1733	28	30	31

				Age at birth of son		
Person	Min	Add	Max	Min	Add	Max
Eber	1689	1724	1764	32	34	35
Peleg	1721	1758	1799	28	30	31
Reu	1749	1788	1830	30	32	33
Serug	1779	1820	1863	28	30	31
Nahor	1807	1850	1894	27	29	30
Terah	1834	1879	1924	128	130	180
Abraham	1962	2009	2109	98	100	101
Isaac	2060	2109	2210	58	60	61
Jacob	2118	2169	2271			

Why might the author of this section of Genesis have rounded these ages to the nearest five years? Possibly this was due to their great age, where a count precise to a single year might not be all that important to the individual when reporting his age, although ratcheting is more likely in this case. Searching for a mathematical reason for the apparent rounding leads us to consider the possibility that the first few generations measured ages in 60-month periods.

Initially, the lunar cycle would have been the most obvious way to track time, especially if Eden and/or its environs did not have significant seasonal variance.

They may have measured longer periods of time in groups of lunar months instead of years. If the first five patriarchs reported ages in 60-month blocks, the ages may have been converted later by multiplying by five, giving us the ages, we have in the biblical record (with one exception) in 12-lunar-month years. There are many possible reasons for the appearance of these numbers, including random chance, but we are obliged to consider both rounding and ratcheting in our calculations because we cannot rule out these possibilities.

Table 23.2. Minimum (Min), maximum (Max), and simple additive (Add) dates for Arphaxad to Jacob from Genesis 11, 21, and 25, accounting for similar potential differences in birthday and rounding convention as in Table 23.1.

CALCULATING THE TIMESPAN AND RANGE

The above imprecision factors come in two categories: 'per-link' and 'overall,' The following calculations will accumulate per-link factors (such as from birthday conventions & rounding), then apply the overall factors (such as calendar conventions) at the end.

Creation to Noah

Table 23.1 lists the minimum, maximum, and simple additive dates for Adam to Noah from Genesis 5, accounting for potential differences in birthday and rounding convention.

NOAH TO ARPHAXAD

Genesis 7:6 and 7:11 state the Flood started in Noah's 600th year, and 8:13 states the Flood ended in his 601st year. This eliminates the possibility of ±5 rounding. Applying the limits of potential birthday conventions and offsets, we find the Flood started between 598 and 601 years after Noah's birth. The simple additive date for the Flood is AM 1656, but it could have been anywhere from AM 1626 to AM 1693. Genesis 11:10 tells us Arphaxad was born two years after the Flood.

This could mean 'in the second year after the Flood started' (just over one year after the Flood ended), 'during the second summer/winter/fall/spring after the Flood ended,' or up to not quite 3 years after the Flood ended. Simply adding up the spans shows Arphaxad was born around AM 1659, with an outside range of 1628 to 1697. Note that we skipped Shem on purpose, because the best links are from Noah to the Flood to Arphaxad, making the ambiguity of Shem's birth year irrelevant.

Arphaxad to Terah

Table 23.2 lists the minimum, maximum, and simple additive dates for Arphaxad to Jacob from Genesis 11, 21, and 25.

A 50-Year Ambiguity from Terah to Abram

The age of Terah when his son Abram was born is ambiguous because we only know Terah was 70 when Abram's oldest brother was born. The narrative from Genesis 11:26–12:5 states that Terah, Abram and family moved from Ur to Haran, lived there a while, and Abram moved on from there to Canaan. That narrative implies (and Acts 7:4 confirms) Abram waited until his father died before leaving for Canaan, and states he was 75 when he left.

If Abram left very soon after Terah died at 205, this would have made Terah 130 when Abram was born. But the text does not exclude the possibility that Abram waited. He may have lived in Haran for decades after his father Terah died before leaving for Canaan. All we know is he was old enough to be married to a wife 10 years younger (Gen. 17:17) before (Gen. 11:31) they moved to Haran.

Terah therefore may have been as old as 180 when Abram was born, assuming Sarai was at least 15 when she married Abram. This is a break from the strict chrono-genealogy and impacts the date of creation by up to 50 years.

Abraham to the Exodus

Genesis 21:5, 25:26 and 47:28 and Exodus 12:40–41 allow us to estimate the number of years from Abram's birth to the Exodus. Assuming a plain reading of Exodus 12, this amounts to 720 years, 430 of which occur between Jacob's move to Egypt and the Exodus. The 400 years of Gen. 15:14 would start in Exodus 1:8 when the Pharaoh who knew Joseph was replaced by one who enslaved the Israelites. Note that although Genesis 21:5 says Abraham was 100 when Isaac was born, this does not allow for ±5 rounding because in Genesis 17:1 we were told he was 99 the year before Isaac was born. Jacob and 11 of his sons moved to Egypt in AM 2299. Simply adding the spans puts the Exodus in **AM 2729** with a range of **2676** to **2834**.

However, Ussher and others have proposed that the 430 years Israel lived in Egypt started with God's promise to Abraham in Genesis 12:1–3 instead of with Jacob's arrival in Egypt. The 400 years of Genesis 15:14 would then start in Genesis 21:8–9, when Ishmael mocked Isaac at his weaning feast.

Rather than attempt to resolve this historic debate here, we acknowledge both positions have strengths and weaknesses, and include the range from both positions for the range of the Exodus: **AM 2461** to **AM**

2834. From this point on, we will use the timespan for the Long-Sojourn view, acknowledging the Short-Sojourn additive, minimum and maximum dates will be 215 years less, Table 23.3.

Table 23.3: Differences between the Masoretic (Mas.), LXX, and Samaritan Pentateuch relevant to the date of Creation.

	Age at birth of son					
	MT	LXX		SP		
Person	**Age**	**Age**	**Effect**	**Age**	**Effect**	**Reference**
Adam	130	230	100	130	0	5:3-5
Seth	105	205	100	105	0	5:6-8
Enosh	90	190	100	90	0	5:9-11
Cainan	70	170	100	70	0	5:12-14
Mahalalel	65	165	100	65	0	5:15-17
Jared	162	162	0	62	-100	5:18-20
Enoch	65	165	100	65	0	5:21-24
Methusaleh	187	167	-20	67	-120	5:21-27
Lamech	182	188	6	53	-129	5:28-31
Noah	500	500	0	500	0	5:32, 9:28-29
Shem	100	100	0	100	0	11:10-11
Pre-Flood Subtotals			**586**		**-349**	
Arphaxad	35	135	100	135	100	11:10-13
Cainan		130	130			11:13 (LXX only)
Salah	30	130	100	130	100	11:12-15
Eber	34	134	100	134	100	11:14-17
Peleg	30	130	100	130	100	11:16-19
Reu	32	132	100	132	100	11:18-21
Serug	30	130	100	130	100	11:20-23
Nahor	29	79	50	**79**	50	11:22-25
Pre-Flood Subtotals			**780**		**650**	
Exodus-Solomon	480	440	-40	0	0	1 Kings 6:1
Grand Totals			**1326**		**301**	

* Some English versions mistakenly translate Nahor's age at Terah's birth as 179 years old, but the Greek manuscripts read 79.

THE EXODUS THROUGH THE BABYLONIAN CAPTIVITY

The books of Kings and Chronicles contain an unbroken chain of timespans from the Exodus to the Babylonian captivity. Simply adding up the years with the maximum length within the implied precision from each link yields a maximum biblically compatible timespan of 437 years from the beginning of Solomon's reign to the Babylonian Captivity.

Thiele claimed regnal years were reported by two different systems: 'accession year' (1-based) and 'non-accession year' (0-based) reckoning. He presented evidence of swaps between conventions in both Judah and Israel, in addition to the two kingdoms using differing conventions simultaneously, which limits the precision of dating simply based on cross-referencing regnal years.

Further complicating the matter, Judah appears to have advanced regnal years in the spring (Nisan), when their new year began, while Israel advanced theirs in the fall (Tishri), when their new year began. Thiele's 383 years from the start of Solomon's reign to the Babylonian Captivity is probably the shortest timespan proposed by a conservative scholar. Additional modifications and discussions of Thiele's work can be found in Kaiser and Kitchen.

Jones accounts for changing regnal year conventions and differing new year months using a more straightforward interpretation than Thiele to arrive at a longer time span of 429 years.

Pierce rejects Thiele completely. And Clarke rejects Austin's, and Ashton and Down's, attempts at linking biblical chronology to Egyptian chronology because they base their ideas on Velikovsky, whom he claims has been thoroughly discredited. All of these authors have a high view of Scripture. Clearly, biblical chronology is a difficult subject, Table 23.4.

Table 23.4: Final Earth age range estimates (all dates BC).

Text	Sojourn	Lunar Min	Min	Add	Max
MT	Short	3822	3909	4005	4124
	Long	4031	4121	4220	4339
SP	Short	4114	4207	4306	5590
	Long	4323	4422	4521	5805
LXX	Short	5108	5232	5331	5450
	Long	**5316**	5447	5546	**5665**

* Minimum with 12-lunar-month years prior to the Exodus.

THE BABYLONIAN CAPTIVITY TO CHRIST

II Kings 23–24 states that the Kingdom of Judah was carried into captivity in three waves, and the extra-biblical historical consensus is that these waves occurred in 597 BC, 587–586 BC, and 582 BC.

The only biblical timespan between then and the New Testament comes from Daniel 9:24–26. This prophecy places a minimum of 7 + 62 'sevens', commonly assumed to mean 483 years from 'the decree to rebuild Jerusalem' until the Crucifixion of Jesus Christ.

Yet, there are multiple such decrees, and we are not sure to which Daniel refers, although Austin argues strongly for one specifically, while at the same time removing a gap of 80–82 years, inserted by Ussher and others, by equating Darius to Artaxerxes.

We must also rely on extra-biblical history to pinpoint the birth of Jesus Christ. This seems to be fairly well established at around **4 BC**, although there are various biblically conservative counter arguments for a variety of dates in that range. The year of Christ's death can be garnered from secular sources, and is attested by Daniel 9. Yet, we chose to peg our age estimate to the start of the Babylonian captivity because it allows for a slightly higher degree of certainty and because there is little dispute after that date.

MASORETIC VS LXX VS SAMARITAN PENTATEUCH

A few hundred years before Christ, Alexandrian Jews produced a Greek translation of the Old Testament called the Septuagint (commonly abbreviated LXX). The authors of the New Testament frequently quoted directly from the LXX when referencing the Old Testament. The Masoretic text is the collection of Hebrew Scriptures collated around AD 700–1000 and is the basis of most modern Old Testament translations.

We have many ancient fragments of Scripture in Hebrew (e.g., the Dead Sea Scrolls), which match the Masoretic very closely, showing the quality of work of the copiers in the intervening years, and supporting the authenticity of the Masoretic.

The LXX puts the earth significantly older than the Masoretic: including 586 additional years before the Flood and 780 additional years from the Flood to Abraham's grandfather, Nahor (Table 23.3). This is mostly due to the LXX including 100 more years in the ages of various Patriarchs at the birth of their son. The LXX also includes a Patriarch named Cainan between Arphaxad and Salah in Genesis 11:13.

This name does not appear at that point in the Masoretic or Samaritan Pentateuch. Most Greek texts of Luke 3:36 agree with the LXX on that point. From Terah forward, the primary date-relevant conflict is 1 Kings 6:1, in which the LXX dates the beginning of Solomon's temple to 440 years after the Exodus vs 480 in the Masoretic. Even though we favor the Masoretic, we cannot know *for certain*, and therefore must acknowledge the possibility of the older dates from the LXX by adding 1,326 years to the maximum age allowed by the Masoretic.

There is another source of differing chronological data, the Samaritan Pentateuch. Written in Hebrew, but with a different etiology, it differs from the Masoretic in several thousand places, sometimes agreeing with the LXX and sometimes not. We do not put much stock in its authority, but see Table 23.3 for details. It subtracts 349 years before the Flood and adds 650 years after it, for a net of 301 years more than the Masoretic.

LIMITED, GAP-FREE IMPRECISION

As detailed above, there are no chronological gaps from Genesis 1:1 to the Babylonian Exile. There is also no place where the text allows the insertion of an unlimited amount of time. In addition, this writing also takes the Genesis 1 narrative literally, leaving no room for a time gap there.

Many have attempted to argue for gaps in the Genesis chrono-genealogies, but, for example, even if Enoch were Jared's great-grandson rather than his son, that would not change the timespan; Jared was still 162 when Enoch was born and this would not change the date of creation. Thus, there is no reason to argue for these gaps.

AMBIGUITIES AND IMPRECISIONS DO NOT EQUATE TO FALSEHOODS

The ambiguities detailed here do not mean the text is untruthful or erroneous. That a modern Western person would use a different number convention to describe age than someone of a different culture or time does *not* mean that either party is mistaken or lying. It merely means that a proper time convention translation is necessary.

In the absence of complete information, the number should be understood to imply a range of possible ages. Our interpretation needs to allow for various possible implications of the original text, resulting in a range of possible ages. A range narrower than intended by the Bible could conflict with valid outside evidence, and influence people to (incorrectly) disbelieve the Bible.

However, the Bible does make historical claims that can be used to estimate the age of the earth, so we should not pretend the earth could be any age. These claims can and should be used by Christians to evaluate the accuracy of extra-biblical historical claims.

RESULTING DATE RANGES

From creation to the Babylonian Captivity, we calculated a per-link imprecision of 219 years (including the 50-yr ambiguity concerning how long Abram remained in Haran), plus an overall systemic imprecision of 89 years. It is not possible to date creation with any more accuracy using just the genealogical data. We should allow for the possibility of ±10 years of imprecision from calendar system changes, and the possibility of up to 3% less solar years before the Exodus if the ancients used 12-lunar-month years or longer blocks of lunar months which would later be converted to 12-lunar-month years.

We must also consider the possibility of 1326 additional years if the LXX chrono-genealogies represent the original wording, 301 additional years if the Samaritan Pentateuch is correct, 215 less years for the 'Short Sojourn' view, and 46 fewer or 8 more years due to the ambiguities in the king lists of Judah and Israel. This yields an outside range of 3236 to 5078 years from Creation to the Babylonian Captivity. If the traditional historic date of 587 BC or 586 BC for the Captivity is correct, the earth cannot be more than **7,680** years old (Table 23.4), having been created between 5665 BC and 3822 BC. The date of the Flood is more significant to the evaluation of extra-biblical history than is the date of creation. The Flood probably occurred between ***2600*** BC and *2300* BC, but certainly between 3386 BC and 2256 BC (**Table 23.5**).

Table 23.5: Date Estimates for the Flood (all dates BC).

Text	Sojourn	Lunar Min	Min	Add	Max
MT	Short	2256	2280	2349	2431
	Long	2464	2495	2564	2646
SP	Short	2886	2930	2999	3081
	Long	3095	3145	3214	3296
LXX	Short	2972	3020	3089	3171
	Long	3181	3235	3304	3386

* Minimum with 12-Lunar-month years prior to the Exodus.

Note that the only way to get a 'traditional' date of creation of approximately ***4000 BC*** is to use the Short Sojourn calculation and minimal to simple-additive adjustment parameters. This makes it likely that the earth is several hundred years older than most biblical creationists expect.

However, we reject the idea that there are 'missing generations' that might increase the age to as much as 10,000 years, as Whitcomb and Morris did in their seminal and influential book *The Genesis Flood. Soli deo gloria.*

JESUS ON THE AGE OF THE EARTH

The standard secular timeline, from an alleged 'big bang' some 15 billion years ago to now, is accepted by most people in the evangelical Christian world, even though many would deny evolution. Some would even say that to dispute billions of years is to place an unnecessary stumbling block in the way of any scientifically-minded potential converts, Figure 23.5.

Fig.23.5: In The Beginning God Created ... !

This is in contrast to the teaching of the Lord Jesus Christ, the Creator made flesh, as well as several of the biblical authors, which makes it plain that this is wrong—people were there *from the beginning* of creation. But in the evolutionary timeline, people have only been around for 4.55 million years—this puts them toward the *end* of the timeline. This means that He is most definitely claiming that the world *cannot* be billions of years old.

For example, dealing with the doctrine of marriage, Jesus says in Mark 10:6:

"But **from the beginning of the creation,** God made them male and female."

In Luke 11:50–51, Jesus also says: "That the blood of all the prophets, which was shed **from the foundation of the world**, may be required of this generation; From the blood of Abel to the blood of Zacharias … ". And in Romans 1:20, the Apostle Paul says of God: "For his invisible attributes, namely, his eternal power and divine nature, have been clearly perceived, **ever since the creation of the world**, in the things that have been made. So, they are without excuse."

Paul is plainly saying that people have been able to perceive these attributes of God in His creation ever since the creation of *the world*. Not ever since people were created.

Comparing the appearance of people on the timelines below, which are both to scale, is instructive. Jesus, speaking around 4,000 years after creation, was correct to say that Day 6, when humans were created, was effectively 'the beginning of creation' as seen from thousands of years later. By contrast, a creation fifteen billion years ago on the secular timescale would put humans at the *end* of the time scale. It shows clearly how the acceptance of the secular timeline starkly contrasts with the statements of Jesus.

Today, the vast majority of Christians in not only secular academia, but also theological institutions, Bible colleges, etc. believe—and many teach—that the secular 'billions of years' is fact. When one tries to find out how they deal with these repeated references, responses vary. However, the 'explaining away' that takes place (whenever the problem is not simply ignored) invariably makes it plain that the authority being deferred to is not the Word of God, but rather current secular opinion.

The most striking (and sad) example of this switch in authority source I know of comes from a personal experience. In New York, USA, many years ago, I had arranged to sit down over a hot drink with a distinguished

Columbia university professor, a Christian who was well-known for his active opposition to a straightforward view of Genesis. At that time, he was actually the head of a grouping of Christian academics which had been openly set up to provide opposition to the inroads our ministry was making. Over the years, this group has unfortunately been very effective in persuading most Christian training institutions that compromising on biblical creation in favor of secular thinking (evolution, long ages) is the only 'respectable' position.

This professor himself, in addition to his secular science qualifications, was well regarded in the theological arena as well as being very biblically literate. He had at that time already been a frequent guest lecturer at several leading American evangelical training institutions.

During our courteous exchange, I asked him about the above comments by Jesus in relation to the age of the world. I asked, "Isn't it clear that Jesus taught and believed that the world was young?"

A STUNNING RESPONSE

I expected him to do as other Christian evolutionists have done—to try to find ways to torture the text to escape these obvious implications. Instead, he said that he totally agreed that Jesus believed in a recent creation of all things, Figure 23.6.

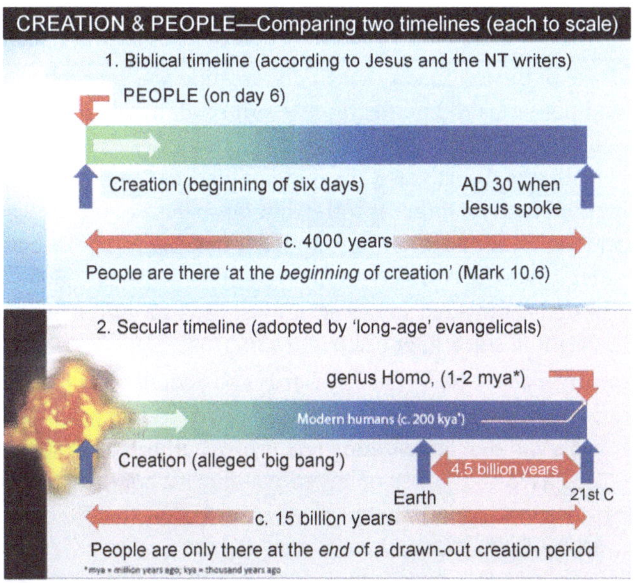

Fig.23.6: Creationists Time Line Vs Big Bang Time Line

Somewhat taken by surprise, I said, "Well, how do you deal with that, then?" (He would of course have assumed, correctly, that I knew of the long-age position of this prominent organization of theistic evolutionists.) His answer simply stunned me, to put it mildly. He said:

"Jesus didn't know as much science as we do today."

His words burned themselves indelibly on my memory, while the recollection of my response has faded somewhat. But I recall saying something about Jesus being the Creator, God made flesh; He was there at creation, He does not lie, that sort of thing. To which his reply was once again unforgettable:

"Ah, but that's where it gets very complex—it has to do with the theology of the Incarnation, where Jesus deliberately laid aside many of the things that had to do with His pre-incarnate divinity."

Our conversation was nearing the end of its allotted period in any case, but I recall being so stunned by this that it took me till well afterwards to fully process the implications.

WHAT IT ALL MEANS

Firstly, and very importantly, the professor's comments were a clear admission that the words of the Lord Jesus Christ Himself, as recorded in the Bible, confirm that He believed that things were recently created.

Remember that this professor was at the time the most prominent of all the professing evangelical academics that were being enthusiastically welcomed into Bible colleges and seminaries—to tell them why it was acceptable to believe in evolution and long ages. He obviously saw it as hopeless to try to claim other than what the Lord is clearly saying in this Bible text. And this is despite many attempts by others to 'explain away' this huge stumbling block for long-agers.

His way of being able to hold onto his theistic evolutionary view was to claim that Jesus was not lying, it was just that He was poorly informed. This was because when He as God the Son became flesh, laying aside aspects of His divinity included divesting Himself of all knowledge about what really happened when He had created all things.

If I had had the presence of mind, an appropriate response might have been to ask something like the following:

"let's assume for the sake of the argument that firstly, creation was by evolution, over millions of years of death and suffering—and that Jesus did perform some sort of lobotomy on Himself, so that He could no longer recall what really took place. So, He just understood Genesis in the most natural straightforward way, not realizing what the real truth was. What you're claiming in that case amounts to this: That God the Father, knowing the *real* truth, permitted not just the Apostles, but His beloved Son, while on Earth, to believe and teach things that were utter falsehoods. Furthermore, it means that the Father permitted these false teachings to appear—repeatedly—in His revealed Word. With the result that for some 2,000 years, the vast majority of Christians were seriously misled about such things as not just the time and manner of creation, but gospel-crucial matters such as the origin of sin, and of death and suffering."

The Lord Jesus repeatedly made it clear that His words and actions were on the Father's authority, in all respects. Some examples are firstly John 8:28: So, Jesus said to them, "When you have lifted up the Son of Man, then you will know that I am he, and that I do nothing on my own authority, but speak just as the Father taught me". And John 12:49–50: "For I have not spoken on my own authority, but the Father who sent me has himself given me a commandment—what to say and what to speak. And I know that his commandment is eternal life. What I say, therefore, I say as the Father has told me."

One thing is very clear from all this. Namely, that the erroneous belief that 'science' insists that evolution and long ages are 'fact' is the most serious challenge to biblical authority, and thus to the faith in general, that Christendom has ever faced. If even Jesus' words in Scripture can't be trusted on some issues, how are we supposed to trust anything in the Bible at all.?

Other leading theistic evolutionists have similarly made plain their belief that Jesus was mistaken. For example, on the American theistic evolutionary site 'BioLogos,' led by Francis Collins, there appeared the following:

"If Jesus as a finite human being erred from time to time, there is no reason at all to suppose that Moses, Paul, John wrote Scripture without error. Rather, we are wise to assume that the biblical authors expressed themselves as human beings writing from the perspectives of their own finite, broken horizons."

THEOLOGICALLY CRIPPLED

This is all the more serious because Jesus and the apostles used the history they taught to back up the theology that they taught. The Resurrection (1 Corinthians 15), marriage (Mark 10:1–12), atonement (Romans 5:12–21), and Heaven (Revelation 21–22:5) are only a few of the areas in which compromising Christians are theologically crippled, because they don't have the same strong stand on Genesis that Jesus and the apostles did when they taught about these areas.

What a tragedy that so many Christian leaders have been bluffed and intimidated into assuming that secular interpretations of the evidence should dictate their understanding of God's Word. And right at a point in history when there are more scientific reasons than ever to confirm the utter rationality of trusting the Bible, not evolutionary conclusions.

THEISTIC EVOLUTION AND THE KENOTIC HERESY

This error from many leading theistic evolutionists is not a new idea. It was rejected by the Church in general as the *kenotic heresy* in the 4th Century already, but has been revived in modern times, and for reasons as shown in the main text.

This asserts that in the Incarnation, Jesus emptied Himself of divine attributes, which is a misunderstanding of Philippians 2:6–7:

"[Jesus] Who, being in very nature God, did not consider equality with God a thing to be grasped; rather, he emptied Himself by taking the very nature of a servant, being made in human likeness."

This does indeed talk about 'emptying' (kenosis), but what does it actually *say*? "He emptied Himself by taking … ." That is, He didn't empty anything *out of* Himself, such as divine attributes; rather, His emptying of Himself was by *taking*. That is, it was a subtraction by means of adding—*adding human nature* to His divine nature, not taking away anything divine.

This is what makes our salvation possible: he "shares our humanity" (Hebrews 2:14–17), and is our "kinsman-redeemer" (Isaiah 59:20); but He is also fully divine so He can be our Savior (Isaiah 43:11) and can bear the infinite wrath of God for our sins (Isaiah 53:10), which no mere creature could withstand.

Nevertheless, on Earth, Jesus *voluntarily* surrendered the *independent* exercise of divine powers like omniscience without His Father's authority. But Jesus never surrendered such *absolute* divine attributes as His perfect goodness, mercy, and (for our purposes), *truth*, so He would never teach something false. Furthermore, Jesus preached with the authority of God the Father (John 5:30, 8:28), who is always omniscient. Accordingly, these theistic evolutionists really must charge God the Father with error as well.

CREATION WEEK - EX *NIHILO*

God made the world and everything in it in consecutive six ~24-hour days, Figure 23.7. The Scriptures are clear and emphatic on this—especially what God wrote with His finger in Exodus *20:8–11*. But what exactly happened during Creation Week? Can we use science to show that what God made during Creation Week was made recently? Or do we need to appeal to some sort of 'apparent age' apologetic? Or perhaps we should just be satisfied with calling it a miracle, and leaving it at that?

The question of how science and natural processes interface with the creation miracle raises too much of important issues in the origins debate.

Fig.23.7: In the Beginning God Created – *Curtsy YMI*

Ken Coulson, a Ph.D. creation geologist (from Loma Linda University, specializing in Cambrian microbialite formations) has written *Creation Unfolding: A New Perspective on Ex Nihilo* to address these questions.

YOUNG EARTH CREATIONISTS APPROACHES TO CREATION WEEK

In explaining the origin of things in Creation Week, "***God spoke, and it happened***" is a fair answer, as far as it goes. It was clearly a miracle. However, young-earth creationists typically don't leave the answer there. The world is full of process, which we study through science. How do we interface that with "Creation Week"?

Coulson sees "some glaring problems within mainstream young-Earth creationism [YEC] that stem from what I believe is an overly suspicious approach to conventional science."

YEC websites run by people with little or no scientific or theological expertise have promoted the misguided belief that "a YEC interpretation of nature is *obvious*, and only a fool would fall for the secular view that believes the Earth is 4.6 billion years old." This has incited evangelicals in the church at large to say things like, "there is not a shred of evidence in support of an evolving universe," or "how could anyone believe in millions of years of Earth history?" Put simply: science *proves* the young age of the earth and universe, and it's only the foolish naturalistic bias that blinds the scientific establishment to the obvious.

However, there are processes at play in creation that we know can form much of what we see, ***given enough time***. But to do so they would require far more time than Creation Week allows. As Coulson explains:

"As we shall see, there are many aspects of creation that, without deferring to special revelation, will only lead to an evolutionary and/or old-Earth perspective."

For instance, Coulson cites the example that new continental crust is being formed at a certain rate per year through processes we see happening today, Figure 23.2. However, if we extrapolate those processes back at the rates we currently observe, they would give an age for the crust much older than the Bible allows for. And yet, geophysicists can successfully predict the composition of the continental crust by assuming it formed through those natural processes. The geophysics of continental crust formation is *successful* science, both as operational science *and* historical science. So, if natural processes suffice to create the continental crust, and the rates of the natural processes through which it forms indicate a million/billion-year age, Figure 23.8, where is there room for supernatural creation?

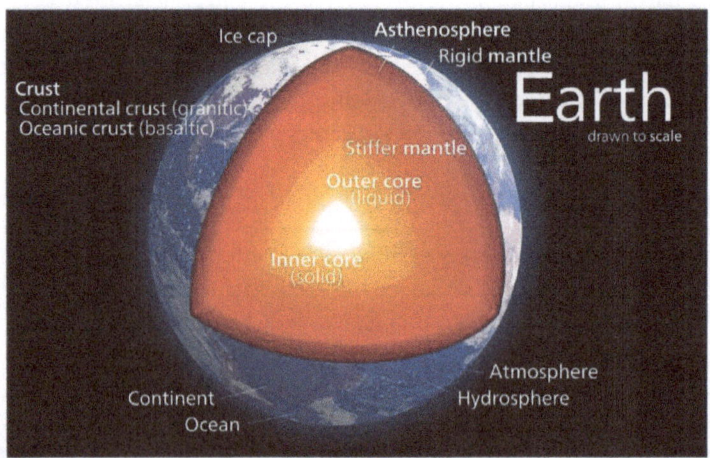

Fig.23.8: *If natural processes successfully describe the evolution of the earth's interior, did God use them in making the earth's interior? Curtsy - Image: Kelvinsong/CC BY-SA 3.0 (modified)*

When a Christian geologist or geophysicist sees this, how does he respond? How can he integrate what he knows from the Bible, that God created the world in six days, with the success of his scientific discipline in predicting so much about the nature, history, and origin of Earth's interior by the operation of natural processes? This is the question that drives Coulson's view:

"This view presents a scientific and theological synthesis that both affirms a six calendar-day creation while at the same time providing a solid scientific methodology from which the Christian layperson, educator, and scientist can approach the origin of the universe."

A NEW CREATION APOLOGETIC

Coulson stated there is only one way to deal with this: "The presence of vast processes in conjunction with short time frames can, therefore, only be understood by appealing to some kind of mature creation apologetic."

However, what sort of "mature creation apologetic"? There have been different forms since they first arose in the 19th century, largely in response to the growing acceptance of deep time geology. Coulson surveys their history, starting with scriptural geologist Granville Penn and the *Omphalos* theory of Philip Gosse. He then highlights the approaches of Henry Morris and Ken Ham, which were much more modest approaches than Gosse. Morris allowed for apparent history in inorganic traces (such as radioisotope abundances and starlight from distant stars) but rejected the application of apparent history after Creation Week or to explain the fossil record, which he (rightly) held to be inconsistent with a 'very good' pre-Fall world. Ham, however, had the 'thinnest' form of mature creation, embracing 'functional maturity'— 'apparent history' only so far as needed for the immediate function of the biosphere.

Coulson then identifies what he regards as the main objection to mature creation apologetics: what has been dubbed 'non-essential apparent age'.

For instance, SN1987A:

"Why would God create in-transit starlight representative of a fake supernova that itself becomes manifest only thousands of years later, and then only for a few months before fading into the stellar background? It seems fair to say that SN1987A and its related cosmological phenomena do not seem to be required for the immediate functionality of a mature universe".

Coulson suggests compartmentalizing matters. Consider Adam's skin. Coulson suggests it was rapidly 'aged' in anticipation of the natural laws and conditions that would naturally age human skin to match his functional 'age' (of a 20–30-year-old). In that case, it would have freckles, blemishes, and even wrinkles. But it would of course have been the skin of a *perfectly healthy* young man; no scars or defects that would affect the function of the skin. As an analogy, he considers Moses' snake (*Exodus 4:3*). Was it genetically perfect without defect or signs of aging? Probably not. It was probably a snake the Egyptians would've recognized. In other words, it was made *in anticipation of the conditions it would be exposed to*. Which means it would've looked like an ordinary snake—mature, healthy, *and* with phenomena consistent with a fallen world. However, is that deceptive? Obviously, Coulson thinks not. Rather, it goes back to purpose: what was the snake made *for*? To convince the Egyptians that God was the true Creator. The message was thus something recognizable to the audience.

In fact, Coulson says the removal of such age and growth-process phenomena (e.g., tree rings) has problems of its own:

"An across-the-board, blanket-like removal of all such age and growth-related phenomena by God during Creation Week would change the very way nature was supposed to look, grow, feel and sound. Such an anomaly would also make it difficult, if not impossible, for man to complete his God-given mandate to have dominion over the entire Earth. To have dominion over something means to *understand* how it works. If Adam suddenly decided to dump his newly acquired theological training in exchange for a career in paleontology, archaeology or osteology, he would soon find himself up against some contradictory and anachronistic data."

In truth, though, Coulson offers a slightly different approach. He seeks to avoid the language of 'mature creation,' 'appearance of age,' and 'apparent history.' For him, the problem is that these speak in static terms to describe *dynamic* events. They focus only on the effect and ignore the cause. To do this, Coulson suggests two concepts: ***supernatural formative processes*** (**SFPs**) and a *conceptual universe*, which he spends a great deal of writings elucidating. With these, he believes he has found a way for mature creation apologetics to interface fruitfully with secular geophysics and astronomy while remaining faithful to the biblical text.

One of the weaknesses of traditional mature creation apologetics is its lack of attention to the *processes* that govern creation. Plus, God appears to have supernaturally formed at least some things through processes during Creation Week, such as the land appearing out of the waters below and the vegetation sprouting on Day 3. But how did God govern those SFPs? Coulson uses examples of post–Creation Week miracles as a template for understanding Creation Week.

For instance, Jesus turned water into wine. If it were submitted to scientific analysis, it would've reflected the typical natural history for producing wine in that time and place. The same would be true for the almonds that budded from Aaron's staff and the snake that formed from Moses' staff.

Jesus Himself was unremarkable in appearance—an ordinary-looking human. And, of course, Jesus created bread and fish more than once, which would've reflected the specimens of each recognizable to the locals. In each case, Coulson argues that "God's supernatural creative strategy involved a *commitment to existing natural processes*."

Coulson also points out that, in all these cases, a commitment to science as the only way to know things would lead us to *wrong* conclusions. The physical characteristics of Aaron's almonds, Moses' snake, Jesus' DNA, and the bread, fish, and wine he made would all lead us to *inevitably* conclude that they formed through natural processes. After all, experience teaches us that time-dependent processes are *needed* for these things to come into existence. But this would be wrong. When God says He created something supernaturally, we're no longer given the option to interpret its origin in terms of natural processes. If we try, we *will* conclude a false origin.

However, Coulson argues this approach can be pushed back into Creation Week as well. For instance, Snelling has suggested that kilometer-thick strata were deposited on Day 3 as the land appeared out of the water. Faulkner has suggested that cratering occurred rapidly for the moon on Day 4.

These clearly must be *accelerated* processes; a rapid *maturing* of the elements under consideration *from an undeveloped or embryonic state*. Nonetheless, Coulson adds to that the notion of the *constancy of relative rates of process*. For instance, if two plants grew at different rates relative to each other, God would've sped them up *so that their rates of growth remained the same relative to each other*. This would create a sort of 'time lapse' effect.

Fig.23.9: *Coulson's 'time lapse creation' has God hitting 'fast forward' on natural processes to produce the land and plants on Day 3 of Creation Week.*

Plants would all grow normally *relative to each other*, but the system *as a whole* would be sped up. Indeed, in rapidly *maturing* the plant world like this, some plants may have gone through cycles of life *and death*. That may even be crucial for establishing a *functional* plant ecosystem. We can call this the *time-lapse creation* model, Figure 23.9.

Coulson then applies this to geology. The land *appeared out of the sea* on Day 3. Several creationists have noted this, and posited *catastrophic* explanations for the Precambrian rock relationships akin to Noah's Flood (indeed, even larger!). However, Coulson points out a problem for this interpretation: stromatolites are present all over the world throughout much of the Precambrian record. They are mound-like structures that typically accrete because of the work of photosynthetic bacteria. However, there are many stromatolite horizons in some of the earliest rocks on Earth to form. But that would imply that communities of cyanobacteria grew and died multiple times, and thus, under ordinary circumstances, it would have required substantial time to form. Moreover, a catastrophic upheaval of the land would create conditions inimical to the formation of stromatolites. Coulson explains the import of his solution:

"This might mean that the Day 3 regression of Earth's oceans was not catastrophic at all. Something is geologically catastrophic when a *single rate*, say erosion due to a flood, is accelerated many times relative to other geologic rates—e.g., deposition of sediment, growth of plants, production of soil. Since *all the rates* were accelerated and remained constant with each other, an observer would witness the development of a genuine, shallow-water environment proceeding at break-neck, yet time-lapsed speed."

However, Coulson says we can't apply SFP theory to arrive at anything empirically consistent with big bang history of the universe. In the Bible, the earth was made on Day 1 and the celestial objects on Day 4. The big bang history says the sun and most of the stars were formed before the earth. To address this issue, Coulson introduces his 'conceptual universe' idea: "a fully functional, conceptual universe, already 'existed' *in the mind of God* prior to Genesis 1:1."

GOD'S MIND FORMED THE MATURATION OF THE UNIVERSE

In essence, Coulson posits that God had a whole maturation process for the universe *in His mind* before *Genesis 1:1*. However, when God manifested the physical product, He didn't manifest the whole maturation process He had in His mind. Instead, He only manifested the *end product* of the maturation process in His mind. SFP theory still applies in this scenario, since God conceived not simply the final product, but the whole maturation *process*. However, Coulson doesn't require the maturing process *in the universe* to have had a physical manifestation, like he suggests for Earth processes on Day 3. As such, events such as SN1987A occurred *in the mind of God* as part of God *mentally* maturing the universe.

But why create the earth before the sun? Of course, God could've done whatever he wanted; the question is in one sense rather moot. However, Coulson explains:

"Contrary to popular opinion, the pinnacle of God's creation is not the universe, *it is man*. Man is God's creative, crowning glory. In order to underscore this reality, God purposefully placed the planet upon which man would dwell as first in the created order."

Assessment

What are we to make of Coulson's mature creation perspective? Coulson's project is an interesting one, and also likely to be quite controversial.

A Successful Mature Creation Apologetic

The crucial question is: does Coulson succeed in overcoming the standard challenges to mature creation apologetics? I think some aspects of his model do this better than others.

TIME-LAPSE CREATION

I think the most successful aspect of Coulson's view is his 'time-lapse creation' idea. It gives us a reason to affirm the *reality* of large-scale processes during Creation Week, since God basically *accelerated* the maturation *processes* of the various systems He created. On the surface, the only feature of this history that seems 'apparent' is the *age*.

However, there is a major challenge for this perspective. How could *seasonal* plants grow all over the earth when they were exposed only to one daylight period and a single day of weather (of whatever season)? Indeed, without any sun? The same problem affects geological processes in Creation Week, with the possible presence of, for example, tidal rhythmites in Precambrian rocks, as well as the need for photoperiodism to explain the existence and growth of stromatolites. It was one day before the existence of the sun and moon; how could many days, months, and years of biological processes happen without photoperiodism and the existence of the sun and moon? This is basically a special instance of the old 'apparent histories make God a deceiver' objection. After all, much of how the rocks and plants developed in this scenario occurred *as if*, for example, the sun and moon were present, though we know they were not.

I think Coulson's appeal to the budding of Aaron's staff in *Numbers 17:8* as a template is a good response. The staff budded, blossomed, and produced almonds *in one night*. However, almonds require not simply more time than one night to grow, but also seasonal variations and photoperiodism (as well as soil, a root system, and nutrients). As such, the almonds rapidly matured *as though* all the needed natural conditions were present. Other examples of biblical miracles (such as Jesus' human body, and the miraculous wine, fish, and bread) would all correspond to this pattern, too.

Applied to the creation of vegetation and the appearance of the land on Day 3 of Creation Week, it would suggest that all the processes involved in the system (including organism growth) were accelerated proportional to each other *and* the organisms grew as though all the needed natural conditions were present.

Nonetheless, how could all this happen in one period of daylight? Coulson has a good answer for this, too: "The periods of day, evening, and morning served to fix the reader's temporal frame of reference, *not to cause the growth of plants* [or cyanobacteria in stromatolite formations]." The growth of plants was an *inside* consideration for the time-lapsed system, not a concern for the absolute temporal frame of reference.

A CONCEPTUAL UNIVERSE

I think Coulson's 'conceptual universe' idea is more problematic. The biggest problem, I think, revolves around giving a *merely conceptual* maturation process the status of 'real.' It seems to me implausible, even if God is the one doing the conceptualizing. **It blurs the distinction between conceptualization and creation**; a distinction we experience every day. Whatever sort of ontological status we give to thoughts and concepts, there is clearly a fundamental lack of concreteness to them that exists for causal objects such as we see all around us in the physical universe. And this is not to promote materialism, either; God and spiritual agents are just as concrete and causal as the physical world, though *we* lack any physical sense experience of them (in general).

However, to the extent that God's conceptualization of any sort of maturing cosmos lacks concreteness, it seems it lacks *genuineness* as a process *history*. As such, the specter of God as a deceiver is not so easy to avoid.

Moreover, I'm not sure how much Coulson needs to rely on the 'conceptual universe' idea to achieve his aims. I don't think it's needed to make Day 4 work in Coulson's perspective. All God needs to do is cordon off Earth from any effects of a rapid maturation of the cosmos on Day 4, and such a process can be allowed to proceed *in physical history* and not merely in concept. In this case, SN1987a was thus the product of not just the post-Creation Week history, but that and the *real* maturation of the cosmos on Day 4. Indeed, Coulson acknowledges this possibility:

"It is true, God could have brought an *immature* universe into our existence, causing it to develop in time and space like the 'sprouting' plants and trees of *Genesis 1:11–12* or the Earth. ... Doing so, however, doesn't change the fact that God has every developmental facet of our universe, including its future state, fully planned out in His mind."

Coulson's response, though, assumes that God merely having the plan in His mind makes it count as 'real.' That, however, is precisely the problem—is it 'real' if the plan is *just* in God's mind?

Additionally, Coulson is treating God's thoughts and ways in a human-like manner, while God's thoughts and ways immensely superior than human type of thinking and formulating any creative thought. It is distinctively separate and above human perceptions. It is stated in Isaiah 55:8-9 "For my thoughts are not your thoughts, neither are your ways my ways," declares the LORD. "As the heavens are higher than the earth, so are my ways higher than your ways and my thoughts than your thoughts."

APOLOGETICS, FAITH, AND AMBIGUITY

Coulson thinks we *can't* prove the universe is young. Indeed, he says God designed things that way: "God's design is not to prove that the earth or the universe is 6,000 to 10,000 years old." Instead, he says that God is clear enough from creation to justifiably condemn us for ignoring general revelation, but ambiguous enough to allow people the freedom to ignore Him. Mature creation and the mismatch of scientific conclusions and scriptural declarations are instances of God veiling Himself.

This is an interesting approach. There is little problem with God inputting some ambiguity into creation. And God does indeed save through the foolishness of the Cross rather than through the mere application of human intellect. Nonetheless, I don't think this provides much support for Coulson's mature creation apologetic.

First, what of *Romans 1:19–20*? I think Coulson rightly points out that this doesn't have anything to do with proving the world is young: "*Psalm 19, 104* and other such passages are mainly concerned with a broad theological fact—God is powerful, and this power is displayed in the created cosmos." Still, if *Romans 1:19–20* has nothing to do with proving the earth is young, can we be sure the pattern God used in general revelation applies to scientific process-age arguments? Perhaps God providentially embedded some supernatural signatures in the rocks, knowing they would only become relevant in modern times, and did so to encourage the faithful and provide a challenge to skeptics.

None of this suggests scientific process-age arguments would ever prove *beyond a shadow of a doubt* that the world is young. But it does attenuate the link between ambiguity in creation and the sort of general mature creation apologetics Coulson advocates. Why? It shows that process-age arguments are also consistent with the balance of ambiguity and clarity God has deployed in general revelation. After all, such process-age arguments have some convincing power, but they are based on large-scale extrapolations, and these, in turn, are based on assumptions. As such, the naturalistic mindset can ignore or sidestep them with auxiliary hypotheses. After all, if they can do it with abiogenesis, they can do it with anomalous process ages in the rocks. At the very least, either approach is justifiable.

Nonetheless, if there is a lot of ambiguity in creation, it may lend more support to the sort of mature creation approach Coulson employs. However, ***I think Coulson overplays*** the ambiguity embedded in creation. He says ambiguity is so pervasive that, despite the fact everyone must exercise some sort of faith, "I do believe it is the Christian who must exercise *more* faith" than the materialist. Is this really consistent with saying that general revelation is clear enough in what it reveals about God to hold everyone accountable for ignoring or rejecting God? It doesn't seem so. Plus, I don't think God's ambiguity in creation is that severe. If it were, materialism and atheism would have been broadly popular views throughout history. And yet Isaac Newton, only 300 years ago, quite justifiably said this: "Atheism is so senseless and odious to mankind that it never had many professors." We can acknowledge that it's much easier *today in the West* than it ever has been elsewhere to be an atheist or materialist. But that suggests the issue is not something intrinsic in faith or general revelation, but rests largely in the particulars of our culture and recent history.

TWO APPROACHES

This brings us to two ways to view Creation Week, as Coulson observes. He says it revolves around the question: is there an apparent age for the Earth (and universe)? He outlines the implications of each idea:

1. "If there is *no* apparent age for the Earth, this would mean that various processes *currently* at work within the earth cannot be extrapolated backwards in time for the purpose of determining a 'beginning'. In this scenario, some of the Earth's *internal* processes can be extrapolated backwards millions or billions of years (partial melting, fractionation, gravitational differentiation, radioisotope decay), others can be extrapolated backwards hundreds of thousands or even thousands of years (the decay of the earth's magnetic field), while others cannot even be correlated at all, since they were clearly supernatural acts (polonium halos). This would mean that God used other, as yet unknown, mechanisms in conjunction with known processes when forming the planet.
2. "If there *is* an 'apparent age' of the earth, this would mean that various processes at work within the earth *can* be extrapolated backwards in time for the purpose of determining a 'beginning.' In this scenario, *all* the earth's internal processes can be extrapolated backwards into Creation Week because, although these processes were accelerated, they were operating at the same rates, *relative to each other*, as they do in the present (SFP theory)."

Coulson favors the latter approach, and I think he has laid out its strengths relatively well. However, I think he misconstrues the true strength of the 'no apparent age' (NAA) approach.

Coulson says:

"Choosing the first option has the benefit of underscoring a few processes for the purpose of 'proving' that the Earth's internal structure was supernaturally created about 6,000 to 10,000 years ago."

For instance, the magnetic field of the earth seems to match the c.-6,000-year timescale of the Bible and is therefore *positive scientific support* for the biblical timescale.

However, if the earth has no apparent age, the existence of processes that roughly match the biblical timescale is *inconsequential* for its reasonableness. Instead, it's what Coulson thinks the proponent of this approach must *admit* that is the true *strength* of this approach:

"This means scientists cannot rely on current natural rates for the purpose of understanding anything that was made during Creation Week."

The advantage of this approach is that, if true, it reveals the futility of trying to form a comprehensively cogent naturalistic account for the history of the universe. Since different processes give different process ages, how could such a system have formed naturalistically? The resulting conditions are complex in very specific ways; ways that would be too improbable for it to have formed spontaneously within the history of the observable universe. But such conditions are only ever associated with intelligent activity. In other words, anomalous processes with a system are an argument for ***intelligent design*** (or **supernatural creation**).

But does this make the NAA approach better than the AA approach? Not necessarily.

First, what of those supposedly 'discrepant' process ages? Some seem more secure than others. But we run into a serious problem of underdetermination: i.e., the physical data seems open to multiple conflicting, yet empirically equivalent, interpretations. The data by itself is rarely, if ever, enough to justify belief in one interpretation of the evidence over all others. Indeed, perhaps the starkest examples of underdetermination among scientific disciplines concern the historical sciences of the deep past. After all, think about the million-fold or more extrapolations that are required to obtain process ages! Are we sure that nothing happened in between now and when the system started 'ticking' to affect the 'age' the system gives? And what of the radical 'gappiness' of the rock record—it is more gap than record at *multiple different scales!* These issues are a major problem even *after* Creation Week, but they become unavoidable *during* Creation Week.

As such, there *are* scientific reasons to adopt the NAA approach, but they are not so epistemically secure as to rule out the AA approach. The AA approach also has several scientific and theological advantages, as Coulson points out. In other words, there is enough ambiguity in the data *and* the theology for creationists to develop *both* approaches.

I think this is a good thing. It forces us to be tentative about our conclusions in favor of our preferred approach and flexible in our apologetic strategy. For those who feel the weight of the concerns that motivate Coulson, his 'apparent age' approach provides a fruitful way forward. For creationists who are more skeptical of secular science's ability to understand the history of the cosmos, and those who think there are genuine process-age discrepancies, there is a fruitful way forward. Most importantly, though, this flexibility emphasizes that our approach is founded on *Scripture* rather than our scientific extrapolations and theological implications.

Coulson has presented an interesting case for a mature creation apologetic. It is wide-ranging, well thought out, and achieves many of its aims. There are, however, some important weaknesses, neither is it the only viable approach. Indeed, a variety of approaches is justifiable and healthy. Nonetheless, I think Coulson has presented a useful scientific/theological synthesis for understanding Creation Week.

THE AGE OF THE EARTH - WESTERN CULTURE

I enjoyed Cherry Lewis's fascinating biography of Arthur Holmes (1890–1965), the English geologist famous for his work on radioactive dating and the age of the earth. It traces how ideas about the age of the earth changed over one man's lifetime. Dr Lewis is a geologist/geochemist, currently working as Research Communications Manager at Bristol University, and is Secretary of the History of Geology Group (HOGG) there.

Arthur Holmes was 'the only child of staunchly Methodist parents' (p. 7) and he 'well remembered his parents' Bible, and the magic fascination of the date of Creation, 4004 BC' (p. 27). In later years Holmes reminisced 'that the Earth has grown older much more rapidly than I have—from about six thousand years when I was ten, to four or five billion years by the time I reached sixty.' I would like to learn more about Holmes' own reasons for his apostasy from his Christian heritage, because his story sadly is all too common.

A short, interesting book, *The Dating Game* is filled with photos and human interest. Cherry Lewis has researched her topic thoroughly, and quotes widely from diaries and letters. Clearly, Holmes was a perceptive and independent thinker, a dynamic lecturer and diligent worker.

IT IS REALLY A GAME

Those who think that science has proved the earth is billions of years old should find this book disturbing. Lewis clearly shows that the quest for the age of the earth is not objective science but a subjective, arbitrary and erratic pursuit.

Even her name for the book illustrates that point. Some reviewers must have urged her to use a different title, but what could be more fitting than *The Dating Game*? Lewis refused to change it, but apologized to any readers who found the book 'on the "Romance" shelves' (p. 242).

My dictionary defines 'game' as 'a contest for amusement in the form of a trial of chance, skill or endurance according to a set of rules.' She vividly paints the characters of the players in the 'dating game,' and tracks the progress of the score for a hundred years. To the uninitiated, a history like hers is one of the best ways to understand a subject.

In a game, the score is determined, not by impersonal scientific measurement, but by the strength, skill and creativity of the players. The rules of play are not laws of nature, but arbitrarily agreed by the players, and sometimes changed during play. We see this acutely demonstrated in the events she describes.

Arthur Holmes' interest in the game was aroused in his teens one summer holiday. This is when the great physicist William Thomson (Lord Kelvin, 1824–1907) instigated the dating game in *The Times*. Lewis described how Arthur and his friend 'were on the edge of their seats with the excitement of it all, for not only did they become familiar with all the arguments, they also got to know all the big names in science at that time—William Ramsay, Ernest Rutherford, Frederick Soddy and Robert Strutt' (p. 12). Watching *The Times* exchange influenced Holmes to take up the sport.

THE KELVIN AFFAIR

Arthur Holmes began his career at a most interesting time, as Cherry Lewis describes. For forty years Lord Kelvin had completely demolished all opposition. But by the early 1900s, Kelvin was gradually losing his dominance, Figure 23.10. The upcoming generation had a new weapon and were about to dislodge him. Holmes would soon be a key player.

Fig.23.10: *Lord Kelvin was a major player in the dating game. Curtsy Photo by Andrew Snelling*

By the end of the 1800s, Lord Kelvin was saying that the earth was between 40 million and 20 million years old, with a 'personal preference' for the lower value (p. 39). Of course, Kelvin had not measured the age (otherwise he would not talk about 'personal preference'), but he had calculated it (or rather, logically his methods could calculate at best an *upper limit* for the age). And before he could start his calculations, he had to make *assumptions* about the past. In particular, he had to assume a history for the earth.

In fact, *every age calculation is based on an assumed history*—assumed because, without an eyewitness report, we cannot travel back in time to observe what happened. This reality is not generally recognized—that it is impossible to measure the age of something scientifically—impossible. The numbers quoted for the age of the earth (or the age of the dinosaurs or the age of a volcano) are the outworking of personal beliefs, made to look authoritative by much technical equipment and complicated calculations. It is not until we understand this fundamental fact that the antics of the players in the dating game make sense.

In Kelvin's case, he assumed the earth was initially a molten blob and calculated how long it would take for the blob to cool (assuming numbers for all the relevant parameters such as its initial temperature, conductivity and reflectivity).

Was Kelvin's answer, right? Scientifically, it is impossible to say. How could anyone check? What would you compare it with? You could only say whether it seemed reasonable, and Kelvin certainly thought so. For a start, it agreed with similar calculations of the age of the sun by Hermann von Helmholtz (1821–1894).

But for the geologists (and evolutionary biologists as shown below), 20 Ma was far too short because they envisaged the earth was unimaginably old. Kelvin knew their long ages flowed from their assumption that geological processes have always operated slowly, like they do at present. They were *not* based on experimentally established laws such as the laws of heat transfer. Kelvin was blunt, 'A great reform in geological speculation seems now to have become necessary' (p. 35). In other words, Kelvin challenged them to assume a different history. He knew the geologists could easily harmonize their age for the earth with his result by imagining a bit of catastrophe.

The geologists did not take kindly to Kelvin's suggestion. 'Although incensed,' Lewis explained, 'most geologists were clearly intimidated by Kelvin's authority and felt obliged to heed his arguments' (p. 35).

The problem was not his 'authority.' The problem was that they agreed with his assumed history for the planet, and could find no mistake with his chosen parameters. After that it was just a matter of cranking the mathematical handle and the age popped out.

But T.C. Chamberlin (1843–1928), head of the Department of Geology at the University of Chicago was not prepared to concede defeat. He *speculated* that there might yet be discovered new sources of energy within the particles of matter (unknown to him then) that would allow more time than Kelvin had calculated. In geological circles this response is regarded as heroic but it really shows that the age issue cannot be resolved scientifically. Clearly, he was simply defending his belief in long-ages as a matter of faith, without any observational basis whatever.

The evolutionary biologists such as Charles Darwin and T.H. Huxley were not happy with Kelvin either. 20 Ma was nowhere near enough time for evolution to occur. Darwin was particularly dissatisfied, describing Kelvin as his 'sorest trouble' (p. 35). That suited Kelvin because he opposed evolution. Loren Eiseley writes:

'It can be observed from Darwin's letters that this development in physics gravely troubled him. He refers to Lord Kelvin as an "odious specter", and in a letter [of 1869] … he writes: "Notwithstanding your excellent remarks on the work which can be effected within a million years, I am greatly troubled at the short duration of the world according to Sir W. Thomson [Lord Kelvin] for *I require for my theoretical views a very long period* before the Cambrian formation." … Painfully and doubtfully he [Darwin] wrote to Wallace in 1871, "I have not as yet been able to digest the fundamental notion of the shortened age of the sun and earth."'

Christians often regard *Kelvin as a great creationist apologist* because of his opposition to evolution. They credit him with keeping Darwin at bay for 40 years. But Kelvin did serious damage to the credibility of the Christian worldview because he publicly promoted an *earth history that contradicted the Bible*. The Bible says the earth was originally covered with water; Kelvin said it was a molten blob. The Bible indicates the earth is 6,000 years old; Kelvin said 20 million was acceptable. If Christians won't stand on the plain teaching of their own book, why should anyone else accept the Bible as authoritative?

RADIOACTIVITY CHANGED THE GAME

Fig.23.11: Arthur Holmes

Therefore, the dating game was set for a fascinating turn when Holmes, Figure 23.11, began his career. The dynamics changed dramatically when (Antoine-) Henri Becquerel (1852–1908) discovered radioactivity in 1896. Heat generated by the radioactive decay of elements within the earth was quickly invoked to explain cooling of the earth over long ages. Thus, radioactivity allowed a different history for the earth to be proposed, one that could extend 'for as long as geologists and biologists might need' (p. 49)—notice the word '*need*'! It is widely paraded that the discovery of radioactivity solved the heat problem but that is not so. Empirical evidence still favor's the view that the earth is much younger than presently believed.

Most importantly, radioactivity allowed age calculations to be applied to individual rocks and minerals, and this is where Holmes became famous. It was Ernest Rutherford (1871–1937), Lewis explains, who was the 'very first person ever to date the true age of a rock' (p. 54). Her word *true* is curious because her explanations demonstrate that ages are not *found* but *assumed*. (Remember, every age calculation is based on assumptions about the past.) Two pages later, Lewis says that Robert Strutt 'recognized the flaw in the method' (p. 56). So much for Rutherford's *true* age.

Apart from the 'flaw,' Lewis reveals that in those early days they did not know there were two different uranium decay chains or different isotopes of uranium and lead. In fact, they did not know isotopes existed—this had to wait till Frederick Soddy (1877–1956) in 1913 (pp. 112–115). This illustrates another vital fact about the dating game: *assumptions are always made in ignorance*. That's why every age result is always tentative, just waiting for a new finding to knock it over, as Lewis illustrates again and again.

Holmes was just 21 when he published his first uranium-lead result for a rock from Norway (*still before the discovery of isotopes*). He also recalculated ages from data previously published by Boltwood, the oldest result being 1,640 million years (pp. 63–64). That was a vast increase on any numbers previously published. The response of the scientific community was stunned incredulity. Geologists 'had been given vast time scales to fill with sediments of which there was no evidence' (p. 65).

METHODS AND DATES ARE *SELECTED*

One interesting episode Lewis describes involved 'dating' two rocks from northern England using the helium method (pp. 149–150). The 'dates' Holmes obtained were 182 million years for the igneous sill, and 26 million years for the dyke. Holmes was delighted. He considered the results 'to be in excellent agreement with the geological evidence.' But then, how would anyone know?

But today the sill is considered to be 295 million years old, not 182 million, and the dyke 60 million years compared with 26 million. The helium method has been blamed for the discrepancy because it is held that helium leaks from the rocks, giving too low an age.

But why did Holmes so quickly accept a faulty method? Lewis explains,

'So strong was the desire to find a successful dating technique, he convinced himself that although the helium results were "slightly low," they concurred "quite satisfactorily with the scanty results based on lead ages"' (p. 151).

So much for objective scientific methods.

Interestingly, the helium method continued to be used on meteorites and some very ancient dates were obtained (p. 221). It was argued that, unlike terrestrial samples, meteorites did not lose helium (again, how would you know?). However, when ages were obtained that were anomalously *high* (i.e., did not agree with what was expected), it was suggested that meteorites gained helium by being bombarded by cosmic radiation as they cruised the heavens. This is another example of an *ad hoc* assumption invoked to dismiss anomalous results.

These stories illustrate how in the dating game, the methods used, and dates obtained, are selected after the event according to whether the results agree with what is already believed and desired to be correct. Recent creationist work is particularly relevant to the helium method. Helium retained in zircons from granite actually provides strong evidence *against* the idea of millions of years, and *for* the idea that accelerated nuclear decay occurred in the past. A main pillar of the argument is precisely the rapid leakage of helium noted above—yet much helium still remains in the zircons!

THE HUBBLE AFFAIR

Lewis describes a curious complication that emerged in the late 1940s. As the age of the earth gradually crept up towards 3000 million years, and beyond, the earth eventually became twice as old as the universe (p. 191). As with Kelvin, the issue became a battle of wills across scientific disciplines.

The age of the universe was calculated from the Hubble constant, assuming the big bang history for the universe. Edwin Hubble (1889–1953) had such standing that no-one seriously questioned his value for the constant, Figure 23.12. As recently as 1936, Hubble had concluded that any further revision of the constant would only be of minor importance.

So the blame was levelled at radioactive dating. Even in 1949 it was considered highly improbable that observational changes in the value of the Hubble constant would resolve the timescale problem. Some astronomers were again suggesting that the radioactive decay rate had changed with time (yet modern creationists are castigated for the same suggestion!).

Fig.23.12: *In the late 1940s the age of the earth became twice as old as the age of the universe! Curtsy Image by NASA, ESA, J. Blakeslee and F. Ford (JHU)*

But the astronomers eventually gave in. In the 1950s new measurements of the Hubble constant extended the age of the universe and at last it was 'safely older than the age of the Earth' (p. 191). This dramatic episode again illustrates that the age issue is a battle of wills and beliefs, and *not* a scientifically measurable parameter.

PATTERSON TAKES THE PRIZE

As readers would expect, Lewis reveals the answer to the age question just before the end of her book. She relates that Clair Patterson (1922–1995) 'goes down in the history books as the man who finally dated the true age of the Earth. Wild miracle, finally achieved' (p. 225).

But here we see an ironic twist. Patterson did not date the earth primarily using earth rocks. His key evidence came from meteorites! What have meteorites to do with the age of the earth?

Remember that before anyone can calculate an age for anything, they have to assume its history, namely how it formed and what has happened to it from that time to the present. Early in the 20th century, T.C. Chamberlin had developed the idea that the earth had formed by the accumulation of cold, solid particles and rocks he called 'planetesimals.' By the 1950s this explanation was widely accepted, and meteorites were considered to be junk left over from when the earth formed.

The number calculated from the meteorite data based on these assumptions gave an age of 4.55 ± 0.07 Ga, the age Patterson announced in 1956, and which is still accepted today.

At first Holmes was not enthusiastic with the method

'to use the isotopic composition of lead from iron meteorites as part of the basic data for calculating the age of the earth or its crust, is unsound in principle ... the correct procedure is to use terrestrial materials' (p. 227).

That of course raises one very obvious problem. As Lewis explains,

'If there was no genetic relationship and the Earth and meteorites had not formed at the same time from the same material, then the primeval lead of meteorites would not be that of the Earth; thus, there would be no point of trying to determine the age of Earth from meteorites, and everyone would be back to square one' (p. 225).

To answer this challenge, Patterson produced a graph in 1956 showing the isotopic composition of lead from four meteorites and lead from modern ocean floor sediment. Because the ocean floor sample plotted on the same line as the meteorites, Patterson argued that they all formed from the same cosmic material (p. 226).

That settled the matter, and the age of 4.55 Ga is universally quoted. Yet, as more ocean floor sediment has been analyzed, it has been found that they do not all fall on the straight line but plot all over the place.

Another problem concerns a view developing 'that the lead isotope clock of the Earth may have been reset by the formation of the Earth's core' (p. 227). In other words, the consensus history of the earth is different now from what Patterson assumed, yet his result is still accepted as the *true* age of the earth. As Lewis muses, 'Patterson's results were more fortuitous than was realized fifty years ago' (p. 228). This raises a question: if we know Patterson's *assumptions* are wrong, why should we believe that his *answer* is right?

Clearly the age of the earth is not a scientific issue but a religious and philosophical one.

WHY NO MORE CHANGES?

Lewis's book highlights another fascinating insight into how science works. In the first fifty years of the 20th century, the age of the earth increased from 20 million years to 4,550 million. But in the second fifty years the age has not changed at all. Why?

Some would argue it's because scientists have discovered the correct answer. But how would anyone check? It could also be argued that the changing age was driven by changing cultural and philosophical values in the West. Holmes lived through a period that saw the progressive development of an all-encompassing naturalistic philosophy in Western thought. All supernatural actions by a Creator God were ruled out; only naturalistic explanations were allowed.

The key parameter for every naturalistic explanation is time:

'Time is in fact the hero of the plot ... given so much time the "impossible" becomes possible, the possible probable and the probably virtually certainly certain. One has only to wait: time itself performs miracles.'

In the first fifty years, every academic discipline was developing its naturalistic models. The age of the earth was the crucial parameter in every case: in geology, biology, astronomy, cosmology, geography, archaeology, anthropology, history and so on. Enough time for one discipline was often too little for another—hence the Kelvin and Hubble conflicts. With the present state of play, anywhere between 3 and 7 billion years would probably be suitable. So, 4.55 billion is a happy choice—and it looks precise and authoritative.

4.55 billion years is comfortably less than the age of the universe, allowing enough time for the big bang, stellar evolution and the origin of our solar system. It is also allegedly old enough for geological evolution, for the chemical evolution of the first living cell, for the evolution of life, and for landscape evolution, etc. So. by the mid-20th century the jostling between disciplines had settled down—the different naturalistic models appeared to be meshing together. Everyone had enough time to work with, and there was nothing to gain by changing the number.

IT AFFECTS ME PERSONALLY

The age of the earth is not just an academic issue. As Lewis states,

'By knowing the age of Earth rocks, Moon rocks and rocks from other planets we ... are more able to understand our place in the order of things, our relationship with other celestial bodies. It helps us to navigate our way around the Universe and build up a picture of why we are here at all' (p. 4–5).

That, of course, is a religious question involving the meaning of life. In Mozambique in his early twenties, Arthur Holmes wrote home about the stars: 'I felt somehow what a fearful meaningless tragedy the whole Universe appeared to be' (p. 93). If naturalism is true, then Holmes was right—there is no meaning to this Universe.

However, the Bible reveals the true history of the world and why we are here. There is a purpose for this universe, and for every human life. That's why the age of the earth is a critical issue for the Christian worldview. Long ages destroy the credibility and message of the Bible.

I enjoyed Lewis's book because she so vividly demonstrated that the billion-year age of the earth is subjective and arbitrary. Thus, it is perfectly valid scientifically to start with the biblical data on the age of the earth and interpret the scientific evidence accordingly. In reality, the only sure way of knowing the age of anything is by reliable eyewitnesses.

CAN ALL THOSE SCIENTISTS BE WRONG?

This is understandable. Most popular books, magazines, TV programs, movies and even ordinary conversation seem constantly to confirm that the big bang, the natural origin of life from primeval ooze, and the evolution of all living things from some original organism, are simply accepted by the scientists. It is believed that the only people to question these things are religious fanatics or the scientifically illiterate. So, can "all those scientists" be wrong? History certainly says they can, Figure 23.13.

Fig.23.13: When creationists suggest to the average person that evolution is not scientifically viable, a common response is: "How can all those scientists be wrong?" Curtsy - istockPhoto

Observe that, without confirming data from experiment, or attempts at falsifying a scientific theory by antagonists' observations and alternative theories, a scientist's ideas can be strongly colored by philosophical bias. This is especially so with *interpretations* of 'evidence' rather than direct observation of phenomena in the present, and applies particularly to theories about historical events such as the concept of evolution. Indeed, as we will see, not only one, but a whole body of scientists can see the world through a paradigm that is wrong at its root. That is because a scientist is like any other person in that one can hold a belief very strongly even in the face of strongly opposing evidence.

THE CASE OF ASTRONOMY

Perhaps the best-known scientists who went 'against the trend' are Galileo and Copernicus. The 'majority of scientists', who were their contemporaries, believed the earth was the center of the universe, and all the heavenly bodies revolved around it. As with modern scientists and evolution, their belief was based on a philosophical idea, not observation. And **they were wrong**.

Galileo's famous 'fight' with the church was not with the Bible, but with church leaders who followed what the scientists of their day held as scientific truth, and thus with the scientific community as a whole. Scientists held this belief even though continuously improving observations and calculations showed that there must be a flaw in the universally-accepted idea of 'epicycles' (heavenly bodies moving in circles within circles). It took a long time, and much published observational evidence from the newly-developed telescopes before the scientific community began to accept that they had believed in a faulty system—the earth was not the absolute rotational center of the heavenly bodies.

Further observation through improved telescopes dismantled another universally-held belief of the time: that the heavenly bodies were perfect spheres, and moved in perfect circles. Irregularities were observed on the moon, indicating it was not a perfect sphere. Alarm! The earth's orbit around the sun was an ellipse. More horror! "All those scientists" had been wrong. The very basis of their view of the universe was false.

Today scientists tell us that our universe burst into existence from nothing for no reason with a big bang. Is it not possible that all those scientists could also have a false view of our universe and its origin?

THE CASE OF CHEMISTRY

'Phlogiston' was used in the late 17th and early 18th centuries to explain how substances burned or rusted. It was believed (by 'most scientists') to be a substance contained in combustible materials, which came out when the object burned. It took the persistent work of several leading scientists of the day, including Antoine-Laurent de Lavoisier, to demonstrate that burning was a chemical reaction, usually with oxygen. Substances that burned usually got heavier because of the added oxygen, rather than lighter from losing phlogiston. The majority were wrong. Later, Lavoisier was executed during the fanatically anti-Christian 'reign of terror' in France. One story goes that the sentencing judge said, "The Republic needs neither scientists nor chemists."

Today most scientists believe that the basic chemicals of life (such as proteins) put themselves together in defiance of experimentally established chemical probabilities. Is it possible these scientists may also be wrong?

Alchemy is the idea that base metals (such as lead) could be turned into gold. This concept persisted for hundreds of years, and, although experiments directed at this goal led to the discovery of many interesting chemical substances, proper experiment proved it impossible (by chemical methods). Much money and time (and whole careers) were wasted on this wrong scientific idea, which blinded so many to other, more useful, possibilities.

Is it possible that scientists searching natural phenomena for the origin and variety of life are also wasting their time and energy on a futile exercise?

THE CASE OF MEDICINE

That wrong ideas can persist pervasively for hundreds of years is evident in the theory of 'Humours.' The basic concept goes all the way back to Aristotle (384–322 BC), but was clarified and popularized by the famous physician, Hippocrates (who originated the code of practice incorporating the 'Hippocratic oath' traditionally sworn by beginning doctors).

The concept was that the body has four basic fluids—bile (Greek *chole*), phlegm, black bile (Greek *melanchole*), and blood (Latin *sanguis*). These were supposed to correspond to four traditional temperaments: choleric, phlegmatic, melancholic, and sanguine. Under the theory, these four must be kept in balance for good health.

Mostly the recommended treatment for imbalance involved good diet and exercise, but sometimes laxatives and enemas were administered to help purge the unwanted 'humour' from the body. Similarly, if one had a fever, it was put down to an excess of blood, so the 'cure' was 'bleeding' of the patient (commonly by leeches), called bloodletting. Obviously, this 'cure' was often worse than the disease. Nevertheless, doctors persisted with it through the Middle Ages because no one was prepared to question 'Galen,' the first-century physician, writer and philosopher who publicized the idea in his popular and authoritative writings. In spite of Galen's example and teaching of observation and experiment, and mounting evidence that there was something wrong, it was common medical practice up to the late 19th century.

Again, they were wrong! Their whole view of the cause of disease was wrong, and all because they believed another scientist's theories without question. This is like many scientists today who believe in evolution for no better reason than that other trusted scientists believe it.

THE CASE OF BIOLOGY

Where do vermin come from? Do cockroaches, rats, and maggots just 'appear' out of rotting vegetable matter and animal waste, or even from rocks? For a long time, it was believed that they did, even by famous thinkers such as Aristotle (4th century BC). The idea was called 'spontaneous generation' and regarded as a fact into the mid-19th century. It took a creation scientist, Louis Pasteur (1822–1895), to prove that life comes only from life, a process called 'biogenesis.' Those who believed in spontaneous generation were wrong.

Today, in spite of Pasteur's proof, and our continuing observations, many scientists still believe in abiogenesis (that *all* life has come from non-living chemicals). How that could happen is called (by evolutionists) a 'mystery', because it defies chemistry, but they still believe it. Why?

Science Decided by Majority Vote!

Actually, a major reason most scientists believe in evolution is that most scientists believe in evolution! This is a type of 'confirmation bias': the alleged scientific consensus was reached by counting heads, which themselves reached their conclusion by counting heads. If most of them were asked for actual evidence, they would likely give very weak answers outside their field of expertise.

For example, one of the world's leading experts on fossil birds—and a staunch critic of the dino-to-bird dogma, is Dr Alan Feduccia, Professor Emeritus at the University of North Carolina. He remains an evolutionist, however, yet when challenged, his prime 'proof' was corn changing into corn!

As the famous author Michael Crichton (1942–2008), who had a previous career in medicine and science, said:

"Let's be clear: the work of science has nothing whatever to do with consensus. Consensus is the business of politics. Science, on the contrary, requires only one investigator who happens to be right, which means that he or she has results that are verifiable by reference to the real world. In science consensus is irrelevant. What is relevant is reproducible results. The greatest scientists in history are great precisely because they broke with the consensus."

"There is no such thing as consensus science. If it's consensus, it isn't science. If it's science, it isn't consensus. Period."

Nevertheless, like the believers in *epicycles*, and *phlogiston*, and *humours*, and *spontaneous generation*, many scientists today believe in evolution. Can so many be wrong? History says 'yes.' Mounting evidence in genetics, molecular biology, information theory, cosmology and other areas all say 'yes.' These scientists believe in the dominant paradigm, naturalism, in spite of the evidence against it. They don't wish to confront the idea of a Creator, but, as in the past, honest appraisal of the evidence of operational science will prove them wrong; the Creator will be vindicated (*Romans 1:18–22*), God's Wrath on Unrighteousness For the wrath of God is revealed from heaven against all ungodliness and unrighteousness of men, who by their unrighteousness suppress the truth. For what can be known about God is plain to them, because God has shown it to them. For his invisible attributes, namely, his eternal power and divine nature, have been clearly perceived, ever since the creation of the world, in the things that have been made. So, they are without excuse. For although they knew God, they did not honor him as God or give thanks to him, but they became futile in their thinking, and their foolish hearts were darkened. Claiming to be wise, they became fools

APPENDIX

i. In Australia, as in most British systems, 'professor' means head of a department. In the US, professor simply means someone who teaches at tertiary level, which could apply to someone, for example, who would be called a 'junior lecturer' in a British system.

ii. From Greek λοβός *lobos* = lobe (of the brain), and τομή tomē = slice/cut. A serious and irreversible operation that cuts certain connections to the cerebral cortex, the 'thinking' part of the brain.

iii. The extended passage cites Genesis 1:27 and 2:24 as real history, and about the same man and woman. In the parallel passage in Matthew 19:4–5, Jesus attributes Genesis 2:24 to the One who created them, i.e., to God himself.

iv. Thiele, E.R., *The Mysterious Numbers of the Hebrew Kings*, Zondervan, Grand Rapids, MI, 1951. [note: An article that addresses independent evidence supporting Thiele's dating method appeared online after the publication of the current article, see: Young, R.C., Evidence for inerrancy from a second unexpected source: the Jubilee and Sabbatical cycles, Bible and Spade 21(4), 2008;biblearchaeology.org/post/2015/07/25/Evidence-for-Inerrancy-from-a-Second-Unexpected-Source-The-Jubilee-and-Sabbatical-Cycles.aspx.]

v. One might be tempted to apply something like a simple Pearson Chi-Squared test to the data (Adam to Jacob). Doing so strongly rejects ($p = 7 \times 10^{-8}$, with 9 degrees of freedom) the null hypothesis that these numbers form a random distribution, but knowing that this is not a random distribution is not the same thing as understanding why the numbers are reported in this way.

vi. Radiometric dating can actually be used to falsify long ages. It has been used on rocks and fossils that we actually know the date of, and dating tests give the wrong dates. See articles under creation.com/dating.

REFERENCES

1. Williams, P., Some remarks preliminary to a biblical chronology, *J. Creation* (formerly *CEN Tech J.*) **2**(1):98–106, 1998; creation.com/chronology.
2. Osgood, J., The times of the Judges, *J. Creation* (formerly *CEN Technical J.*) 1(1):141–158, 1984.
3. Note that recent discussions on the length of the solar day do not come into play here. The length of the tropical and sidereal year are unaffected by the earth's spin. See Carter, R.W., 'Ancient' coral growth layers, *J. Creation* 26(3):50–53, 2012; creation.com/ancient-coral; Jackson, D., Letter to the Editor; reply by Carter, R., *J. Creation* 28(2):55–56, 2014.
4. Sarfati, J., Biblical chronogenealogies, *J. Creation* (formerly *TJ*) 17(3):14–18, 2003; creation.com/biblical-chronogenealogies.
5. Ussher, J., *Annales veteris testamenti, a prima mundi origine deducti*, 1650, in which he calculated creation to have occurred on the "nightfall preceding 23 October 4004 BC". A modern English translation, Annals of the World (Master Books, Green Forest, AR, 2006), is available from various sources, including creation.com.
6. Jones, F.N., *The Chronology of the Old Testament*, Master Books, Green Forest, AR, 1993.
7. Beechik, R., Sojourn of the Jews, *J. Creation* (formerly *TJ*) 15(1):60–61, 2001; creation.com/sojourn-of-the-jews.
8. Austin, D., Chronology of the 430 years of Exodus 12:40, *J. Creation* 21(1):67–68, 2007.
9. Austin, D., Reply to Brenton Minge, *J. Creation* 22(1):59–60, 2008.
10. Viccary, M., Reply to Brenton Minge, *J. Creation* 22(1):60, 2008.
11. Kaiser, W.C., *A History of Israel from the Bronze Age through the Jewish Wars*, Broadman and Holman, Nashville, TN, 1998.
12. Kitchen, K.A., *On the Reliability of the Old Testament*, Eerdmans, Grand Rapids/Cambridge, 2003.
13. Jones, F., *Chronology of the Old Testament*, King's Word Press, The Woodlands, TX, 1999.
14. Pierce, L., Evidentialism—the Bible and Assyrian chronology, *J. Creation* (formerly *TJ*) **15**(1):62–68, 2001.
15. Ashton, J. and Down, D., *Unwrapping the Pharaohs*, Master Books, Green Forest, AR, 2006.
16. Clarke, P., Why Pharaoh Hatshepsut is not to be equated to the Queen of Sheba, *J. Creation* 24(2):62–68, 2010; See also Habermehl's letter to the editor and Clarke's reply in *J. Creation* 25(1):44–46, 2011.
17. Austin, D., Is Darius, the king of Ezra 6:14–15, the same king as the Artaxerxes of Ezra 7:1?, *J. Creation* 22(2):46–52, 2008; creation.com/darius.
18. Whitcomb, J.C. and Morris, H.M., *The Genesis Flood*, Presbyterian and Reformed Publishing, Phillipsburg, NJ, 1961.
19. Clarke, P., The Queen of Sheba and the Ethiopian problem, *J. Creation* 27(2):55–61, 2013.
20. Here he references Austin, D.,'The Queen of the South' is the 'the Queen of Egypt', *J. Creation* 26(3):79, 2012.
21. Clarke, P., The Queen of Sheba and the Ethiopian problem, *J. Creation* 27(2):55–61, 2013.
22. Here he references Austin, D.,'The Queen of the South' is the 'the Queen of Egypt', *J. Creation* 26(3):79, 2012.
23. See Sarfati, J., The Incarnation: Why did God become Man?, December 2010; creation.com/incarnation.

24. See Sarfati, J., Why Bible history matters, *Creation* 33(4):18–21, 2011, as well as creation.com/nt and creation.com/gen-hist.
25. See Sarfati, J., Biblical chronogenealogies, *J. Creation* 17(3):14–18, December 2003; creation.com/chronogenealogy.
26. ISCAST (Institute for the Study of Christianity in an Age of Science and Technology); see Sarfati, J., The Skeptics and their 'Churchian' Allies, November 1998; creation.com/iscast.
27. Sparks, K., "After Inerrancy, Evangelicals and the Bible in the Postmodern Age, part 4" Biologos Forum, 26 June 2010. See also Cosner, L., Evolutionary syncretism: a critique of Biologos, 7 September 2010; creation.com/biologos.
28. A concept elucidated by Rusbult, C., Apparent age: part I, asa3.org/ASA/education/origins/aa-cr.htm, 2003; Apparent age: part II, asa3.org/ASA/education/origins/aa2-cr.htm#i, 2004.
29. Snelling, A.A., Thirty miles of dirt in a day, *Answers* 3(4):28–30, 2008.
30. Faulkner, D.R., Interpreting craters in terms of the Day 4 Cratering Hypothesis, ARJ 7:11–25, 2014.
31. Byl, J., Rapidly matured creation, bylogos.blogspot.com/2019/02/rapidly-matured-creation.html, 6 February 2019.
32. Newton, I., A short Schem of the true Religion, www.newtonproject.ox.ac.uk/view/texts/normalized/THEM00007, 2002.
33. For example, Smith, C., *The Secular Revolution: Power, interests, and conflict in the secularization of American public life*, University of California Press, Berkeley, 2003; reviewed by Weinberger, L., Secularizing America, *J. Creation* 20(3):43–45, 2006.
34. Gordon, B., Constrained Integration view; in; Copan, P. and Reese, C.L. (Eds.), *Three Views on Christianity and Science*, Zondervan Academic, Kindle Edition, p. 146, 2021.
35. Reed, J., Changing paradigms in stratigraphy—"a quite different way of analyzing the record", *J. Creation* 30(1):83–88, 2016.
36. Sarfati, J., *Refuting Evolution*, ch. 1, 4th ed., Creation Book Publishers, 2008; creation.com/refutingch1.
37. Walker, T., Challenging dogmas: Correcting wrong ideas, *Creation* 34(2):6, 2012; creation.com/challenging-dogmas.
38. Sarfati, J., Galileo Quadricentennial: Myth vs fact, *Creation* 31(3):49–51, 2009; creation.com/galileo-quadricentennial.
39. phlogiston, *Encyclopædia Britannica*, *Encyclopædia Britannica Online*, 2012; Britannica.com/EBchecked/topic/456974/phlogiston.
40. Alchemy, answers.com/topic/alchemy.
41. From Greek χυμός (*chumos*) meaning juice or sap; Humours, Science Museum; sciencemuseum.org.uk.
42. What is spontaneous generation? allaboutscience.org. Spontaneous Generation; allaboutthejourney.org/spontaneous-generation.htm.
43. Discover Dialogue: Ornithologist and evolutionary biologist Alan Feduccia plucking apart the dino-birds, *Discover* 24(2), February 2003; see also creation.com/4wings.
44. Crichton, M., *Aliens cause global warming*, 17 January 2003 speech at the California Institute of Technology; s8int.com/crichton.html.
45. See, Batten, D., *'It's not science'*, 2002. Return to text.
46. Williams, A., *The Universe's Birth Certificate*, *Creation* 30(1):31, 2007, Sarfati, J., *Biblical chronogenealogies*, *Journal of Creation* 17(3):14–18, 2003. Return to text.

47. Oard, M., *Aren't 250 million year old live bacteria a bit much?*, 2001. Return to text.
48. Sanford, J., *Genetic entropy and the mystery of the genome*, Ivan Press, 2005; see: *Plant geneticist: 'Darwinian evolution is impossible'*, *Creation* 30(4):45–47, 2008. Realistic modelling shows that genomes are young, in the order of thousands of years. See Sanford, J., *et al.*, Mendel's Accountant: A biologically realistic forward-time population genetics program, *SCPE* 8(2):147–165, 2007; www.scpe.org/vols/vol08/no2/SCPE_8_2_02.pdf. Return to text.
49. Wieland, C., *Dinosaur soft tissue and protein—even more confirmation!*, 2009. Return to text.
50. Allen, D., *Warped earth, Creation* 25(1):40–43, 2002. Return to text.
51. Walker, T., *Coal: memorial to the Flood, Creation* 23(2):22–27, 2001; Wieland, C., *Forests that grew on water, Creation* 18(1):20–24, 1995. Return to text.
52. *'Millions of years' are missing* (interview with Dr Ariel Roth), *Creation* 31(2):46–49, 2009. Return to text.
53. Williams, A., *World's oldest salt lake only a few thousand years old, Creation* 17(2):5, 1995. Return to text.
54. Pierce, L., *Niagara Falls and the Bible, Creation* 22(4):8–13, 2000. Return to text.
55. Wieland, C., *Caving in to reality, Creation* 20(1):14, 1997. Also Q&A on limestone caves; creation.com/caves. Return to text.
56. *What about carbon dating? Creation Answers Book* chapter 4. Return to text.
57. Humphreys, D.R., Young helium diffusion age of zircons supports accelerated nuclear decay, in Vardiman, L., Snelling, A. and Chaffin, E. (eds.), *Radioisotopes and the Age of the Earth*, ICR and CRS, 848 pp., 2005. Return to text.
58. Gentry, R., *et al.*, Differential lead retention in zircons: Implications for nuclear waste containment, *Science* 216(4543):296–298, 1982; DOI: 10.1126/science.216.4543.296. Return to text.
59. DeYoung, D.B., *Transient lunar phenomena: a permanent problem for evolutionary models of Moon formation, Journal of Creation* 17(1):5–6, 2003; creation.com/tlp; Walker, T., and Catchpoole, D., *Lunar volcanoes rock long-age timeframe, Creation* 31(3):18, 2009. Return to text.
60. See *creation.com/magfield#planets*. Return to text.
61. Anon., *Saturnian surprises, Creation* 27(3):6. Return to text.
62. Bernitt, R., *Fast stars challenge big bang origin for dwarf galaxies, Journal of Creation* 14(3):5–7, 2000. Return to text.
63. McIntosh, A., and Wieland, C., *'Early' galaxies don't fit, Creation* 25(2):28–30, 2003. Return to text.
64. Faulkner, D., *Comets and the age of the solar system, Journal of Creation* 11(3):264–273, 1997. Return to text.
65. See also Morris, H.M., *Games some people play: the supreme rule of this game is to stifle arguments against evolution—any way you can, Creation* 23(4):35, 2001; Wieland, C., *The rules of the game: as the 'rules' of science are now defined, creation is forbidden as a conclusion—even if true, Creation* 11(1):47–50, 1988. Return to text.
66. Press, F. and Siever, R., *Earth*, 4th Ed., W.H. Freeman and Co., New York, p. 40, 1986. Return to text
67. Eiseley, L., *Darwin's Century: Evolution and the Men who Discovered It*, pp. 234, 235, 240, Doubleday, Anchor, New York, 1961. Return to text
68. Woodmorappe, J., *Lord Kelvin revisited on the young age of the earth, J. Creation* **13**(1):14, 1999. Return to text
69. Humphreys, D.R., Austin, S.A., Baumgardner, J.R. and Snelling, A.A., Helium diffusion rates support accelerated nuclear decay, in: Ivey, R.L., *Proceedings of the 5th International Conference on Creationism*, Creation Science Fellowship, Pittsburgh, PA, pp. 175–196, 2003. Return to text.

70. Patterson, C.C., Age of meteorites and the Earth, *Geochimica et Cosmochimica Acta* 10:230–237, 1956. Return to text.
71. Williman, A.R., *Long-age isotope dating short on credibility, J. Creation* 6(1):2–5, 1992. Return to text
72. Wald, G., The origin of life; in: *Physics and Chemistry of Life*, p. 12, 1955; quoted in: Morris, J.D., *The Young Earth*, Creation-Life Publishers, Colorado Springs, CO, p. 41, 1994. Return to text.
73. See Mortenson, T., *The Great Turning Point: The Church's Catastrophic Mistake on Geology—Before Darwin*, Master Books, Green Forest, AR, 2004. Return to text.

24

LIFE IN A FINELY TUNED COSMOS

NATURALISM AND THE AGE OF THE EARTH

Contemporary concern over the negative impact of theories of biological evolution is justified, but many Christians do not understand the stranglehold that philosophical naturalism has on geology and astronomy. The historical roots of philosophical naturalism reach back into the sixteenth century in the works of Galileo Galilei, Figure 24.1 and Francis Bacon.

Fig.24.1: Galileo Galilei - *1564 - 1642*

Evolutionary and naturalistic theories of the earth's creation based on uniformitarian assumptions and advocating old-earth theories emerged in the late eighteenth century. In the early nineteenth century, many Christians sought to harmonize biblical teaching with old-earth geological theories such as the gap theory and a tranquil or local Noachian flood. However, many evangelicals and High Churchmen still held to the literal view of Genesis 1–11.

Two Enlightenment-generated philosophical movements in the eighteenth century, deism and atheism, elevated human reason to a place of supreme authority and took an anti-supernaturalistic view of the Bible, holding it to be just another human book. The two movements, with their advocacy of an old-earth and their effect on astronomy and geology, preceded Darwin and supplied him with millions of years needed for his naturalistic theory of the origin of living things. From this lineage it is clear that geology is not an unbiased, objective science and that old-earth theories, naturalism and uniformitarianism are inseparable.

Intelligent-design arguments usually used to combat evolution fail to account for the Curse imposed by God in *Genesis 3* and are therefore only partially effective. Intelligent-design advocates should recognize that the naturalism represented in evolutionary theories began much earlier than Darwin. A return to the Scriptures and their teaching of a ***young earth*** is the great need of the day.

NEGATIVE IMPACT OF EVOLUTION

Many are concerned about the negative impact of evolution on today's world. Some see the consequences in terms of moral and spiritual chaos in society and the church. Others see the damage that the brainwashing of evolution is causing in academic and intellectual arenas. They correctly argue that Neo-Darwinism (or any related theory of biological evolution, such as 'punctuated equilibrium theory') is not pure science, but largely philosophical naturalism masquerading as scientific fact. Many such critics of evolution are part of what is called the 'Intelligent Design' (hereafter ID) movement. But many are also within the '**young-earth creationist**' (hereafter YEC) movement.

I strongly agree with and appreciate a great deal of what leaders in the ID movement are writing, not only about the scientific problems with all theories of biological evolution, but especially about the stranglehold that philosophical naturalism (hereafter simply 'naturalism') has on science.

However, from my reading of ID books and articles and listening to lectures by some of those leaders, I do not think that they see clearly enough the extent to which science is dominated by naturalism. The reason for this observation is that many ID leaders have made oral or written statements something like this: 'We are not going to deal with the question of the age of the earth because it is a divisive side issue. Instead, we want to address the main issue, which is the control of science by naturalism.'

The implication of such statements is that the age of the earth is unrelated to naturalism. Many Christians have not even considered the arguments for young-earth creationism because they think that the ID movement has the right view and is dealing with evolution correctly. But this disjunction of naturalism and the age of the earth is incorrect, as I hope to show.

As I read their writings, the ID people do not seem to understand the historical roots of the philosophical control of science. Or, perhaps, they do not appear to have gone back far enough in their historical investigations. A closer look at history, especially the history of the idea of an old earth, provides abundant evidence that the originators of the idea of an old earth and old universe interpreted the physical evidence by using essentially naturalistic assumptions. Similarly, a closer look at the way modern old-earth geologists and old-universe cosmologists reason shows that both geology and astronomy are controlled by the same naturalism that dominates the biological sciences, and indeed nearly all of academia.

I submit, therefore, that the age of the earth strikes at the very heart of naturalism's control of science and that fighting naturalism only in the biological sciences amounts to fighting only one-third of the battle. Worse still, many of the people involved at the highest levels in the ID movement (e.g., Hugh Ross, Robert Newman, Walter Bradley) are not neutral regarding the age of the earth (as the recognized leader of the ID movement, Phillip Johnson, attempts to be), but are actively and strongly opposed to the young-earth view.

Although the ID movement is fighting naturalism in biology, it is actually tolerating or even promoting naturalism in geology and astronomy—which is not a consistent strategy—thus undermining its potential effectiveness.

I. HISTORICAL ROOTS

The idea of an old earth really began to take hold in science in the late eighteenth and early nineteenth centuries, before Darwin's controversial theory appeared on the scene. Prior to this, in Europe and North America (where science was born and developed under the influence of Christianity and assumptions about physical reality were rooted firmly in the Bible), the dominant, majority view was that God created the world in six literal days about 6,000 years earlier and judged it with a global, catastrophic Flood, Figure 24.2.

Fig.24.2: *Noah's Flood – 5000 BC*

How, then, did the old-earth idea arise?

Two important people in the sixteenth century greatly influenced the development of old-earth thinking at the end of the eighteenth and beginning of the nineteenth centuries. Those two were Galileo Galilei and Sir Francis Bacon. As is well known, Galileo (1564–1642) was a proponent of Copernicus's theory that the earth revolves around the sun, not vice versa. Initially the Roman Catholic Church leadership had no problem with this idea, but for various academic, political and ecclesiastical reasons, in 1633 the pope changed his mind and forced Galileo to recant his belief in heliocentricity on threat of excommunication. But eventually heliocentricity became generally accepted and with that many Christians absorbed two lessons from the so-called 'Galileo affair.' One was from a statement of Galileo himself. He wrote, '***The intention of the Holy Ghost is to teach us how to go to heaven, not how heaven goes.***' In other words, the Bible teaches theology and morality, but not astronomy or science. The other closely related lesson was that the church will make big mistakes if it tries to tell scientists what to believe about the world.

Galileo's contemporary in England, Francis Bacon (1561–1626), Figure 24.3, was a politician and philosopher who significantly influenced the development of modern science. He emphasized observation and experimentation as the best method for gaining true knowledge about the world. He also insisted that theory should be built only on the foundation of a wealth of carefully collected data. But although Bacon wrote explicitly of his belief in a recent, literal six-day creation, he like Galileo insisted on not mixing the study of what he called the two books of God: creation and the Scriptures. He stated,

Fig.24.3: Sir Francis Beacon (1561 – 1626)

But some of the moderns, however, have indulged in this folly, with such consummate carelessness, as to have endeavored to found a natural philosophy on the first chapter of Genesis, the book of Job, and other passages of holy Scripture— 'seeking the dead among the living.' And this folly is the more to be prevented and restrained, because, from the unsound admixture of things divine and human, there arises not merely a fantastic philosophy, but also a heretical religion.

As a result of the powerful influence of Galileo and Bacon, a strong bifurcation developed between the interpretation of creation (which became the task of scientists) and the interpretation of Scripture (which is the work of theologians and pastors). With the advent of the nineteenth century, the old-earth geologists, whether Christian or not, often referred to Bacon and Galileo's dictums to silence the objections of the 'scriptural geologists,' a group of Christian clergy and scientists writing from about 1820 to 1850 who raised biblical, geological and philosophical arguments against old-earth theories and for the literal truth of Genesis—a literal six-day creation about 6,000 years ago and a global catastrophic Flood at the time of Noah, which they believed was responsible for most of the geological record.

The warning of the old-earth proponents was powerful in its effect on the minds of the public. The message was that defenders of a literal interpretation of Genesis regarding Creation, Noah's Flood and the age of the earth were repeating the same mistake the Roman Catholic Church made three centuries earlier in relation to the nature of the solar system. And just look at how that retarded the progress of science and exposed the church to ridicule, said the old-earth advocates.

II. NEW THEORIES ABOUT THE HISTORY OF CREATION

In contrast to the long-standing young-earth creationist (YEC) view, different histories of the earth began to be developed in the late eighteenth century, which were evolutionary and naturalistic in character. Three prominent French scientists were very influential in this regard. In 1778 Georges-Louis Comte de Buffon (1708–1788), Figure 24.4, postulated that the earth was the result of a collision between a comet and the sun and had gradually cooled from a molten lava state over at least 75,000 years (a figure based on his study of cooling metals).

Fig.24.4: Georges-Louis Comte de Buffon (1708–1788)

Buffon was probably a deist or possibly a secret atheist. Pierre Laplace (1749–1827), an open atheist, published his nebular hypothesis in 1796. He imagined that the solar system had naturally and gradually condensed from a gas cloud during a very long period of time. In his *Zoological Philosophy* of 1809, Jean Lamarck (1744-1829), Figure 24.5, who straddled the fence between deism and atheism, proposed a theory of biological evolution over long ages, with a mechanism known as the inheritance of acquired characteristics.

New theories in geology were also being advocated at the turn of the nineteenth century as geology began to develop into a disciplined field of scientific study. Abraham Werner (1749–1817) was a German mineralogist and probably a deist. Although he published very little, his impact on geology was enormous, because many of the nineteenth century's greatest geologists had been his students. He theorized that the strata of the earth had been precipitated chemically and mechanically from a slowly receding universal ocean. According to Werner's unpublished writings, the earth was at least one million years old. His elegantly simple, oceanic theory was quickly rejected (because it just did not fit the facts), but the idea of an old earth remained with his students.

Fig.24.5: *Jean Lamarck (1744-1829)*

The Scotsman, James Hutton (1726–1797), Figure 2.46, was trained in medicine but turned to farming for many years before eventually devoting his time to geology. In his *Theory of the Earth*, published in 1795, he proposed that the continents were gradually and continually being eroded into the ocean basins. These sediments were then gradually hardened and raised by the internal heat of the earth to form new continents, which would be eroded into the ocean again. With this slow cyclical process in mind, Hutton could see no evidence of a beginning to the earth, a view that precipitated the charge of atheism by many of his contemporaries, though he too was most likely a deist.

Fig.24.6: *James Hutton (1726–1797)*

Neither Werner nor Hutton paid attention to the fossils in rocks. But another key person in the development of old-earth geological theories who did, was the Englishman, William Smith (1769–1839). He was a drainage engineer and surveyor and helped build canals all over England and Wales, which gave him much exposure to the strata and fossils. He is called the 'Father of English Stratigraphy' because he produced the first geological maps of England and Wales and developed the method of using fossils to assign relative dates to the strata. As a vague sort of theist he believed in many supernatural creation events and supernaturally induced floods over the course of much more time than indicated in the Bible.

The Frenchman, Georges Cuvier (1768–1832), was a famous comparative anatomist and paleontologist. Although he was a nominal Lutheran, recent research has shown that he was an irreverent deist. Because of his scientific stature, he was most influential in popularizing the catastrophist theory of earth history. By studying fossils found largely in the Paris Basin he believed that over the course of untold ages there had been at least four regional or nearly global catastrophic floods, the last of which probably was about 5,000 years ago. This obviously coincided with the date of Noah's Flood, and some who endorsed Cuvier's theory made this connection. However, in his published theory, Cuvier himself never explicitly equated his last catastrophe with Noah's Flood.

Finally, Charles Lyell (1797–1875), Figure 24.7, a trained lawyer turned geologist and probably a deist (or Unitarian, which is essentially the same), began publishing his three-volume *Principles of Geology* in 1830. Building on Hutton's uniformitarian ideas, Lyell insisted that the geological features of the earth can, and indeed must, be explained by slow gradual processes of erosion, sedimentation, earthquakes, volcanism, etc., operating at essentially the same average rate and power as observed today. By the 1840s his view became the ruling paradigm in geology. So, at the time of the scriptural geologists (ca. 1820–50), there were three views of earth history.

Fig.24.7: Charles Lyell (1797–1875)

It should be noted that two very influential geologists in England (and in the world) at this time were William Buckland (1784–1856) and Adam Sedgwick (1785–1873). Buckland became the head professor of geology at Oxford University in 1813 and Sedgwick gained the same position at Cambridge in 1818. Both were ordained Anglican clergy and both initially promoted old-earth catastrophism. But under the influence of Lyell, they both converted to uniformitarianism with public recantations of their catastrophist views in the early 1830s. Buckland is often viewed as a defender of Noah's Flood because of his 1823 book, *Reliquiae Diluvianae*. But this apparent defense of the Flood was actually a subtle attack on it, as scriptural geologists accurately perceived. Because of their powerful positions in academia and in the church, Sedgwick and Buckland led many Christians in the 1820s to accept the new geological theories about the history of the earth and to abandon their faith in the literal interpretation of Genesis and in the unique and geologically significant Noachian Flood.

One more fact about geology at this time deserves mention. The world's first scientific society devoted exclusively to geology was the London Geological Society (LGS), founded in 1807. From its inception, which was at a time when very little was known about the geological formations of the earth and the fossils in them, the LGS was controlled by the assumption that earth history is much older than and different from that presented in Genesis. And a few of its most powerful members were Anglican clergy. Not only was very little known about the geological features of the earth, but at that time there were no university degrees in geology and no professional geologists. Neither was seen until the 1830's and 1840's, which was long after the naturalistic idea of an old earth was firmly entrenched in the minds of those who controlled the geological societies, journals and university geology departments.

III. CHRISTIAN COMPROMISES WITH OLD-EARTH GEOLOGICAL THEORIES

During the early nineteenth century many Christians made various attempts to harmonize these old-earth geological theories with the Bible. ***In 1804, the gap theory began*** to be propounded by the 24-year-old pastor, Thomas Chalmers (1780–1847), who after his conversion to evangelicalism in 1811 became one of the leading Scottish evangelicals. It should be noted that Chalmers began teaching his gap theory before the world's first geological society was formed (in London in 1807), and before Cuvier's catastrophist theory appeared in French (1812) or in English (1813) and over two decades before Lyell's theory was promoted (beginning in 1830). In part because of Chalmers' powerful preaching and writing skills, the gap theory quickly became the most popular reinterpretation of Genesis among Christians for about the next half-century.

However, the respected Anglican clergyman, George Stanley Faber (1773–1854), Figure 24.8, began advocating the day-age theory in 1823. This was not widely accepted by Christians, especially geologists, because of the obvious discord between the order of events in Genesis 1 and the order according to old-earth theory. The day-age view began to be more popular after Hugh Miller (1802–1856), the prominent Scottish geologist and evangelical friend of Chalmers, embraced and promoted it in the 1850s after abandoning the gap theory.

Also, in the 1820s the evangelical Scottish zoologist, Rev. John Fleming (1785–1857), began arguing for a tranquil Noachian deluge (a view which Lyell also advocated, under Fleming's influence). In the late 1830s the prominent evangelical Congregationalist theologian, John Pye Smith (1774–1851), advocated that Genesis 1–11 was describing a local creation and a local flood, both of which supposedly occurred in Mesopotamia. Then, as German liberal theology was beginning to spread in Britain in the 1830s, the view that Genesis is a myth, which conveys only theological and moral truths, started to become popular.

Fig.24.8: George Stanley Faber (1773–1854)

Accordingly, from all this it should be clear that by 1830, when Lyell published his uniformitarian theory, most geologists and much of the church already believed that the earth was much older than 6,000 years and that the Noachian Flood was *not* the cause of most of the geological record. Lyell is often given too much credit (or blame) for the church's loss of faith in Genesis. In reality, most of the damage was done before Lyell, often by Christians who were otherwise quite biblical, and this compromise was made at a time when geologists knew very little about the rocks and fossils of the earth.

Nevertheless, many evangelicals and High Churchmen still clung to the literal view of Genesis because it was exegetically the soundest interpretation. In fact, until about 1845 the majority of Bible commentaries on Genesis taught a recent six-day creation and a global catastrophic Flood. So, in the early nineteenth century competing old-earth geological theories and competing old-earth interpretations of the early chapters of Genesis existed, and the scriptural geologists fought against all these theories and interpretations.

IV. PHILOSOPHICAL DEVELOPMENTS

As a prelude to this Genesis-geology controversy, the eighteenth century also witnessed the spread of two competing but largely similar worldviews: deism and atheism. These two worldviews flowed out of the Enlightenment, in which human reason was elevated to the place of supreme authority for determining truth. This enthroning of human reason not only challenged the authority of the church in society, but also led to all kinds of anti-supernatural attacks on the Bible, undermining its authority as a source of historical, as well as moral and theological truth. Deism and atheism were slightly different ways of packaging an anti-supernatural view of history.

Apart from the deists' belief in a rather vaguely defined Creator God and a supernatural beginning to the creation, they were indistinguishable from atheists in their views of Scripture and the physical reality. *In deism, as in atheism, the Bible is merely a human book*, containing errors, and not the inspired Word of God, and the history and function of the creation can be totally explained by the properties of matter and the 'inviolable laws of nature' in operation over a long period of time. Deists and atheists often disguised their true views, especially in England where they were not culturally acceptable. Many of them gained influential positions in the scientific establishment of Europe and America, where they subtly and effectively promoted what is today called *naturalism*. Brooke comments on the subtle influence of deistic forms of naturalism when he writes,

Without additional clarification, it is not always clear to the historian (and was not always clear to contemporaries) whether proponents of design were arguing a Christian or deistic thesis. The ambiguity itself could be useful. By cloaking potentially subversive discoveries in the language of natural theology, scientists could appear more orthodox than they were, but without the discomfort of duplicity if their inclinations were more in line with deism.

But the effects of deistic and atheistic philosophy on biblical studies and Christian theology also became widespread on the European continent in the late eighteenth century and in Britain and America by the middle of the nineteenth century. As Reventlow concluded in his massive study,

We cannot overestimate the influence exercised by Deistic thought, and by the principles of the Humanist world-view which the Deists made the criterion of their biblical criticism, on the historical-critical exegesis of the nineteenth century; the consequences extend right down to the present. At that time a series of almost unshakeable presuppositions were decisively shifted in a different direction.

Therefore, the biblical worldview, which had dominated the Western nations for centuries, was rapidly being replaced by a naturalistic worldview. And it was into the midst of these revolutions in worldview and the reinterpretation of the phenomena of nature and the Bible that the scriptural geologists expressed their opposition to old-earth geology in the first half of the nineteenth century.

In summary, deism (which is a slightly theologized form of naturalism) flourished briefly in the early eighteenth century and then went underground as it spread into liberal biblical scholarship and in the nineteenth century into science. Atheism (naked naturalism) became increasingly popular and aggressive in the eighteenth and nineteenth centuries, especially on the European continent. So, naturalism first affected astronomy and geology and then only later did it gain control of biology. Many old-earth geologists (e.g., Sedgwick) vigorously opposed Darwin's theory in 1859. But they failed to realize that Darwin simply applied the same naturalistic thinking to his theory of the origin of living creatures that the geologists had applied to their theories about the origin of the earth and geological record of strata and fossils. Their naturalistic geological theories laid the foundation for naturalistic biology.

Clearly, Buffon's theory that the earth was the result of a collision of a comet and the sun and then cooled from a molten state over at least 75,000 years was a naturalistic theory. His deism led him to try to separate science from religious and metaphysical ideas and to reject teleological reasoning and the idea of any supernatural, divine intervention in nature. It is therefore no surprise that he firmly rejected the biblical Flood (along with its implications for the history and age of the earth).

Laplace's nebular hypothesis for the origin of the solar system over much more than 75,000 years (which became the seedbed of the big bang theory) was atheistic and therefore naturalistic. So was Werner's deistic geological theory of a slowly receding ocean producing the geological record over one million years. So were Hutton's and Lyell's deistic uniformitarian theories. William Smith's and Georges Cuvier's deistic catastrophist theories were also quite naturalistic in that they too ignored Scripture and considered only natural causes for the geological record (though they had a supernaturalistic view of the origin of biological life).

V. GEOLOGY—AN OBJECTIVE SCIENCE?

These developers of old-earth theory were hardly objective, unbiased, let-the-facts-speak-for-themselves interpreters of the physical evidence, as is so often supposed. Regarding early nineteenth-century geology, a respected historian of science has noted:

Most significantly, recent work in cultural anthropology and the sociology of knowledge has shown that the conceptual framework that brings the natural world into a comprehensible form becomes especially evident when a scientist constructs a classification [of rock strata].

Previous experience, early training, institutional loyalties, personal temperament, and theoretical outlook are all brought to bear in defining particular boundaries as 'natural.'

It would be misleading to think that all these factors influenced all scientists to the same degree.

Furthermore, a major component of anyone's theoretical outlook is his religious worldview (which could include atheism or agnosticism). Worldview had a far more significant influence on the origin of old-earth geology than has often been perceived or acknowledged. A person's worldview not only affects the interpretation of the facts but also the observation of the facts.

Another prominent historian of science rightly comments about scientists, and non-scientists, '[M]en often perceive what they expect, and overlook what they do not wish to see.' In his enlightening description of the late-1830s controversy over the identification of the Devonian formation in the geology of Britain, Rudwick wrote:

'Furthermore, most of their recorded field observations that related to the Devonian controversy were not only more or less 'theory laden,' in the straightforward sense that most scientists as well as historians and philosophers of science now accept as a matter of course, but also 'controversy laden.' The particular observations made, and their immediate ordering in the field, were often manifestly directed toward finding empirical evidence that would be not merely relevant to the controversy but also *persuasive*. Many of the most innocently 'factual' observations can be seen from their context to have been sought, selected, and recorded in order to reinforce the observer's interpretation and to undermine the plausibility of that of his opponents.

In his covert promotion of Scrope's uniformitarian interpretations of the geology of central France, Lyell had similarly said in 1827, 'It is almost superfluous to remind the reader that they who have a theory to establish, may easily overlook facts which bear against them, and, unconscious of their own partiality, dwell exclusively on what tends to support their opinions.' However, many geologists, then and now, would say that Lyell was blind to this fact in his own geological interpretations.

So, the influence of worldview on the observation, selection and interpretation of the geological facts was significant, especially given the limited knowledge of people individually and collectively in the still infant stage of early nineteenth-century geology. As the philosopher of science, Thomas Kuhn, has noted,

'Philosophers of science have repeatedly demonstrated that more than one theoretical construction can always be placed upon a given collection of data. History of science indicates that, particularly in the early developmental stages of a new paradigm, it is not even very difficult to invent such alternatives.'

Just as the catastrophist felt irresistibly driven by the 'obvious' evidence to believe in great regional or global catastrophes, so also the uniformitarian 'saw' equally undeniable evidence that they had never happened. In the same way, scriptural geologists, like Rev. Henry Cole (with virtually no geological knowledge) or Rev. George Young (with excellent geological competence), felt that all the opposing geologists were 'blind' to the plain evidence for a recent supernatural creation and a unique global Flood.

Not only did various influences bias the developers of old-earth theory, they were in fact either blatantly or subtly hostile toward Scripture. We get a glimpse of the anti-scriptural attitudes of old-earth geologists from the writings of Charles Lyell. Writing to Roderick Murchison (a fellow old-earth geologist) in a private letter dated August 11, 1829, just months before the publication of the first volume of his uniformitarian *Principles of Geology* (1830), Lyell reflected,

"I trust I shall make my sketch of the progress of geology popular. Old [Rev. John] Fleming is frightened and thinks the age will not stand my anti-Mosaical conclusions and at least that the subject will for a time become unpopular and awkward for the clergy, but I am not afraid. I shall out with the whole but in as conciliatory a manner as possible."

About the same time Lyell corresponded with his friend, George P. Scrope (another old-earth geologist and MP of British Parliament), saying, 'If ever the Mosaic geology could be set down without giving offense, it would be in an historical sketch.' Why would Lyell want to rid geology of the historically accurate (inspired) record of the Flood? Because as a Unitarian ***he was living in rebellion against his Creator, Jesus Christ,*** and he wanted geology to function with naturalistic presuppositions, just like his uniformitarian forefather, James Hutton, who wrote,

'The past history of our globe must be explained by what can be seen to be happening now. ... No powers are to be employed that are not natural to the globe, no action to be admitted except those of which we know the principle.'

So contrary to what people in the ID movement and many Christians influenced by the ID movement seem to think, naturalism (with its attendant anti-Bible, especially anti-Genesis, attitude) took hold of geology and astronomy in the late eighteenth and early nineteenth centuries. And this spread of the infection of naturalism in science was concurrent with the development of the same critical naturalistic approach to Genesis in biblical scholarship. In other words, it was reasoned, *Moses did not write Genesis under divine inspiration.* Rather, Genesis is no different from any other fallible human book and was in fact the purely natural product of many human authors and redactors working many centuries after Moses.

Although some of the catastrophists and uniformitarians believed in a Creator and some even professed to be Christians, the old-earth theories were developed by applying naturalistic philosophical assumptions in their interpretations of geological and astronomical evidence. Many old-earthers were *not* 100 percent philosophical naturalists. But all of them were operating largely with naturalistic assumptions, whether they realized it or not. In other words, they reconstructed their histories of the earth and solar system by appealing *only* to the presently observed laws and processes of nature plus time and chance (i.e., excluding the supernatural interventions of God at the Fall and the Flood, which disrupted or altered at least some of the laws and processes of nature).

It was on the basis of this anti-biblical naturalistic thinking that, fifty years later, Darwin promoted his naturalistic uniformitarian theory in biology to explain the incredible design in living things. Old-earth geological theories and old-universe astronomical theories are nothing but naturalistic philosophy (or really religion) masquerading as scientific fact, just like the evolutionary biological theories of Neo-Darwinism and punctuated equilibrium are.

VI. NATURALISM AND UNIFORMITARIANISM

Much more needs to be explored regarding this subject of naturalism and uniformitarianism. There has been some shallow and even incorrect thinking and writing on this subject by YECs as well as by their old-earth Christian and non-Christian critics. John Reed has written two very helpful articles.

I want to state clearly that naturalistic assumptions do not *necessarily* mean that a scientific conclusion is wrong. For example, a person with naturalistic assumptions as his starting point could conceivably deduce the law of inertia from his observations. Or, in the matter of actualities, Francis Crick, who is an atheist, was a co-discoverer of the structure of the DNA molecule. But these examples have to do with what I like to call *operation* science. This research uses the so-called 'scientific method' of observation of repeatable experiments in a controlled environment to determine how the present creation, or an individual entity in the creation, operates. For example, medical research, engineering research, and much research in biology, chemistry and physics fall into the category of operation science. This is the kind of science which put a man on the moon, a refrigerator in almost every kitchen, and finds cures for diseases. But operation science does not have any significant bearing on any doctrine of Scripture, and it is rarely affected by a scientist's religious worldview.

However, the matter of the *origin* of the law of inertia or of the DNA molecule or of the origin, age and history of the earth and universe (and everything in them) is a distinctly different question. These questions fall into the domain of what is often called *origin* science. This kind of research does not use the 'scientific method' of experimentation (except sometimes to propose *possible* causes of past events). Rather, to determine the *actual* past cause for some present effect that was produced in the unobservable past (e.g., a fossil or Grand Canyon), origin scientists use the legal-historical method of consideration of any relevant eyewitness testimony of the past event and careful investigation of the existing circumstantial evidence of the past event. Sciences such as archeology, paleontology and historical geology fit into this category of origin science. Origin science is like criminal investigation—by studying the evidence which exists in the present, researchers are trying to 'discover the past.' Origin scientists, then, are reconstructing history, which has direct and significant bearing on many important doctrines of Scripture. Here, naturalistic and uniformitarian assumptions strongly influence the observation, selection and interpretation of the physical data and can lead to very erroneous conclusions. In this case, Jesus' warning that bad trees cannot produce good fruit (*Matt. 7:18*) and Paul's warnings about deceptive philosophy (*Col. 2:8*) and 'arguments of what is falsely called "knowledge"' (*1 Tim. 6:20*) are very relevant. Old-earth geological theories were theories about *history*. Since they started with anti-biblical presuppositions, it is no surprise that they ended up concluding that the history in the Bible was wrong.

Fig.24.9: *Creation Day 5*

Naturalistic, and even uniformitarian, thinking of sorts is not to be totally excluded from Christian thinking. From roughly the end of the post-Flood, Ice-Age period (about 500–700 years after the Flood) to the present time, physical processes (e.g., volcanoes, earthquakes, wind and water erosion and sedimentation, meteor impacts, etc.) have been operating essentially as they do today and at the same average rate and intensity presently observed. Furthermore, although some different starting conditions for the processes and laws of nature prevailed in the interval between Creation Week and the Flood, there was a uniformity of natural processes then, too.

Some of the laws of nature started functioning during Creation Week after God made particular things (e.g., laws governing the growth and reproduction of plants did not commence until God supernaturally made the first kinds of plants on Day 3, laws related to the movements of the heavenly bodies commenced when God made those bodies on Day 4, and certain laws affecting animal life began to take effect on Day 5, Figure 24.9, when God made the first birds and sea creatures). Certainly, by the time God made Adam all the laws of nature were operational.

Sin Corrupted the World

However, it is likely that some of the laws of nature were altered in some way by God's Curse on the whole creation in *Genesis 3*, resulting in the bondage to corruption that Paul speaks of in *Rom. 8:19–23*. This present world is similar to, but significantly different from, the perfect world that God originally created during the six literal days of Creation Week. We now live in, and scientists' study, a creation damaged by human sin and divine judgment, Figure 24.10. Today all old-earth geologists and astronomers (whether professing Christians or not) deny the cosmic impact of the Fall, just as their predecessors did in the early nineteenth century.

Fig.24.10: *Sin Brought Death and Destruction*

Such a denial is an obvious implication of a non-Christian's worldview. Many old-earth Christians explicitly deny this cosmic impact of the Fall. Others unconsciously reject it. That is, they explicitly affirm that the Fall affected the whole creation, but because they accept the evolutionary view of history (even if they reject Darwinism to explain the origin of the various kinds of life), they unwittingly imply that the Curse of *Genesis 3* had no discernable impact on the non-human creation.

Furthermore, although many laws continued to operate during the Flood (e.g., water still flowed downhill and with sufficient speed could erode and carry silt, sand, rocks and boulders but with reduced speed would drop and sort its load, as it does today), there was a significant divinely induced disruption in the 'normal' course of nature during that year-long event, due to several supernatural acts of God (e.g., the Flood began exactly seven days after God said it would, God brought the animals to Noah in the Ark, the floodgates of heaven and fountains of the deep broke open simultaneously on a global scale, etc.).

In light of these considerations, biblically informed students of God's creation should invoke supernatural explanations only when there is an explicit biblical indication that God has done supernatural things (e.g., Creation Week, the Fall, the Flood and the Tower of Babel). Otherwise, Christians should seek to explain what they see in creation by the processes and laws of nature. The laws of nature describe not what God *must* do, but what He *normally does* to uphold his creation providentially.

God does not have to obey the laws of nature. Rather, nature must obey God. Put another way, the laws of nature reflect the customs of God as He works in creation, and miracles are simply God acting in His creation in an uncustomary manner for a special purpose.

What all YECs (both the scriptural geologists in the early nineteenth century and the YECs in the last 50 years) have always argued is that Genesis 1–11 is inspired, inerrant history given to us by the Creator. One cannot correctly interpret the physical evidence of His acts in creation (either the customary 'natural' acts or the uncustomary supernatural acts) if he ignores His written revelation about those acts. Even more problematic is the use of *naturalistic* interpretations of the present physical evidence to reinterpret the plain meaning of God's Word. But that is what the ID movement and most Christian leaders and Bible scholars have been doing and advocating in varying degrees (explicitly or implicitly, consciously or unconsciously) for almost the past 200 years, as they have tried to accommodate millions of years (and sometimes Darwinian evolution) in their interpretation of Scripture.

VII. 'INTELLIGENT DESIGN' ARGUMENTS OF AN EARLIER TIME

One more observation about the early nineteenth century is necessary. As atheism was advancing in the late eighteenth century, Christians and others expended much effort to defend the existence of a Creator God. To do this they developed arguments from design, especially in living creatures. The most famous design argument at this time was developed by the Anglican minister, Rev. William Paley (1743–1806), in his *Natural Theology: Evidence of the Existence of and Attributes of the Deity Collected from the Appearances of Nature* (1802). It was very popular, going through 20 editions by 1820 and continuing in use as a set text at Cambridge University into the early twentieth century. Darwin and all his old-earth mentors studied and knew the book well.

But there were other such writings, including a work by one of the scriptural geologists and a fellow Anglican clergyman, Rev. Thomas Gisborne (1758–1846), who in 1818 published *Testimony of Natural Theology to Christianity*. Gisborne said that Paley's work was very good as far as it went, but it was weak because of its omissions. Paley's argument only vindicated God's so-called positive attributes, such as goodness, wisdom, eternity and omnipotence. But it failed to point to God's holiness and justice as well as his mercy, as witnessed in nature. Paley, in other words, had ignored the cosmic impact of sin and God's Judgment on His once perfect creation. Gisborne sought to rectify this weakness by illuminating the witness of nature to these neglected divine attributes.

Then in the 1830s the celebrated 8-part series of 'Bridgewater Treatises' appeared. These presented design arguments from

1. the moral and intellectual nature of man,
2. the physical nature of man,
3. astronomy and physics,
4. animal and plant physiology,
5. the human hand,
6. chemistry, meteorology and digestion,
7. geology (written by the old-earth geologist, William Buckland),
8. the history, habits and instincts of animals (the only one of the eight treatises written by a young-earth creationist). Robson correctly identifies two important weaknesses of these efforts to defend the existence of God. First, because they largely divorced themselves from divine revelation (the Bible), the natural theology that was produced failed to deal with one of the greatest difficulties in theology, namely the existence of evil.

To put it simply, by arguing for a Designer without incorporating the Fall, they raised the obvious question of what sort of Designer would create some of the pathological features of this world. Second, argued Robson, contrary to the intent of the authors of the 'Bridgewater Treatises,' their arguments had an inherent tendency toward deism or even pantheism.

Regarding the impact of the Fall, a consideration of the following subsequent criticisms of the design argument is necessary. The famous atheist, Bertrand Russell, told why he was an atheist. One reason was that:

'When you come to look into this argument from design, it is a most astonishing thing that people believe that this world, with all the things that are in it, with all its defects, should be the best that omnipotence and omniscience have been able to produce in millions of years. I really cannot believe it.

More recently, the evolutionist philosopher, David Hull, argued in a similar way in his review of Phillip Johnson's *Darwin on Trial* (InterVarsity, 1991), which essentially launched the ID movement. Hull wrote,

'The problem that biological evolution poses for natural theologians is the sort of God that a Darwinian [*sic*] version of evolution implies ... . The evolutionary process is rife with happenstance, contingency, incredible waste, death, pain and horror .... Whatever the God implied by evolutionary theory and the data of natural history may be like, He is not the Protestant God of waste not, want not. He is also not a loving God who cares about His productions. He is not even the awful God portrayed in the book of Job. The God of the Galápagos is careless, wasteful, indifferent, almost diabolical. He is certainly not the sort of God to whom anyone would be inclined to pray.'

This line of reasoning applies even if one rejects neo-Darwinian evolution and instead believes that God supernaturally created new forms of life occasionally over the course of millions of years of death, bloodshed, and extinction.

The early nineteenth-century design arguments, while enthusiastically received by the already 'converted' of that day, failed to stem the rising tide of atheism and other forms of anti-biblical (and therefore anti-God) skepticism. In fact, history shows that the unrecognized assumptions of naturalism, which were buried in the foundations of the old-earth, 'the-age-of-the-earth-doesn't-matter' design arguments, actually paved the way for Darwin's theory, which would demolish the force of those design arguments in most people's minds.

VIII. MODERN COMPROMISE WITH OLD-EARTH NATURALISM

Phillip Johnson and the other old-earthers in the ID movement have not gone back far enough in their historical studies. Johnson appears to think that naturalism took control of science only after Darwin, or maybe even at the time of the 100th anniversary of Darwin's book. Speaking about a famous international celebration of about 2,000 scientists in Chicago in 1959, Johnson writes,

'What happened in that great triumphal celebration of 1959 is that science embraced a religious dogma called naturalism or materialism. Science declared that nature is all there is and that matter created everything that exists. The scientific community had a common interest in believing this creed because it affirmed that in principle there is nothing beyond the understanding and control of science. What went wrong in the wake of the Darwinian triumph was that the authority of science was captured by an ideology, and the evolutionary scientists thereafter believed what they wanted to believe rather than what the fossil data, the genetic data, the embryological data and the molecular data were showing them.

Nancy Pearcey likewise seems historically short-sighted. In her excellent discussion of the victory of Darwin's theory, she speaks of the Christians who tried to make peace with Darwinian evolution. She states, 'Those who reformulated Darwin to accommodate design were hoping to prevent the takeover of the idea of evolution by philosophical naturalism. They sought to extract the scientific theory from the philosophy in which it was imbedded.' But those Christians and many before them had for over 50 years allowed and even advocated (albeit unknowingly) the takeover of geology and astronomy by naturalism, and then advocated the day-age theory or gap theory and local-flood theory to save old-earth theory. I attended the ID movement conference in 1996, where Pearcey originally gave this paper. When in the comment period after the presentation I remarked about philosophical naturalism taking control of science decades before Darwin through old-earth geology and referred to my just-completed Ph.D. work on this matter, I had no response from anyone, either publicly or privately. It seemed that the old-earthers did not want to know about naturalism's involvement in the development of the idea of millions and billions of years of history.

The above-mentioned conference was sponsored by the Christian Leadership Ministry (hereafter CLM), a ministry of Campus Crusade for Christ which is focused on university professors and is very supportive of the ID movement and of such old-earth proponents at Hugh Ross and Walter Bradley. Through its link to the Origins website, CLM targets 'top scientists and philosophers on issues concerning intelligent design and theism.' That site linked to CLM states confidently,

'For Christians, the date of creation is not a primary issue of faith and should not be regarded as such, because the Bible does not specifically state a date of creation. This fact can be easily confirmed by reviewing sources such as The NIV Study Bible, The Believers Study Bible, The New Geneva Study Bible and evangelical commentaries. ... Therefore, *we believe Christians are free to follow the scientific evidence, minus hostile philosophical assumptions like naturalism.*'

For starters, what most Christian scholars believe *today* on this issue is no confirmation of the correct interpretation of Scripture, because popular scholarly vote does not determine truth. If it does, then the Protestant Reformation was wrong (which is not the case), for the Reformers were definitely in the minority for many decades. But note the emphatic statement in italics. These old-earth proponents do not understand that the 'scientific evidence' for billions of years is really only a ***naturalistic*** interpretation of the observed geological and astronomical evidence. Remove the 'hostile philosophical assumptions' of naturalism from geology and astronomy, and there is *no* scientific evidence for millions and billions of years.

'Scientific' Theories – Deceptively in the Church

Another example of people who say they are fighting naturalism's stranglehold on science, while at the same time promoting naturalistic 'scientific' theories in the church, is the new book by Hugh Ross and Fazale Rana, *Origins of Life* (2004). Their Reasons To Believe website advertisement for the book says, 'For years naturalistic theories have monopolized academia as the only possible scientific explanation for the origin of life. ... Rana and Ross explode the myth that scientific evidence supports naturalistic theories. ... ' The subtle implication is that the origin of life is the only topic in which *naturalism* reigns. But it also reigns in billions-of-years theories of geology and astronomy, which Ross and Rana - sadly and effectively persuade Christian laymen, pastors and scholars to accept and use as they interpret their Bibles. Undoubtedly, Ross and Rana are deceiving themselves and other Christians by this opposition to *naturalism* in the area of the origin of life while they simultaneously promote the big bang and billions of years.

YEC may not See Clearly

Even a few young-earth creationists do not seem to see things very clearly. Nelson and Reynolds state in their debate with old-earthers, 'Our advice, therefore, is to leave the issues of biblical chronology and history to a saner period.

Christians should unite in rooting out the tedious and unfruitful grip of naturalism, methodological and otherwise, on learning.'

But there never will be a saner period, because sin will continue to darken the minds of people who do not want to submit to their Creator and His Word. Nelson and Reynolds are mistaken when they say that 'the key thing is to oppose any sort of attempt to accommodate theism and naturalism.' No, the key is to oppose the accommodation of biblical revelation with naturalistic interpretations of the creation, which is what all old-earth reinterpretations of Genesis are.

Adulterous Union of *Biblical Teaching* and Naturalism

The issue is not a vaguely defined *theism's* marriage with naturalism but rather the adulterous union of *biblical teaching* and naturalism. Thus, fighting naturalism only in biology will not work. Ignoring the Bible—especially Genesis—and its testimony to the cosmic impact of sin and God's judgments at the Fall, the Flood and the Tower of Babel, even though arguing for design in living things (and even *God's* designing activity), ***will not lead people to the true and living God***, but rather away from Him and His holy Word. Nor will fighting naturalism only in biology, while tolerating or even promoting naturalism in geology and astronomy, break the stranglehold of naturalism on science.

Be Aware of ID

Therefore, the 'wedge' of the ID movement is not a wedge (leading to more truth) at all. It is simply a nail, which will not split the log open. It will not lead the scientific establishment to embrace the biblical view of creation, nor will it lead most people to the true God, the Creator who has spoken in only one book, the Bible.

In his book about his 'wedge strategy,' Johnson explains how Christians should proceed in what he thinks is the coming public dialogue between religion and science (actually, it has been going on for years before the ID movement was born, as a result of the efforts of young-earth creationists and others). He says, 'The place to begin is with the Biblical passage that is most relevant to the evolution controversy. It is not in Genesis; rather, it is the opening of the Gospel of John.' He then quotes and discusses John 1:1–3, *In the beginning was the Word, and the Word was with God, and the Word was God. He was in the beginning with God. All things were made through him, and without him was not anything made that was made.* followed by Rom. 1:18–20, *For the wrath of God is revealed from heaven against all ungodliness and unrighteousness of men, who by their unrighteousness suppress the truth. For what can be known about God is plain to them, because God has shown it to them. For his invisible attributes, namely, his eternal power and divine nature, have been clearly perceived, ever since the creation of the world, in the things that have been made. So, they are without excuse.* Though those passages are certainly relevant, they do not directly address the creation-evolution and age-of-the-earth debates as Genesis does. Furthermore, John and Paul clearly believed Genesis was literal history and based their teaching on Genesis, as Jesus did. More recently, in a 2001 interview, Johnson also stated,

"I think that one of the secondary issues [in the creation-evolution debate] concerns the details of the chronology in Genesis. ... So, I say, in terms of biblical importance, that we should move from the Genesis chronology to the most important fact about creation, which is John 1:1. ... It's important not to be side-tracked into questions of biblical detail, where you just wind up in a morass of shifting issues."

On what basis does Johnson assert that the most important fact about creation is John 1:1? He has never provided a theological or biblical argument to defend this assertion. It is difficult to see how his comments indicate anything but a very low view of and indifference to the inspired inerrant text of Genesis 1–11. I suggest that Johnson's failure to see (or to explain to his listeners, if he does see) that the idea of billions of years of geological and cosmic history is nothing but ***philosophical naturalism masquerading as scientific fact***, is the reason that he avoids the text of Genesis.

The Ill Influence of Naturalism

This failure to see the influence of naturalism, even by a person warning about the danger of naturalism, is further illustrated in a paper by one of America's greatest evangelical philosophers, Norman Geisler. In 1998 Geisler was president of the Evangelical Theological Society and gave the presidential address at the November annual meeting of the ETS. In it he warned of a number of dangerous philosophies that are assaulting the church and having considerable influence. The first one he discusses is *naturalism* (both methodological and philosophical naturalism), which he says has been one of the most destructive philosophies. Therefore, he devotes more space to it than any of the other dangerous philosophies that he discusses. As far as it goes, it is a very helpful warning about the dangers of naturalism. He even says that 'James Hutton (1726–1797) applied [David] Hume's anti-supernaturalism to geology, inaugurating nearly two centuries of naturalism in science.'

What is terribly ironic and very disappointing is that *Geisler has endorsed the writings of Hugh Ross*, who aggressively but subtly (whether consciously or not) *promotes naturalistic assumptions and thinking in the church by persuading Christians to accept billions of years and the big bang as scientific fact*.

Also, in Geisler's own *Encyclopedia of Christian Apologetics*, published the year *after* his ETS presidential address, he tells his readers, 'Most scientific evidence sets the age of the world at billions of years.' But as I have shown, it was not the *evidence* that set the age at billions of years, but rather the naturalistic *interpretation* of the evidence. Because of the confusion of evidence and interpretation of evidence, Geisler rejects the literal-day interpretation of Genesis 1 and believes that the genealogies of Genesis 5 and 11 have gaps of thousands of years, even though he says that '*prima facie* evidence' in Genesis supports literal days and no genealogical gaps in Genesis. After laying out the various old-earth reinterpretations of Genesis (all of which are based on naturalistic interpretations of the scientific evidence, have serious exegetical problems and have been refuted by YECs), he mistakenly concludes, 'There is *no necessary conflict between Genesis and the belief that the universe is millions or even billions of years old*.'

Nonetheless, Geisler is not the only evangelical philosopher who is highly trained to spot philosophical naturalism and yet has missed it in the issue of the earth's age. I am not aware of any leading evangelical philosopher who is a convinced YEC. If our greatest Bible-believing and Bible-defending philosophers cannot see naturalism's control of geology and astronomy, how will the rest of the church see it?

Bewitching Influence of Old-Earth Thinking

Herein is the bewitching influence of old-earth thinking. The fact is that we all (from the intellectually lowest to the most brilliant) have been brainwashed. 'Brainwashed' is a strong word, so let me explain. As we saw earlier, soon after Lyell published his *Principles of Geology* (1830–33), geology came under the control of the dogma of uniformitarianism, and catastrophism essentially passed off the scene. Reflecting this fact, in 1972 the following definition of 'catastrophism' appeared in a geological dictionary written by two of the leading geologists and academics of the day: '*Catastrophism*: The hypothesis, now more or less completely discarded, that changes in the earth occur as a result of isolated giant catastrophes of relatively short duration, as opposed to the idea, implicit in *Uniformitarianism*, that small changes are taking place continuously.'

However, at about the same time a very unexpected thing was occurring in geology—the birth of 'neo-catastrophism.' All the neo-catastrophists were evolutionists and believed in the billions of years of earth history. But they believed that much of the geological record was formed quickly and catastrophically, as the early nineteenth-century catastrophists had believed. One of the leading neo-catastrophists was Derek Ager, a British geologist who had conducted geological investigations in about 50 countries of the world. In one of his books, he reviewed the early nineteenth-century development of catastrophism and uniformitarianism and made this revealing comment:

"My excuse for this lengthy and amateur digression into history is that I have been trying to show how I think geology got into the hands of the theoreticians [i.e., the uniformitarians, in Ager's view] who were conditioned by the social and political history of their day more than by observations in the field .... In other words, we have allowed ourselves to be brain-washed into avoiding any interpretation of the past that involves extreme and what might be termed 'catastrophic' processes."

Ager admits that he was brainwashed through his geological education and early years in geological work, so that he could not see the evidence for catastrophe. The evidence was staring him in the face, but a mind-controlling set of assumptions made him blind to it. However, what he failed to see was that he had not only been brainwashed with assumptions coming from nineteenth-century social and political philosophy; he had been blinded by a whole philosophical-religious worldview called *naturalism* (he was a willing victim, however, for his writings give sufficient indication that he was a sinner in rebellion against God and his Word). So, as far as I am aware, until the day of his death a few years ago he was blinded (by naturalism) from seeing the overwhelming evidence in the rocks and fossils for Noah's Flood. If the geologists themselves were (and most geologists, even most Christian geologists, still are) ***brainwashed*** with the assumptions of philosophical naturalism, think of other Christians (including the most brilliant evangelical philosophers and OT Bible scholars), who through education, museums, national-park tours, TV science programs, etc., have been led to believe that the geologists have proven that the earth is billions of years old and that the global, catastrophic, year-long Flood never happened.

IX. THE WEDGE BETWEEN SCRIPTURE AND SCIENCE

The source of naturalism's control of science goes further back than Darwin, back to the old-earth and old-universe theories of the late eighteenth and early nineteenth centuries and even back to the writings of Galileo and Bacon (to whose dictums about Scripture and science the early nineteenth-century old-earth geologists frequently referred), who drove the first wedge between Scripture and science.

The Age of the Earth Matters

The age of the earth matters enormously if one wants to fight naturalism in science effectively and if he wants to be faithful to the inspired, inerrant Word of the Creator of heaven and earth, who was there at the beginning of creation and at the Flood, and has faithfully and clearly told us what happened, Figure 24.11.

Fig.24.11: *Young Earth*

ID Movement Mixture of Agnostics/Theists -Lacks Realism to the Word of God

But the ID movement is such a mixture of agnostics and theists of great theological variety that it can never be concerned about faithfulness to the true God and His Word. As noted earlier, there really is no wedge in Johnson's strategy. It is rather a nail strategy that will not split the log. A vaguely defined intelligent designer (not even necessarily divine) is as far as a Scripture-less approach can reach. Having deliberately ignored the biblical teaching given by the Creator—especially in Genesis—the ID arguments will not open the door to the true God.

If Johnson and the other Christian ID participants want eventually to bring Genesis into the origins debate, I predict:

- they will be accused of having been deceptive (a suspicion that many evolutionists have already expressed) during all the years that they have distanced themselves from YEC and ignored Genesis, and
- they will scare away most of their old-earth bedfellows in the ID movement who for various reasons do not want to live under the authoritative Word of God.

The lack of faithfulness to Scripture in the ID movement should be a concern to every Bible-believing Christian. Christians do not help God or help the revolutionized world by ignoring His holy Word.

This is a call to my Christian brothers in the ID movement to return to the Word of God, especially to the book of Genesis, which opens eyes to see the naturalism that controls geology and astronomy and leads people to think mistakenly that science has proven that the creation is billions of years old. I urge them to use their considerable mental powers and speaking and writing abilities to expose the lie of the naturalistic interpretations of old-earth geology and old-universe astronomy and to defend the clear truth of Genesis, both in the church and in the secular world.

PROGRESSIVELY POLLUTING THE CHURCH

The evidence is abundant and clear. The enemy has invaded the holy citadel. Naturalistic (atheistic) ways of thinking have increasingly polluted the church over the last 200 years through old-earth 'scientific' theories and through liberal theology. Who will take up the sword of the Spirit (*Eph. 6:17*), *and take the helmet of salvation, and the sword of the Spirit, which is the word of God,* —especially Genesis 1–11—and help expel the enemy of naturalism? The only alternative is to ignore the invasion and pollution and further abet it by compromise with the evolutionary belief in millions of years. Early Nineteenth-Century Views of Earth History

Biblical View (Scriptural Geologists)

SC---F---------------P----------SE

(Time to Present: ca. 6,000 years)

- God supernaturally created the world and all the basic 'kinds' of life in six literal days **(SC)** and
- then judged the world with a global Flood **(F)** at the time of Noah, which
- produced most of the geological/fossil record, and all present-day **(P)** processes have
- continued essentially since the Flood. This will continue until God supernaturally brings the world to
- an end **(SE)**.

Catastrophist View (e.g., Cuvier, Smith)

SB------------C----------C----------C----------C ----------------------------P -------C?---NE?

(Time to Present: 'untold ages')

During the earth's long history (millions of years at least) since

- God supernaturally began a primitive earth **(SB)**, there
- have been many natural regional or
- global catastrophic floods, which
- produced most of the geological/fossil record and
- current geography of the earth.
- After each catastrophe **(C)**
- God supernaturally created some new forms of life.
- Since the past catastrophes were natural events, there may be another in the future on earth, which may also have a
- natural (or supernatural) end **(NE)**.

Uniformitarian View (e.g., Hutton, Lyell)

SB?-- -------------------------P -------------NE?

(Time to Present: 'untold ages')

All geological processes on the earth (perhaps)

- had a beginning **(SB)** millions of years ago on a primitive earth. These processes (e.g., erosion, sedimentation, volcanoes and earthquakes) continued into the present and will continue into the future
- at the *same* rate and intensity
- as observed today **(P)**.
- No one knows whether there will be an end to the current natural processes (NE?).

SIX-DAYS CREATION

Fallacious Argument

It is commonly claimed by secular scientists that creationism is a "science stopper." The contention is that to ascribe anything (e.g., the origin of living organisms) to the direct action of God is to cut off all scientific inquiry. This seems such simple common sense that it has been very persuasive. Nevertheless, it is not difficult to show that the argument is fallacious, Figure 24.12.

Fig.24.12: *6-Day Creation*

A number of general points can be made.

- First, the argument is based on ignorance of all the different ways in which Christian faith can enter into science and of how fruitful these have been. After all, many of the great scientists of the past were committed Christians and many of those were consciously exploring the implications of their Christian faith for science.
- Second, whereas the direct action of God may cut off one type of explanation, others will remain and may even be enhanced. To say that God created the different kinds of animals and plants certainly cuts off explanation in terms of evolutionary continuity. However, it leaves wide-open scientific investigation of every other pattern of relationship (ecological, developmental, etc.) between these kinds. Scientists have been so indoctrinated in the belief that all patterns can only be explained historically in terms of the happenstances of Darwinian evolution that many wouldn't even know how to look for explanations in other terms.
- Third, there is abundant documentation of the fact that evolutionary naturalism has often stopped scientific research. To take just one example, the evolutionary assumption that certain organs or features are vestigial has often long delayed the (fruitful) research into their functions.

More specifically, one can appeal to experience and this is where this writing becomes a personal testimony to the scientific fruitfulness of a commitment to creation.

During the undergraduate days of my room-mate when his "heretical" views became known, his professor (Otto Lowenstein, Professor of Zoology) made a point of telling him that no creationist would be allowed to do research in his department!

However, he did allow him to do research. From the pressure that was put on him, he can only assume that it was thought that he could be convinced of the error of his ways. If that was the intention, then it badly backfired. Many a visiting scholar was brought into his laboratory to convince him, from their area of expertise, that evolution was indisputably true.

Of course, hardly knowing their field, he never had an answer at the time, but after they had gone, he would look up the relevant research and carefully analyze it. He always found that the evolutionist case was much weaker than it had seemed and that alternative creationist interpretations were available which were just as or more convincing. His position was further strengthened by the results of his own research.

He had decided to tackle the issue of the identity and nature of the created kinds. This was in response to a common evolutionist challenge that always seemed to him to be a reasonable one. If there are created kinds then they should be identifiable. He wanted to investigate the processes of variation within a kind, and gain some handle on the limits to that variation. He needed to be able to keep and breed large numbers of species. His background was in vertebrate studies, so that meant fish. His supervisor was a fan of the cichlid aquarium fish, so that was quickly settled! Those years of research were fascinating. For all the diversity of species, he found the cichlids to be an unmistakably natural group, a created kind.

The more he worked with these fish the clearer his recognition of "cichlidness" became and the more distinct they seemed from all the "similar" fishes he studied. Conversations at conferences and literature searches confirmed that this was the common experience of experts in every area of systematic biology.

Distinct kinds really are there and the experts know it to be so. Developmental studies then showed that the enormous cichlid diversity (over 1,000 "species") was actually produced by the endless permutation of a relatively small number of character states: 4 colors, ten or so basic pigment patterns and so on. The same characters (or character patterns) appeared "randomly" all over the cichlid distribution.

The patterns of variation were "modular" or "mosaic"; evolutionary lines of descent were nowhere to be found. This kind of adaptive variation can occur quite rapidly (since it involves only what was already there) and some instances of cichlid "radiation" (in geologically "recent" lakes) were indeed dateable (by evolutionists) to within timespans of no more than a few thousand years. On a wider canvas, fossils provided no comfort to evolutionists.

All fish, living and fossil, belong to distinct kinds; "links" are decidedly missing. Incidentally, creationists have no reason to be committed to any particular classification scheme, nor to any particular taxa above the kind level. "Orders," "classes," and "phyla" must not be allowed to become hallowed by tradition. They may be correctly identified (higher taxa *are* real), but there again they may not. Some "missing links" have been artifacts of bad classification systems. Morphology (and now biochemistry) have dominated classification, but ecology may yet prove to be a better guide.

His fish were supposedly strictly freshwater, but were found in the tropical fresh waters of three continents—from the Americas through Africa to Asia. He hypothesized that all, or at least most, fish kinds that survived the Flood must be able to survive both seawater and fresh, and much mixing of the two. After the post-Flood diversification within the kinds, we should still find that, in marine kinds, there are some species that can tolerate much fresher water and, in freshwater kinds, some species that can tolerate much saltier water. With his cichlids he found that this was indeed the case.

He was able to keep some species in pure seawater for more than two years with no harmful effects—they lived and reproduced normally. Literature searches again revealed that this was a common pattern throughout the fish classes.

He was also looking at heredity and already becoming skeptical of the dogma that "DNA is all" (so linked to reductionistic and evolutionary schemes). He discovered there is substantial evidence that there is more to heredity than genes and genic processes.

Indeed, it is clear that the whole cell system is a minimum unit of organism heredity. Genic processes have much to do with variation within kinds, but probably little to do with the distinction of kinds. Genes are best regarded as triggers in complex developmental systems rather than as creators or causes of organic structures. In this regard he found that there had been a vibrant creationist research program in developmental biology before Darwin that has been partly taken up again by the modern "structuralist" biologists (e.g., Stuart Kauffman and Brian Goodwin). Not surprisingly, the latter evolutionists are anti-Darwinian and anti-Dawkins. However, their work can readily be interpreted in creationist terms. It may, of course, ultimately prove wrong (our science is always approximate and liable to error), but it at least makes the point that creationism is not a science stopper. In my view, evolutionary explanations turn out to be fatally inconsistent.

REFUTING EVOLUTION

Many evolutionary books, including *Teaching about Evolution and the Nature of Science,* contrast religion/creation opinions with evolution/science facts. It is important to realize that this is a misleading contrast. Creationists often appeal to the facts of science to support their view, and evolutionists often appeal to philosophical *assumptions* from *outside* science. While creationists are often criticized for starting with a bias, evolutionists also start with a bias, as many of them admit. The debate between creation and evolution is primarily a dispute between two worldviews, with mutually incompatible underlying assumptions, Figure 24.13.

Fig.24.13: Refuting Evolution

This writing takes a critical look at the definitions of science, and the roles that biases and assumptions play in the interpretations by scientists.

THE BIAS OF EVOLUTIONARY

It is a fallacy to believe that facts speak for themselves—they are always *interpreted* according to a framework. The framework behind the evolutionists' interpretation is ***naturalism***—it is assumed that things made themselves, that no divine intervention has happened, and that God has not revealed to us knowledge about the past.

Evolution is a deduction from this assumption, and it is essentially the idea that things made themselves. It includes these unproven ideas: nothing gave rise to something at an alleged 'big bang,' non-living matter gave rise to life, single-celled organisms gave rise to many-celled organisms, invertebrates gave rise to vertebrates, ape-like creatures gave rise to man, non-intelligent and amoral matter gave rise to intelligence and morality, man's yearnings gave rise to religions, etc.

Professor D.M.S. Watson, one of the leading biologists and science writers of his day, demonstrated the atheistic bias behind much evolutionary thinking when he wrote:

'Evolution [is] a theory universally accepted not because it can be proven by logically coherent evidence to be true, but because the only alternative, special creation, is clearly incredible.

Accordingly, it's not a question of biased religious creationists versus objective scientific evolutionists; rather, it is the biases of the Christian religion versus the biases of the religion of secular humanism resulting in different interpretations of the same scientific data. As the anti-creationist science writer Boyce Rensberger admits:

'At this point, it is necessary to reveal a little inside information about how scientists work, something the textbooks don't usually tell you. The fact is that scientists are not really as objective and dispassionate in their work as they would like you to think. Most scientists first get their ideas about how the world works not through rigorously logical processes but through hunches and wild guesses. As individuals, they often come to believe something to be true long before they assemble the hard evidence that will convince somebody else that it is. Motivated by faith in his own ideas and a desire for acceptance by his peers, a scientist will labor for years knowing in his heart that his theory is correct but devising experiment after experiment whose results he hopes will support his position.'

It's not really a question of who is biased, but which bias is the correct bias with which to be biased! Actually, *Teaching about Evolution* admits in the dialogue on pages 22–25 that science isn't just about facts, and it is tentative, not dogmatic. But the rest of the book is dogmatic that evolution is a fact!

Professor Richard Lewontin, a geneticist (and self-proclaimed Marxist, is a renowned champion of Neo-Darwinism, and certainly one of the world's leaders in promoting evolutionary biology. He recently wrote this very revealing comment (the italics were in the original). It illustrates the implicit philosophical bias against Genesis creation regardless of whether or not the facts support it:

We take the side of science *in spite* of the patent absurdity of some of its constructs, *in spite* of its failure to fulfil many of its extravagant promises of health and life, *in spite* of the tolerance of the scientific community for unsubstantiated just-so stories, because we have a prior commitment, a commitment to materialism. It is not that the methods and institutions of science somehow compel us to accept a material explanation of the phenomenal world, but, on the contrary, that we are forced by our *a priori* adherence to material causes to create an apparatus of investigation and a set of concepts that produce material explanations, no matter how counter-intuitive, no matter how mystifying to the uninitiated. Moreover, that materialism is absolute, for we **cannot allow a Divine Foot** in the door.

Many evolutionists chide creationists not because of the facts, but because creationists refuse to play by the current rules of the game that exclude supernatural creation *a priori*. That it is indeed a **'game'** was proclaimed by the evolutionary biologist Richard Dickerson:

'Science is fundamentally a game. It is a game with one overriding and defining rule:

Rule #1: Let us see how far and to what extent we can explain the behavior of the physical and material universe in terms of purely physical and material causes, without invoking the supernatural.'

In practice, the 'game' is extended to trying to explain not just the behavior, but the *origin* of everything without the supernatural.

Actually, evolutionists are often not consistent with their own rules against invoking an intelligent designer. For example, when archaeologists find an arrowhead, they can tell it must have been designed, even though they haven't seen the designer. And the whole basis of the SETI (Search for Extra Terrestrial Intelligence) program is that a signal from outer space carrying specific information must have an intelligent source. Yet the materialistic bias of many evolutionists means that they reject an intelligent source for the literally encyclopedic information carried in every living cell.

It's no accident that the leaders of evolutionary thought were and are ardently opposed to the notion of the Christian God as revealed in the Bible. Stephen Jay Gould and others have shown that Darwin's purpose was to destroy the idea of a divine designer. Richard Dawkins applauds evolution because he claims that before Darwin it was impossible to be an intellectually fulfilled atheist, as he says he is.

Many atheists have claimed to be atheists precisely because of evolution. For example, the evolutionary entomologist and sociobiologist E.O. Wilson (who has an article in *Teaching about Evolution* on page 15) said:

'As were many persons from Alabama, I was a born-again Christian. When I was fifteen, I entered the Southern Baptist Church with great fervor and interest in the fundamentalist religion; I left at seventeen when I got to the University of Alabama and heard about evolutionary theory.'

Many people do not realize that the teaching of evolution propagates an anti-biblical religion. The first two tenets of *Humanist Manifesto I* (1933), signed by many prominent evolutionists, are:

1. Religious humanists regard the universe as self-existing and not created.
2. Humanism believes that Man is a part of nature and has emerged as a result of a continuous process.

This is exactly what evolution teaches. Many humanist leaders are quite open about using the *public schools to proselytize their faith.* This might surprise some parents who think the schools are supposed to be free of religious indoctrination, but this quote makes it clear:

I am convinced that the battle for humankind's future must be waged and won in the *public-school classroom* by *teachers* who correctly perceive their role as the *proselytizers of a new faith*: a *religion* of humanity that recognizes and respects the spark of what theologians call divinity in every human being. These teachers must embody the same selfless dedication as the most rabid fundamentalist preachers, for they will be *ministers* of another sort, utilizing a classroom instead of a pulpit to *convey humanist values* in *whatever subject they teach*, regardless of the educational level—preschool day care or large state university. The classroom must and will become an arena of conflict between the old and the new—the rotting corpse of Christianity, together with all its adjacent evils and misery, and the *new faith of humanism* … .

It will undoubtedly be a long, arduous, painful struggle replete with much sorrow and many tears, but humanism will emerge triumphant. It must if the family of humankind is to survive.

Teaching about Evolution, while claiming to be about science and neutral on religion, has some religious statements of its own. For example, on page 6:

'To accept the probability of change and to see change as an agent of opportunity rather than as a threat is a silent message and challenge in the lesson of evolution.

However, as it admits that evolution is 'unpredictable and natural,' and has 'no specific direction or goal' (p. 127), this message is incoherent.

The authors of *Teaching about Evolution* may realize that the rank atheism of most evolutionary leaders would be repugnant to most American parents if they knew.

More recently, the agnostic anti-creationist philosopher Ruse admitted, 'Evolution as a scientific theory makes a commitment to a kind of naturalism' but this 'may not be a good thing to admit in a court of law.' *Teaching about Evolution* tries to sanitize evolution by claiming that it is compatible with many religions. It even recruits many religious leaders in support. One of the 'dialogues' portrays a teacher having much success diffusing opposition by asking the students to ask their pastor, and coming back with 'Hey evolution is okay!' Although the dialogues are fictional, the situation is realistic.

It might surprise many people to realize that many ***church leaders do not believe their own book, the Bible***. This plainly teaches that God created recently in six consecutive normal days, made things to reproduce 'after their kind,' and that death and suffering resulted from Adam's sin. This is one reason why many Christians regard evolution as incompatible with Christianity. On page 58, *Teaching about Evolution* points out that many religious people believe that 'God used evolution' (theistic evolution). But theistic evolution teaches that God used struggle for survival and death, the 'last enemy' (1 Cor. 15:26), *The last enemy to be destroyed is death.* as His means of achieving a 'very good' (Gen. 1:31), *And God saw everything that he had made, and behold, it was very good. And there was evening and there was morning, the sixth day.* creation. Biblical creationists find this objectionable.

The only way to assert that evolution and 'religion' are compatible is to regard 'religion' as having nothing to do with the real world, and being just subjective. A God who 'created' by evolution is, for all practical purposes, indistinguishable from no God at all.

Perhaps *Teaching about Evolution* is letting its guard down sometimes. For example, on page 11 it refers to the 'explanation provided in Genesis … that God created everything in its present form over the course of six days,' i.e., Genesis really does teach six-day creation of basic kinds, which contradicts evolution. Therefore, *Teaching about Evolution* is indeed claiming that evolution conflicts with Genesis, and thus with biblical Christianity, although they usually deny that they are attacking 'religion.' *Teaching about Evolution* often sets up straw men misrepresenting what creationists really do believe. Creationists do not claim that everything was created in exactly the same form as today's creatures. Creationists believe in variation *within a kind*, which is totally different from the *information-gaining* variation required for particles-to-people evolution.

More blatantly, *Teaching about Evolution* recommends many books that are very openly atheistic, like those by Richard Dawkins (p. 131). On page 129 it says: 'Statements about creation … should not be regarded as reasonable alternatives to scientific explanations for the origin and evolution of life.' Since anything not reasonable is unreasonable, *Teaching about Evolution* is in effect saying that believers in creation are really unreasonable and irrational. This is hardly religiously neutral, but is regarded by many religious people as an attack.

A recent survey published in the leading science journal *Nature* conclusively showed that the National Academy of Sciences, the producers of *Teaching about Evolution*, is heavily biased against God, rather than religiously unbiased. A survey of all 517 NAS members in biological and physical sciences resulted in just over half responding: 72.2% were overtly atheistic, 20.8% agnostic, and only 7.0% believed in a personal God. Belief in God and immortality was lowest among biologists. It is likely that those who didn't respond were unbelievers as well, so the study probably underestimates the level of anti-God belief in the NAS. The percentage of unbelief is far higher than the percentage among U.S. scientists in general, or in the whole U.S. population.

Commenting on the professed religious neutrality of *Teaching about Evolution*, the surveyors comment:

NAS President Bruce Alberts said: 'There are very many outstanding members of this academy who are very religious people, people who believe in evolution, many of them biologists.' *Our research suggests otherwise.*

THE BASIS OF MODERN SCIENCE

Many historians, of many different religious persuasions including atheistic, have shown that modern science started to flourish only in largely Christian Europe. For example, Dr Stanley Jaki has documented how the scientific method was stillborn in all cultures apart from the Judeo-Christian culture of Europe. These historians point out that the basis of modern science depends on the assumption that the universe was made by a rational creator. An orderly universe makes perfect sense only if it were made by an orderly Creator. But if there is no creator, or if Zeus and his gang were in charge, why should there be any order at all? So, not only is a strong Christian belief not an obstacle to science, such a belief was its very foundation. It is, therefore, fallacious to claim, as many evolutionists do, that believing in miracles means that laboratory science would be impossible. Loren Eiseley stated:

'The philosophy of experimental science … began its discoveries and made use of its methods in the faith, not the knowledge, that it was dealing with a rational universe controlled by a creator who did not act upon whim nor interfere with the forces He had set in operation … . It is surely one of the curious paradoxes of history that science, which professionally has little to do with faith, owes its origins to an act of faith that the universe can be rationally interpreted, and that science today is sustained by that assumption.'

Evolutionists, including Eiseley himself, have thus abandoned the only rational justification for science. But Christians can still claim to have such a justification.

It should thus not be surprising, although it is for many people, that most branches of modern science were founded by believers in *creation*. The list of creationist scientists is impressive. A sample:

1. Physics—Newton, Faraday, Maxwell, Kelvin
2. Chemistry—Boyle, Dalton, Ramsay

3. Biology—Ray, Linnaeus, Mendel, Pasteur, Virchow, Agassiz
4. Geology—Steno, Woodward, Brewster, Buckland, Cuvier
5. Astronomy—Copernicus, Galileo, Kepler, Herschel, Maunder
6. Mathematics—Pascal, Leibniz, Euler

Even today, many scientists reject particles-to-people evolution (i.e., everything made itself). The *Creation Ministries International* (Australia) staff scientists have published many scientific papers in their own fields. Dr Russell Humphreys, a nuclear physicist working with Sandia National Laboratories in Albuquerque, New Mexico, has had over 20 articles published in physics journals, while Dr John Baumgardner's catastrophic plate tectonics theory was reported in *Nature*. Dr Edward Boudreaux of the University of New Orleans has published 26 articles and four books in physical chemistry. Dr Maciej Giertych, head of the Department of Genetics at the Institute of Dendrology of the Polish Academy of Sciences, has published 90 papers in scientific journals. Dr Raymond Damadian invented the lifesaving medical advance of magnetic resonance imaging (MRI). Dr Raymond Jones was described as one of Australia's top scientists for his discoveries about the legume *Leucaena* and bacterial symbiosis with grazing animals, worth millions of dollars per year to Australia. Dr Brian Stone has won a record number of awards for excellence in engineering teaching at Australian universities. An evolutionist opponent admitted the following about a leading creationist biochemist and debater, Dr Duane Gish:

Duane Gish has very strong scientific credentials. As a biochemist, he has synthesized peptides, compounds intermediate between amino acids and proteins. He has been co-authoring of a number of outstanding publications in peptide chemistry.

A number of highly qualified living creationist scientists can be found on the *Creation Ministries International* website. So, an oft-repeated charge that no real scientist rejects evolution is completely without foundation. Nevertheless, *Teaching about Evolution* claims in this Question-and-Answer section on page 56:

Q: Don't many scientists reject evolution?

A: No. The scientific consensus around evolution is overwhelming … .

It is regrettable that *Teaching about Evolution* is not really answering its own question. The actual question should be truthfully answered 'Yes,' even though evolution-rejecting scientists are in a minority. The explanation for the answer given would be appropriate (even if highly debatable) if the question were: 'Is it true that there is no scientific *consensus* around evolution?' But truth is not decided by majority vote!

C.S. Lewis also pointed out that even our ability to reason would be called into question if atheistic evolution were true:

'If the solar system was brought about by an accidental collision, then the appearance of organic life on this planet was also an accident, and the whole evolution of Man was an accident too. If so, then all our thought processes are mere accidents, the accidental by-product of the movement of atoms. And this holds for the materialists' and astronomers' as well as for anyone else's. But if their thoughts, i.e., of Materialism and Astronomy are merely accidental by-products, why should we believe them to be true? I see no reason for believing that one accident should be able to give a correct account of all the other accidents.

THE LIMITS OF SCIENCE

Science does have its limits. Normal (operational) science deals only with repeatable observable processes in the *present*. This has indeed been very successful in understanding the world, and has led to many improvements in the quality of life. In contrast, evolution is a speculation about the unobservable and unrepeatable *past*.

Thus, the comparison in *Teaching about Evolution* of disbelief in evolution with disbelief in gravity and heliocentrism is highly misleading. It is also wrong to claim that denying evolution is rejecting the type of science that put men on the moon, although many evolutionary propagandists make such claims. (Actually, the man behind the Apollo moon mission was the *creationist* rocket scientist Wernher von Braun.)

In dealing with the past, 'origins science' can enable us to make educated guesses about origins. It uses the principles of causality (everything that has a beginning has a cause) and analogy (e.g., we observe that intelligence is needed to generate complex coded information in the present, so we can reasonably assume the same for the past). But the only way we can be really sure about the past is if we have a reliable eyewitness account. Evolutionists claim there is no such account, so their ideas are derived from assumptions about the past.

However, biblical creationists believe that Genesis is an eyewitness account of the origin of the universe and living organisms. They also believe that there is good evidence for this claim, so they reject the claim that theirs is a blind faith.

Creationists don't pretend that any knowledge, science included, can be pursued without presuppositions (i.e., prior religious/philosophical beliefs). Creationists affirm that creation cannot ultimately be divorced from the Bible any more than evolution can ultimately be divorced from its naturalistic starting point that excludes divine creation *a priori.*

GOD'S ORDER IN ALL SUBSTANCE

The Periodic Table of chemical elements

Atoms

The beauty and organization of God's creation can be seen all around us in the macroscopic world. Everything we observe from plant life, stars, animals, rocks, air, and water—virtually everything—is composed of 90 naturally occurring building blocks known as atoms. Order starts with atoms and the subatomic particles that comprise them. This orderliness is not a random or haphazard assemblage of particles happening by accident or spontaneously organizing without an intelligent cause.

Looking into the nature of atoms, creation is clearly seen. When God created, He brought order to the universe even in the smallest things, for God is not the author of confusion.

Fig.24.14: Dmitri Mendeleev (1834–1907). Curtsy - commons.wikimedia.org

Mendeleev's Table

Dmitri Mendeleev (1834–1907), Figure 24.14, a Russian chemist, developed a formal organization of the elements in 1869 after having a dream. *"I saw in a dream a table where all elements fell into place as required.* Awakening, I immediately wrote it down on a piece of paper—only in one place did a correction later seem necessary."

Mendeleev's table is noteworthy because it exhibits the most accurate values for atomic mass allowing one to recognize trends over the entire array of elements. Mendeleev saw that the 65 then-known elements in the table lie at the heart of chemistry but that it was incomplete. There were spaces where elements could be located but no one at that time had discovered them. Mendeleev predicted the discovery of a missing element he called 'ekasilicon' on paper after observing a gap in the ***Periodic Table*** between silicon and tin, Figure 24.15.

When the new element was discovered and named 'germanium,' its properties closely matched Mendeleev's predictions of 'ekasilicon' for atomic mass, density, and melting point.

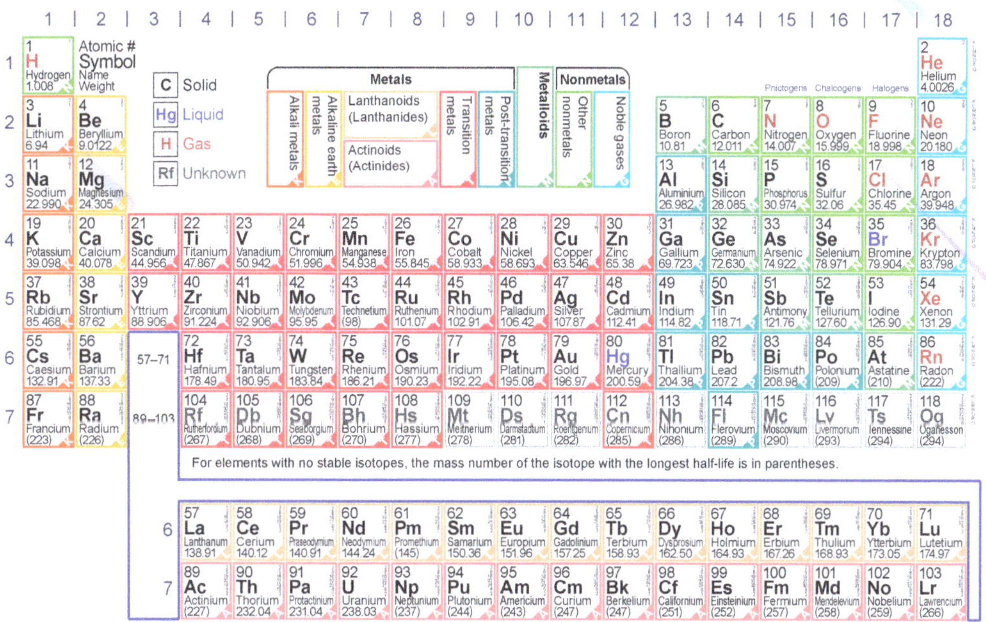

Fig.24.15: *Periodic Table of The Elements*

Information from the Periodic Table

- Elements Names and Symbols
- Atomic Number Z
- Atomic Weight A
- Electron Shell Configuration
- 90 Natural Elements
- 113 Elements Including Synthetic Cones
- Electro-magneticity
- Atomic Radius
- Ionization Energy
- Valence State

THE PERIODIC TABLE

Order and pattern can be clearly seen in the table, one of the most powerful and perhaps least understood tools of science. Atoms' organized structure leads to predictable attributes and chemical behavior, so predictions can be made on how atoms will bond or react with other atoms to form complex, multi-atom chemical compounds such as DNA.

Atoms are periodic because of the way that subatomic particles are arranged and added to each atom; hence they can be systematically placed in a table. Chemists use this table in their work, and everyone can understand structural beauty from a creator God.

The discovery of the Periodic Table tells us many attributes of the mind of God out creator. Most prominent is atom type, distinctively identified by a two-letter symbol, its etymology primarily from abbreviations of Latin and Greek names, or the discoverer of the element, and the **atomic number** of an element **Z** which is equal to the number of protons in the nucleus, Figure 24.16. The **atomic weight** is the total weight of protons plus neutrons found in the nucleus.

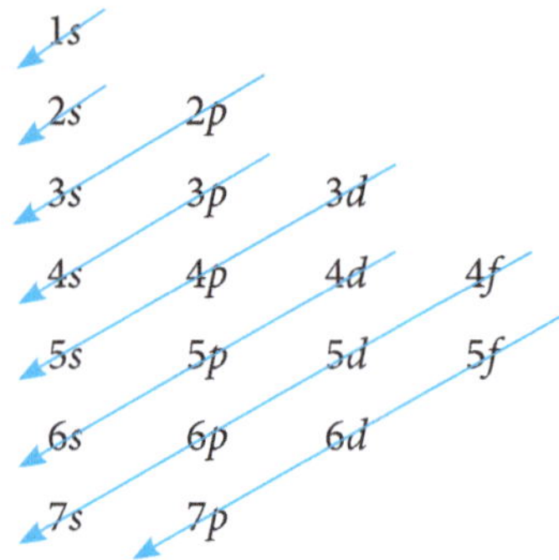

Fig.24.16: Arranging Atomic Table

ATOMIC STRUCTURE

Comprehension of the Periodic Table's organization may seem like a daunting task but a fundamental understanding can be achieved after a description of atomic structure. An atom is the smallest unit of matter that has all the chemical properties of that particular substance. This material, otherwise known as an element, cannot be broken down or changed into another substance by chemical means.

The structure of an atom consists of protons and neutrons in the nucleus, and electrons that move around the nucleus in specific energy levels, known as shells.

A FINE-TUNED COSMOS

The Hand of the Creator

Three millennia ago, an Ancient Near Eastern monarch gazed up at the night sky in wonder, and it stirred his poetic soul.

"The heavens declare the glory of God," he wrote, "and the sky above proclaims his handiwork. Day to day pours out speech, and night to night reveals knowledge."

For believers, it's easy to resonate with King David's awe in the face of cosmic beauty, and to recognize that it points to something beyond itself. The heavens do, indeed, declare the glory of God.

But David goes farther, claiming the skies show God's handiwork and display his knowledge. These statements reflect the direct experience of the senses but also speak to the mind. They almost sound like they might fit under the banner of scientific inquiry, if we may use that modern term in such an archaic context.

And so, they have proven to do, with increasing clarity. The cosmos, as it turns out, is not only beautiful but also exquisitely fine-tuned. Its physical properties are so carefully – and improbably – arranged that they point without mistake to the hand of a cosmic Designer.

Cosmic Fine-Tuning, in Brief

Modern physicists and astronomers have discovered that the odds against the universe existing in its present form by pure chance are, well, astronomical. In fact, they're far, far beyond astronomical, Figure 24.17.

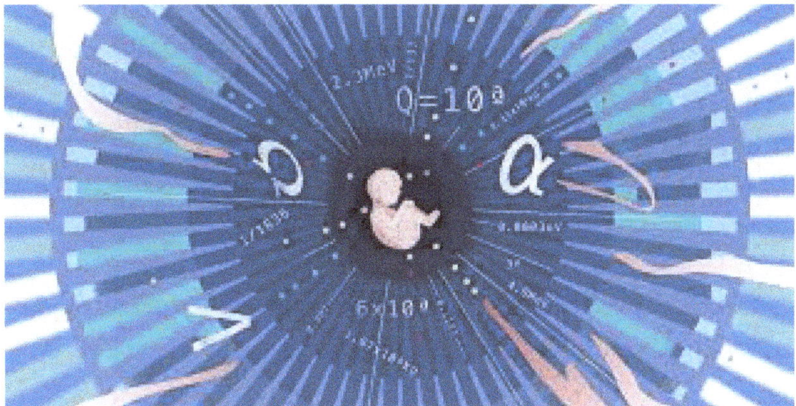

Fig.24.17: The Universe is Finely Tuned

The cosmos is built on a complex set of physical laws and constants, and on relationships between fundamental forces, all of which possess infinitesimally specific values. Even the slightest deviation in any of these values, and the universe as we know it could not exist, much less support life.

The strength and ratios of the four basic universal forces –

- strong nuclear force,
- weak nuclear force,
- electromagnetic force and
- gravitational force –

must be exactly as they are for elements, stars, planets and galaxies to exist. The same goes for the cosmological constant, the mass density of the universe and the expansion rate of the universe, among others.

It's common to use the phrase "one in a million" to describe something so unlikely as to be virtually impossible. But the level of specificity – of fine-tuning – for these various laws and constants ranges from 1 in 10^{37} (that's 10 followed by 37 zeroes) to 1 in 10^{120} (or 10 followed by 120 zeroes).

The numbers become practically meaningless without an analogy. Taking the "best bet" of the lot (1 in 10^{37}, the required degree of precision for both the strong nuclear force and the expansion rate of the universe), astrophysicist Hugh Ross compares it to covering North America with dimes, piling the dimes up to the moon, repeating the process with a billion North Americas, and then finding one specific marked dime, blindfolded. Pick any other dime, and the universe we know and live in ceases to exist.

Or according to physicist-philosopher Robin Collins, imagine a measuring tape stretched across the entire known universe, with a single mark representing the acceptable value of the force of gravity. Move that mark by one inch – on a tape spanning the whole universe – and said universe could never be formed.

As theoretical physicist Michio Kaku sums it up, "It's shocking to find how many of the familiar constants of the universe lie within a very narrow band that makes life possible. If a single one of these accidents were altered, stars would never form, the universe would fly apart, DNA would not exist, life as we know it would be impossible, Earth would flip over or freeze, and so on."

Such precise fine-tuning in the foundational properties of the cosmos offers compelling evidence that it was designed for a purpose. And the evidence for design would indicate the presence of a Designer.

In the words of Nobel Prize-winning physicist Charles Townes, "Intelligent design, as one sees it from a scientific point of view, seems to be quite real. This is a very special universe: it's remarkable that it came out just this way. If the laws of physics weren't just the way they are, we couldn't be here at all. The sun couldn't be there, the laws of gravity and nuclear laws and magnetic theory, quantum mechanics, and so on have to be just the way they are for us to be here."

A PROBLEM FOR MATERIALIST COSMOLOGIES

As far as the evidence from astronomy and physics goes, there's nothing controversial about these observations of cosmic fine-tuning. As physicist Paul Davies notes, "Everyone agrees that the universe looks as if it was designed for life."

Disputes arise, however, not over the science but because of metaphysical assumptions. Materialist cosmologies insist the universe has no design, and certainly no Designer, but came about through sheer chance via random natural processes. But for such cosmologies, the evident fine-tuning in the cosmos poses a serious problem.

One way to deal with the problem is simply to ignore it. Materialists will double down on their belief in chance because for them, it's the only acceptable explanation. They'll claim the appearance of design in the cosmos is merely that – an appearance, an illusion that in no way necessitates a Designer.

But such assertions fly in the face of reason and experience, to say nothing of the evidence. The extreme remote probabilities inherent in cosmic fine-tuning rule out pure chance as an organizing principle for the universe. Moreover, no one would look at a car or smartphone or building and say, "Sure, it may look like it was designed, but that's an illusion. It must've assembled itself by random chance. It's just a coincidence that it serves a useful purpose."

Another way around the fine-tuning problem for materialists is by appealing to necessity. The universe with all of its variables, they'll argue, had to take shape exactly the way it did because of physical laws. It was still due to chance, they'd add, only chance constrained by physical necessity, somehow.

This line of argument, however, makes a fatal error regarding the nature of the cosmos. In metaphysical terms, it fails to recognize that the universe isn't necessary, but contingent. The cosmos has a beginning, which means something – or someone – caused it. As such, the universe doesn't need to exist in any specific form. In fact, it doesn't need to exist at all. There's no reason why its physical constants couldn't have assumed any of a near-infinite number of values, or none at all.

Besides, where did the physical laws governing the cosmos come from? They didn't exist before or outside the universe, and thus couldn't have constrained its formation. The only feasible explanation is that they, along with the cosmos they define, come from the mind and purpose of a Designer.

What about the Multiverse?

By far the most popular option currently in vogue with materialists to explain away cosmic fine-tuning is the multiverse hypothesis. The basic concept is that our universe is merely one of an infinite number of universes, some of which are nearly indistinguishable from our own, others so alien that the laws of physics as we experience them don't apply, Figure 28.18.

Fig.24.18: *Hypothetical Multiverse*

The implications for a fine-tuned cosmos are clear, at least in the mind of multiverse proponents: Given an infinite number of universes, everything becomes not only possible but inevitable. In other words, our universe not only lucked out against unimaginable odds but in fact couldn't avoid doing so.

However, this is nothing more than an exercise in stacking the odds. It's like a child playing a game, growing frustrated at his inability to win and changing the rules so he can't lose. Are the odds insurmountable against our universe arising by chance? No problem. Just assume the existence of a multiverse, in which literally every possibility can and does occur.

It must be kept in mind that there's not one iota of evidence for the existence of other universes, nor indeed can there be, since by definition we're limited to our own universe. And a growing number of scientists, both theist and atheist, have come to criticize the hypothesis for its lack of scientific rigor.

As cosmologist George F.R. Ellis points out, "The trouble is that no possible astronomical observations can ever see those other universes. The arguments are indirect at best. And even if the multiverse exists, it leaves the deep mysteries of nature unexplained."

Theoretical physicist Sabine Hossenfelder is a tad blunter in her assessment: "Universes besides our own are logically equivalent to gods. They are unobservable by assumption; hence they can exist only in a religious sense. You can believe in them if you want to, but they are not part of science."

Even on its own terms, the concept fails to account for why a multiverse would exist. For argument's sake, suppose there are infinite multitudes of universes, each one governed by a divergent set of physical laws. But once again, where did those laws and those universes come from? The multiverse hypothesis doesn't know or care. Its only concern is to deny the necessity for a Creator.

GOOD REASON TO CALL COSMOS

The word "cosmos" comes from the ancient Greeks, originally meaning good order and arrangement, with an added sense of harmony and beauty. The English word "cosmetics" derives from it. And as early as Pythagoras in the 6th century BC, Greek philosophers, astronomers and mathematicians began using it to describe the universe as the created order, a meaning that has come down to the present.

Clearly, even ancient pagan thinkers shared in the universal human intuition that the universe is ordered and arranged by a cosmic Designer. It's doubly ironic, then, that many of their modern counterparts invoke the word that denotes the created order while denying that it's either ordered or created.

"By faith we understand that the universe was created by the word of God," wrote the author to the Hebrews, "so that what is seen was not made out of things that are visible."

The discoveries of modern astronomy and physics only serve to underscore that timeless scriptural truth, echoed in the hearts and intuitions of all humanity. Deep-space imagery reveals new vistas of beauty and grandeur that continue to declare the glory of God. And beneath the surface, exquisitely fine-tuned cosmic properties keep pouring out knowledge of God and displaying his handiwork.

It's a joy and privilege to live at a time when we can watch, listen and learn of such things.

SCIENCE COLLIDES WITH PHILOSOPHY

Fine-Tuning the Universe

When renowned scientists now talk seriously about millions of multiverses, the old question "are we alone?" gets a whole new meaning.

Our ever-expanding universe is incomprehensibly large – and its rate of growth is apparently accelerating – but if so, it's actually in a very delicate balance. It's then incredible that the universe exists at all.

Whether we scientists are inspired, bored, or infuriated by philosophy, all our theorizing and experimentation depends on particular philosophical background assumptions. This hidden influence is an acute embarrassment to many researchers, and it is therefore not often acknowledged. Such fundamental notions as reality, space, time and causality – notions found at the core of the scientific enterprise – all rely on particular metaphysical assumptions about the world.

Fig.24.19: Karl Popper. Curtsy - Wikimedia Commons

This may seem self-evident, and was regarded as important by Einstein, Bohr and the founders of quantum theory a century ago, but it runs against the grain of the views of working scientists in the post-war period.

Indeed, 21st-century mathematicians and scientists seem to have little need of philosophy. The glory days of Karl Popper, Figure 24.19, who argued that falsifiability was a hallmark of good science, and Thomas Kuhn, who noted the phenomenon of paradigm shifts, are long gone — in science, if not in the humanities.

For many years, scientific philosophy as practiced by scientists has languished, punctuated only by lapses such as the Sokal hoax, when NYU physicist Alan Sokal wrote a tongue-in-cheek article with a lot of scientific nonsense that was accepted by a leading journal in the postmodern science studies field (and launched a cottage industry of similar hoaxes).

But maybe the tide is finally turning. Perhaps modern science really needs philosophy after all.

Cosmic Coincidences

The main drivers here are some truly perplexing developments in physics and cosmology. In recent years physicists and cosmologists have uncovered numerous eye-popping "cosmic coincidences," remarkable instances of apparent "fine-tuning" of the universe.

Here are just three out of many that could be listed:

1. **Carbon Resonance and the Strong Force.** Although the abundance of hydrogen, helium and lithium are well-explained by known physical principles, the formation of heavier elements, beginning with carbon, very sensitively depends on the balance of the strong and weak forces. If the strong force were slightly stronger or slightly weaker (by just 1% in either direction), there would be no carbon or any heavier elements anywhere in the universe, and thus no carbon-based life forms like us to ask why.
2. **The Proton-to-Electron Mass Ratio.** A neutron's mass is slightly more than the combined mass of a proton, an electron and a neutrino. If the neutron were very slightly less massive, then it could not decay without energy input. If its mass were lower by 1%, then isolated protons would decay instead of neutrons, and very few atoms heavier than lithium could form.
3. **The Cosmological Constant.** Perhaps the most startling instance of fine-tuning is the cosmological constant paradox. This derives from the fact that when one calculates, based on known principles of quantum mechanics, the "vacuum energy density" of the universe, focusing on the electromagnetic force, one obtains the incredible result that empty space "weighs" 1,093g per cubic centimeter (cc). The actual average mass density of the universe, 10^{-28}g per cc, differs by 120 orders of magnitude from theory.

Physicists, who have fretted over the cosmological constant paradox for years, have noted that calculations such as the above involve only the electromagnetic force, and so perhaps when the contributions of the other known forces are included, all terms will cancel out to exactly zero, as a consequence of some unknown fundamental principle of physics.

But these hopes were shattered with the 1998 discovery that the expansion of the universe is accelerating, which implied that the cosmological constant must be slightly positive.

This meant that physicists were left to explain the startling fact that the positive and negative contributions to the cosmological constant cancel to 120-digit accuracy, yet fail to cancel beginning at the 121st digit.

Curiously, this observation is in accord with a prediction made by Nobel laureate and physicist Steven Weinberg in 1987, who argued from basic principles that the cosmological constant must be zero to within one part in roughly 10^{120} (and yet be nonzero), or else the universe either would have dispersed too fast for stars and galaxies to have formed, or else would have re-collapsed upon itself long ago.

THE ANTHROPIC PRINCIPLE

In short, numerous features of our universe seem fantastically fine-tuned for the existence of intelligent life. While some physicists still hold out for a "natural" explanation, many others are now coming to grips with the notion that our universe is profoundly unnatural, with no good explanation other than the Anthropic Principle — the universe is in this exceedingly improbable state, because if it weren't, we wouldn't be here to discuss the fact.

They further note that the prevailing "eternal inflation" big bang scenario suggests that our universe is just one pocket in a continuously bifurcating multiverse.

Inflation cosmology, by the way, got a significant experimental boost with the March 17, 2014 announcement that astronomers had discovered gravitational waves, signatures of the big bang inflation, in data collected from telescopes based at the South Pole.

In a similar vein, string theory, the current best candidate for a "theory of everything," predicts an enormous ensemble, numbering 10 to the power 500 by one accounting, of parallel universes. Thus, in such a large or even infinite ensemble, we should not be surprised to find ourselves in an exceedingly fine-tuned universe.

But to many scientists, such reasoning is anathema to traditional empirical science. Lee Smolin wrote in his 2006 book The Trouble with Physics:

"We physicists need to confront the crisis facing us. A scientific theory [the multiverse/ Anthropic Principle/ string theory paradigm] that makes no predictions and therefore is not subject to experiment can never fail, but such a theory can never succeed either, as long as science stands for knowledge gained from rational argument borne out by evidence."

And even the proponents of such views have some explaining to do. For example, if there are truly infinitely many pocket universes like ours, as physicists argue is the case, how can one possibly define a "probability measure" on such an ensemble? In other words, what does it mean to talk of the "probability" of our universe existing in its observed state?

But others see no alternative to some form of the multiverse and the Anthropic Principle. Physicist Max Tegmark, in his recent book Our Mathematical Universe, argues that not only is the multiverse real, but in fact that the multiverse *is* mathematics — all mathematical laws and structures actually exist, and are the ultimate stuff of the universe.

MODERN SCIENCE NEEDS PHILOSOPHY

With this backdrop, a growing number of scientists are calling for head-to-head interactions with philosophers. In a recent New Scientist article, cosmologist Joseph Silk reviews these and other issues now faced by the field, and then notes that such problems, probing the meaning of our very existence, are closely akin to those that have been debated by philosophers through the ages.

Thus, perhaps a new dialogue between science and philosophy can bring some badly needed insights into physics and other leading-edge fields such as neurobiology. (Indeed, there is a burgeoning sub discipline of neuro-philosophy.)

Philosophy and Physics

"Drawing the line between philosophy and physics has never been easy. Perhaps it is time to stop trying. The interface is ripe for exploration."

THE INEXPLICABLE FINE-TUNING

The Foundational Forces in Our Universe

The appearance of fine-tuning in our universe has been observed by theists and atheists alike. Even physicist Paul Davies (who is agnostic when it comes to the notion of a Divine Designer) readily stipulates, "Everyone agrees that the universe looks as if it was designed for life." Oxford philosopher John Leslie agrees: "it looks as if our universe is spectacularly 'fine-tuned for life'. By this I mean only that it looks as if small changes in this universe's basic features would have made life's evolution impossible."

The foundational "laws of nature" are amazingly fine-tuned; there is very little room for alteration. The smallest modifications of these laws would completely destroy the possibility of life in the universe. Theoretical physicist, Stephen Hawking says the laws of physics "appear fine-tuned in the sense that if they were altered by only modest amounts, the universe would be qualitatively different, and in many cases unsuitable for the development of life…

The emergence of the complex structures capable of supporting intelligent observers seems to be very fragile. The laws of nature form a system that is extremely fine-tuned, and very little in physical law can be altered without destroying the possibility of the development of life as we know it. Were it not for a series of startling coincidences in the precise details of physical law, it seems, humans and similar life-forms would never have come into being." The universe appears fine-tuned in three specific ways:

Forces Governing the Atom Are Favorable to Life:

The constants and proportions of the strong nuclear force, weak nuclear force, electromagnetic force and the force of gravity must exist within very narrow ranges in order for life to exist in the universe. The ratio of electrons to protons (both in their numbers and mass) must be precariously balanced. Stanford University physicist and cosmologist, Leonard Susskind, says, "The Laws of Physics begin with a list of elementary particles like electrons, quarks, and photons, each with special properties such as mass and electric charge.

These are the objects that everything else is built out of. No one knows why the list is what it is or why the properties of these particles are exactly what they are. An infinite number of other lists are equally possible. But a universe filled with life is by no means a generic expectation…"

If the value of this ratio deviated more than 1 in 10^{37}, the universe, as we know it, would not exist today. If the ratio between the electromagnetic force and gravity was altered more than 1 in 10^{40}, the universe would have suffered a similar fate. The nature of the universe (at the atomic level) could have been different, but even remarkably small differences would have been catastrophic to our existence.

Forces Governing the Matter of the Universe Are Favorable to Life:

On the macro-level, the size of the universe, its rate of growth, and the nature and existence of galaxies, stars, and planets depend largely on the force of gravity. While we sometimes take gravity for granted, the precisely calibrated gravity in the universe is puzzling; Susskind describes it as an "unexplained miracle." If the expansion rate of the universe deviated by more than 1 in 10^{37}, or the mass density of universe varied more than 1 in 10^{59}, there wouldn't be a single habitable galaxy or planet in the universe.

Forces Governing the Creation of Chemicals Are Favorable to Life:

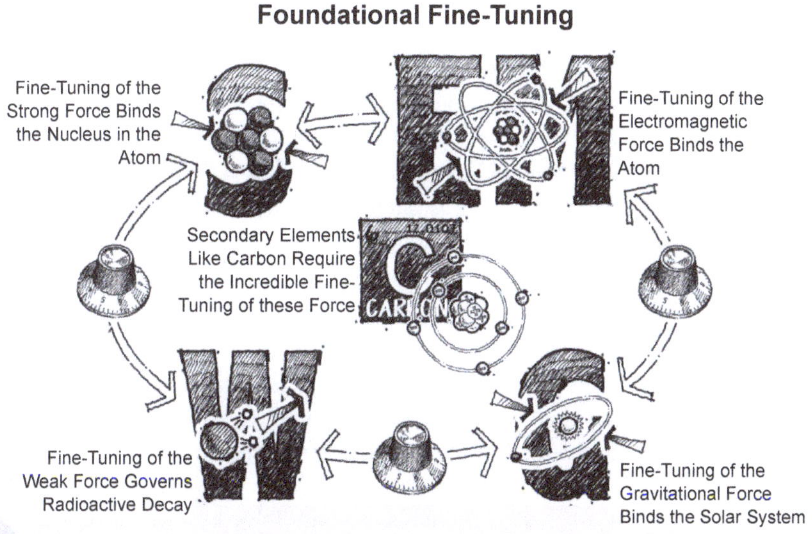

Foundational Fine-Tuning

Fine-Tuning of the Strong Force Binds the Nucleus in the Atom

Fine-Tuning of the Electromagnetic Force Binds the Atom

Secondary Elements Like Carbon Require the Incredible Fine-Tuning of these Force

Fine-Tuning of the Weak Force Governs Radioactive Decay

Fine-Tuning of the Gravitational Force Binds the Solar System

Fig.24.20A: Illustration from God's Crime Scene

The earliest elements in the universe, hydrogen and helium, are insufficient for the existence of carbon-based life forms unless joined by carbon, oxygen and the other necessary elements. These secondary elements were formed in stars, but the process by which these stars converted hydrogen and helium to carbon was an incredibly fine-tuned process. Even small alterations in the laws of physics would have prevented the formation of elements critical to the existence of life. Susskind puts it this way: "In the beginning there were only hydrogen and helium: certainly not sufficient for the foundation of life. Carbon, oxygen, and all the others came later. They were formed in the nuclear reactors in the interiors of stars. But the ability of stars to transmute hydrogen and helium into the all-important carbon nuclei was a very delicate affair. Small changes in the laws of electricity and nuclear physics could have prevented the formation of carbon, Figure 24.20A – Figure 24.B."

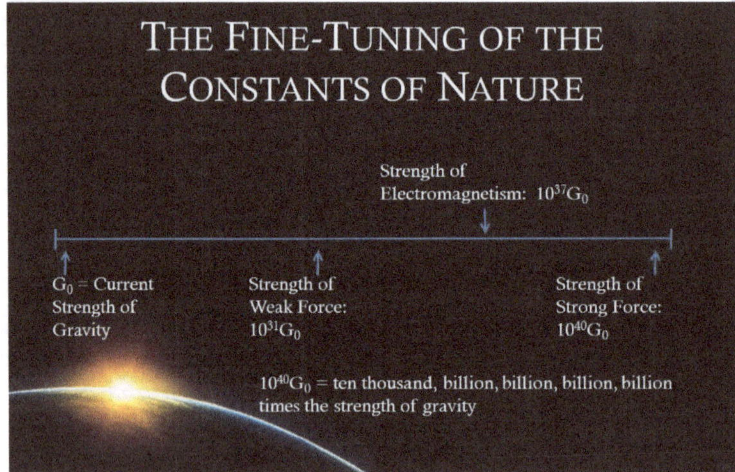

Fig.24.20B: *The Fine-Tuning - Illustration from God's Crime Scene*

- Figure 24.20 A and B: To say that the laws are fine-tuned means that the universe must have precisely the right set of laws in order for life to exist. Gravity – Electromagnetism - Strong Nuclear Force - Weak Nuclear Force – Principle of Quantization
- Electromagnetic Force - Different atomic bonds and thus complex molecules needed for life could not form. No light, no life!
- Strong Nuclear Force - The force that holds the atomic nucleus together. After all, protons are positively charged and like charges repel each other. Thus, shouldn't the nucleus fly apart? If stronger, no hydrogen, an essential element of life. If weaker, only hydrogen.
- Weak Nuclear Force - If stronger, insufficient helium to generate heavy elements in stars. If weaker, stars burn out too quickly and supernova explosions could not scatter heavy elements across the universe

When we use exponential numbers like 1 in 10^{37}, it's easy to underestimate the precision these numbers represent. Let me give you a few illustrations to help you grasp the exactitude of these universal constants:

Astrophysicist Hugh Ross offers the following analogy: Imagine covering the entire North American continent in dimes and stacking them until they reached the moon. Now imagine stacking just as many dimes again on another billion continents the same size as North America. If you marked one of those dimes and hid it in the billions of piles you've assembled, the odds of a blindfolded friend picking out the correct dime is approximately 1 in 10^{37}; the same level of precision required in the ***strong nuclear force*** and the ***expansion rate of the universe.***

Philosopher Robin Collins describes it this way: Imagine stretching a measuring tape across the entire known universe. Now imagine one ***particular mark on the tape*** represents the correct degree of gravitational force required to create the universe we have. If this mark ***were moved more than an inch*** from where it is (on a measuring tape spanning the entire universe), the altered gravitational force would ***prevent our universe from coming into existence***.

Paul Davies credits the following analogy to John Jefferson Davis: ***Imagine trying to fire a bullet at a one-inch target*** on the other side of the observable universe. The accuracy required to accomplish such a feat has been calculated at ***1 in*** 10^{60}. Compare this to the precision required in ***calibrating the mass density of the universe*** (fine-tuned to within ***1 unit in*** 10^{59}).

Hugh Ross also provides the following analogy: Imagine comparing the universe to an aircraft carrier like the USS John C Stennis (measuring 1,092 feet long with a displacement of 100,000 tons). If this carrier were as fine-tuned as the mass density of our universe, subtracting a ***billionth of a trillionth of the mass of an electron from the total mass of the aircraft carrier would sink the ship***. The appearance of fine-tuning in our universe has been observed by theists and atheists alike.

Starting to appreciate the level of fine-tuning in the foundational particles and forces in the universe? A small change in the value of any one particle or force would have a major impact on the larger systems and outcomes.

These fine-tuned relationships (sometimes jokingly referred to as "*happy cosmic accidents*") are critically important to life in the cosmos.

If just one of these parameters were altered, critical difficulties would result at every level of the universe. According to theoretical physicist Michio Kaku, "*...it's shocking to find how many of the familiar constants of the universe lie within a very narrow band that makes life possible. If a single one of these accidents were altered, stars would never form, the universe would fly apart, DNA would not exist, life as we know it would be impossible, Earth would flip over or freeze, and so on.*"

In the book, *God's Crime Scene: A Cold-Case Detective Examines the Evidence for a Divinely Created Universe*, It describes the fine-tuning of the universe in a much more robust manner and examine the explanations given by naturalists who want to stay "inside the room" of the natural universe for an answer. For a much more thorough account of the inadequacy of naturalism in this regard.

THE ACCIDENTAL UNIVERSE

In the fifth century B.C., the philosopher 'Democritus' proposed that all matter was made of tiny and indivisible atoms, which came in various sizes and textures—some hard and some soft, some smooth and some thorny. The atoms themselves were taken as givens.

In the nineteenth century, scientists discovered that the chemical properties of atoms repeat periodically (and created the periodic table to reflect this fact), but the origins of such patterns remained mysterious.

It wasn't until the twentieth century that scientists learned that the properties of an atom are determined by the number and placement of its electrons, the subatomic particles that orbit its nucleus. And we now know that all atoms heavier than helium were created in the nuclear furnaces of stars.

FUNDAMENTAL CAUSES AND PRINCIPLES

The history of science can be viewed as the recasting of phenomena that were once thought to be accidents as phenomena that can be understood in terms of fundamental causes and principles. One can add to the list of the fully explained: the hue of the sky, the orbits of planets, the angle of the wake of a boat moving through a lake, the six-sided patterns of snowflakes, the weight of a flying bustard, the temperature of boiling water, the size of raindrops, the circular shape of the sun. All these phenomena and many more, once thought to have been fixed at the beginning of time or to be the result of random events thereafter, have been explained as *necessary* consequences of the *fundamental laws of nature*—laws discovered by human beings.

HYPOTHESIS OF ENORMOUS NUMBER OF UNIVERSES

This long and appealing trend may be coming to an end. Dramatic developments in cosmological findings and thought have led some of the world's premier physicists to propose that our universe is only one of an enormous number of universes with wildly varying properties, and that some of the most basic features of our particular universe are indeed mere *accidents*—a random throw of the cosmic dice. In which case, there is no hope of ever explaining our universe's features in terms of fundamental causes and principles.

It is perhaps impossible to say how far apart the different universes may be, or whether they exist simultaneously in time. Some may have stars and galaxies like ours. Some may not. Some may be finite in size. Some may be infinite. Physicists call the totality of universes the "*multiverse*."

Alan Guth, a pioneer in cosmological thought, says that *"the multiple-universe idea severely limits our hopes to understand the world from fundamental principles."* And the philosophical ethos of science is torn from its roots. As put to me recently by Nobel Prize–winning physicist Steven Weinberg, a man as careful in his words as in his mathematical calculations, *"We now find ourselves at a historic fork in the road we travel to understand the laws of nature. If the multiverse idea is correct, the style of fundamental physics will be radically changed."*

Theoretical Physics

The scientists most distressed by Weinberg's "fork in the road" are theoretical physicists. Theoretical physics is the deepest and purest branch of science. It is the outpost of science closest to philosophy, and religion.

Experimental Scientists

Experimental scientists occupy themselves with observing and measuring the cosmos, finding out what stuff exists, no matter how strange that stuff may be. Theoretical physicists, on the other hand, are not satisfied with observing the universe. They want to know *why*. They want to explain all the properties of the universe in terms of a few fundamental principles and parameters.

Laws of Nature

These fundamental principles, in turn, lead to the "*laws of nature*," which govern the behavior of all matter and energy. An example of a fundamental principle in physics, first proposed by Galileo in 1632 and extended by Einstein in 1905, is the following: All observers traveling at constant velocity relative to one another should witness identical laws of nature. From this principle, Einstein derived his theory of special relativity.

An example of a fundamental parameter is the mass of an electron, considered one of the two dozen or so "elementary" particles of nature. As far as physicists are concerned, the fewer the fundamental principles and parameters, the better. The underlying hope and belief of this enterprise has always been that these basic principles are so restrictive that only one, self-consistent universe is possible, like a crossword puzzle with only one solution. That one universe would be, of course, the universe we live in. Theoretical physicists are Platonists. Until the past few years, they agreed that the entire universe, the one universe, is generated from a few mathematical truths and principles of symmetry, perhaps throwing in a handful of parameters like the mass of the electron. It seemed that we were closing in on a vision of our universe in which everything could be calculated, predicted, and understood.

HYPOTHESIS ETERNAL INFLATION AND STRING THEORY

However, two theories in physics, *eternal inflation and string theory*, now suggest that the *same* fundamental principles from which the laws of nature derive may lead to many *different* self-consistent universes, with many different properties. It is as if you walked into a shoe store, had your feet measured, and found that a size 5 would fit you, a size 8 would also fit, and a size 12 would fit equally well. Such wishy-washy results make theoretical physicists extremely unhappy.

Evidently, the fundamental laws of nature do not pin down a single and unique universe. According to the current thinking of many physicists, we are living in one of a vast number of universes. We are living in an ***accidental universe***. We are living in a universe incalculable by science.

"Back in the 1970s and 1980s," says Alan Guth, "the feeling was that we were so smart, we almost had everything figured out." What physicists had figured out were very accurate theories of three of the four fundamental forces of nature:

1. the strong nuclear force that binds atomic nuclei together,
2. the weak force that is responsible for some forms of radioactive decay, and
3. the electromagnetic force between electrically charged particles.
4. Einstein's theory of the fourth force, gravity

And there were prospects for merging the theory known as *quantum physics* with Einstein's theory of the fourth force, gravity, and thus pulling all of them into the fold of what physicists called the ***Theory of Everything***, or the Final Theory. These theories of the 1970s and 1980s required the specification of a couple dozen parameters corresponding to the masses of the elementary particles, and another half dozen or so parameters corresponding to the strengths of the fundamental forces. The next step would then have been to derive most of the elementary particle masses in terms of one or two fundamental masses and define the strengths of all the fundamental forces in terms of a single fundamental force.

There were good reasons to think that physicists were poised to take this next step. Indeed, since the time of Galileo, physics has been extremely successful in discovering principles and laws that have fewer and fewer free parameters and that are also in close agreement with the *observed facts of the world.*

For example, the observed rotation of the ellipse of the orbit of Mercury, 0.012 degrees per century, was successfully calculated using the theory of general relativity, and the observed magnetic strength of an electron, 2.002319 magnetons, was derived using the theory of quantum electrodynamics. More than any other science, physics brims with highly accurate agreements between theory and experiment.

Guth started his physics career in this sunny scientific world. Now sixty-four years old and a professor at MIT, he was in his early thirties when he proposed a major revision to the Big Bang theory, something called inflation. We now have a great deal of evidence suggesting that our universe began as a nugget of extremely high density and temperature about 14 billion years ago and has been expanding, thinning out, and cooling ever since. The *theory of inflation* proposes that when our universe was only about a trillionth of a trillionth of a trillionth of a second old, a peculiar type of energy caused the cosmos to expand very rapidly.

A tiny fraction of a second later, the universe returned to the more leisurely rate of expansion of the standard Big Bang model. Inflation solved a number of outstanding problems in cosmology, such as why the universe appears so homogeneous on large scales.

When Guth was visited in his third-floor office at MIT one cool day in May, he was hardly been see above the stacks of paper and empty Diet Coke bottles on his desk. More piles of paper and dozens of magazines littered the floor. In fact, a few years ago Guth won a contest sponsored by the *Boston Globe* for the messiest office in the city. The prize was the services of a professional organizer for one day. "She was actually more a nuisance than a help. She took piles of envelopes from the floor and began sorting them according to size." He wears aviator-style eyeglasses, keeps his hair long, and chain-drinks Diet Cokes. "The reason I went into theoretical physics," Guth tells me, "is that the visitor liked the idea that both could understand everything—i.e., the universe—in terms of mathematics and logic." Guth gives a bitter laugh. Both have been talking about the multiverse.

While challenging the Platonic dream of theoretical physicists, the multiverse idea does explain one aspect of our universe that has unsettled some scientists for years: according to various calculations, if the values of some of the fundamental parameters of our universe were a little larger or a little smaller, life could not have arisen.

For example, if the nuclear force were a few percentage points stronger than it actually is, then all the hydrogen atoms in the infant universe would have fused with other hydrogen atoms to make helium, and there would be no hydrogen left. No hydrogen means no water. Although we are far from certain about what conditions are necessary for life, most biologists believe that water is necessary.

On the other hand, if the nuclear force were substantially weaker than what it actually is, then the complex atoms needed for biology could not hold together. As another example, if the relationship between the strengths of the gravitational force and the electromagnetic force were not close to what it is, then the cosmos would not harbor any stars that explode and spew out life-supporting chemical elements into space or any other stars that form planets.

Both kinds of stars are required for the emergence of life. The strengths of the basic forces and certain other fundamental parameters in our universe appear to be "***fine-tuned***" to allow the existence of life. The recognition of this finetuning led British physicist Brandon Carter to articulate what he called the anthropic principle, which states that the universe must have the parameters it does because we are here to observe it.

Actually, the word ***anthropic***, from the Greek for "man," is a misnomer: if these fundamental parameters were much different from what they are, it is not only human beings who would not exist. No life of any kind would exist.

If such conclusions are correct, the great question, of course, is *why* these fundamental parameters happen to lie within the range needed for life. Does the universe care about life?

INTELLIGENT DESIGN

Intelligent design is one answer. Indeed, a fair number of theologians, philosophers, and even some scientists have used fine-tuning and the anthropic principle as evidence of the existence of God.

For example, at the 2011 Christian Scholars' Conference at Pepperdine University, Francis Collins, a leading geneticist and director of the National Institutes of Health, said, "To get our universe, with all of its potential for complexities or any kind of potential for any kind of life-form, everything has to be precisely defined on this ***knife edge*** of improbability.... [Y]ou have to see the hands of a creator who set the parameters to be just so because the creator was interested in something a little more complicated than random particles."

Intelligent Design - Answer to Fine-Tuning

Intelligent design, however, is an answer to fine-tuning that does not appeal to most scientists. The multiverse offers another explanation. If there are countless different universes with different properties—for example, some with nuclear forces much stronger than in our universe and some with nuclear forces much weaker—then, some of those universes will allow the emergence of life and some will not. Some of those universes will be dead, lifeless hulks of matter and energy, and others will permit the emergence of cells, plants and animals, minds. From the huge range of possible universes predicted by the theories, the fraction of universes with life is undoubtedly small. But that doesn't matter. We live in one of the universes that permits life because otherwise we wouldn't be here to ask the question.

Happy Coincidence, Good Luck, or Act of Providence!

The explanation is similar to the explanation of why we happen to live on a planet that has so many nice things for our comfortable existence: oxygen, water, a temperature between the freezing and boiling points of water, and so on. Is this happy coincidence just good luck, or an act of Providence, or what?

No, it is simply that we could not live on planets without such properties. Many other planets exist that are not so hospitable to life, such as Uranus, where the temperature is –371 degrees Fahrenheit, and Venus, where it rains sulfuric acid.

The multiverse offers an explanation to the fine-tuning conundrum that does not require the presence of a Designer. As Steven Weinberg says:

"Over many centuries' science has weakened the hold of religion, not by disproving the existence of God but by invalidating arguments for God based on what we observe in the natural world. The multiverse idea offers an explanation of why we find ourselves in a universe favorable to life that does not rely on the benevolence of a creator, and so if correct will leave still less support for religion."

Some physicists remain skeptical of the anthropic principle and the reliance on multiple universes to explain the values of the fundamental parameters of physics. Others, such as Weinberg and Guth, have reluctantly accepted the anthropic principle and the multiverse idea as together providing the best possible explanation for the observed facts.

If the multiverse idea is correct, then the historic mission of physics to explain all the properties of our universe in terms of fundamental principles—to explain why the properties of our universe must *necessarily* be what they are—is futile, a beautiful philosophical dream that simply isn't true.

Our universe is what it is because we are here. The situation could be likened to a school of intelligent fish who one day began wondering why their world is completely filled with water. Many of the fish, the theorists, hope to prove that the entire cosmos necessarily has to be filled with water. For years, they put their minds to the task but can never quite seem to prove their assertion. Then, a wizened group of fish postulates that maybe they are fooling themselves. Maybe there are, they suggest, many other worlds, some of them completely dry, and everything in between.

The most striking example of fine-tuning, and one that practically demands the multiverse to explain it, is the unexpected detection of what scientists call dark energy. Little more than a decade ago, using robotic telescopes in Arizona, Chile, Hawaii, and outer space that can comb through nearly a million galaxies a night, astronomers discovered that the expansion of the universe is accelerating. As mentioned previously, it has been known since the late 1920s that the universe is expanding; it's a central feature of the Big Bang model.

Orthodox cosmological thought held that the expansion is slowing down. After all, gravity is an attractive force; it pulls masses closer together. So, it was quite a surprise in 1998 when two teams of astronomers announced that some unknown force appears to be jamming its foot down on the cosmic accelerator pedal. The expansion is speeding up. Galaxies are flying away from each other as if repelled by antigravity. Says Robert Kirshner, one of the team members who made the discovery: "This is not your father's universe." (In October, members of both teams were awarded the Nobel Prize in Physics.)

Physicists have named the energy associated with this cosmological force *dark energy*. No one knows what it is. Not only invisible, dark energy apparently hides out in empty space. Yet, based on our observations of the accelerating rate of expansion, dark energy constitutes a whopping three quarters of the total energy of the universe. It is the invisible elephant in the room of science.

DARK ENERGY

The amount of ***dark energy***, or more precisely the amount of dark energy in every cubic centimeter of space, has been calculated to be about one hundred-millionth (**10^{-8}**) of an erg per cubic centimeter. (For comparison, a penny dropped from waist-high hits the floor with an energy of about three hundred thousand—that is, 3×10^5—ergs.) This may not seem like much, but it adds up in the vast volumes of outer space. Astronomers were able to determine this number by measuring the rate of expansion of the universe at different epochs—if the universe is accelerating, then its rate of expansion was slower in the past. From the amount of acceleration, astronomers can calculate the amount of ***dark energy*** in the universe.

Theoretical physicists have several hypotheses about the identity of dark energy. It may be the energy of ghostly subatomic particles that can briefly appear out of nothing before self-annihilating and slipping back into the vacuum. According to quantum physics, empty space is a pandemonium of subatomic particles rushing about and then vanishing before they can be seen.

Dark energy may also be associated with an as-yet-unobserved force field called the Higgs field, which is sometimes invoked to explain why certain kinds of matter have mass. (Theoretical physicists ponder things that other people do not.) And in the models proposed by string theory, dark energy may be associated with the way in which extra dimensions of space—beyond the usual length, width, and breadth—get compressed down to sizes much smaller than atoms, so that we do not notice them.

These various hypotheses give a fantastically large range for the *theoretically possible* amounts of dark energy in a universe, from something like 10^{115} ergs per cubic centimeter to -10^{115} ergs per cubic centimeter. (A negative value for dark energy would mean that it acts to *decelerate* the universe, in contrast to what is observed.) Thus, in absolute magnitude, the amount of dark energy actually present in our universe is either very, very small or very, very large compared with what it could be. This fact alone is surprising. If the theoretically possible positive values for dark energy were marked out on a ruler stretching from here to the sun, with zero at one end of the ruler and 10^{115} ergs per cubic centimeter at the other end, the value of dark energy actually found in our universe (10^{-8} ergs per cubic centimeter) would be closer to the zero end than the width of an atom.

On one thing most physicists agree: If the amount of dark energy in our universe were only a little bit different than what it actually is, then life could never have emerged. A little more and the universe would accelerate so rapidly that the matter in the young cosmos could never pull itself together to form stars and thence form the complex atoms made in stars. And, going into negative values of dark energy, a little less and the universe would decelerate so rapidly that it would re-collapse before there was time to form even the simplest atoms.

Here we have a clear example of fine-tuning: out of all the possible amounts of dark energy that our universe might have, the actual amount lies in the tiny sliver of the range that allows life. There is little argument on this point. It does not depend on assumptions about whether we need liquid water for life or oxygen or particular biochemistries. As before, one is compelled to ask the question: Why does such fine-tuning occur? And the answer many physicists now believe: The multiverse. A vast number of universes may exist, with many different values of the amount of dark energy. Our particular universe is one of the universes with a small value, permitting the emergence of life. We are here, so our universe must be such a universe. We are an accident. From the cosmic lottery hat containing zillions of universes, we happened to draw a universe that allowed life. But then again, if we had not drawn such a ticket, we would not be here to ponder the odds.

The concept of the multiverse is compelling not only because it explains the problem of fine-tuning. As mentioned earlier, the possibility of the multiverse is actually predicted by modern theories of physics. One such theory, called eternal inflation, is a revision of Guth's inflation theory developed by Andrei Linde, Paul Steinhardt, and Alex Vilenkin in the early and mid-1980s. In regular inflation theory, the very rapid expansion of the infant universe is caused by an energy field, like dark energy, that is temporarily trapped in a condition that does not represent the lowest possible energy for the universe as a whole—like a marble sitting in a small dent on a table. The marble can stay there, but if it is jostled it will roll out of the dent, roll across the table, and then fall to the floor (which represents the lowest possible energy level).

In the theory of eternal inflation, the dark energy field has many different values at different points of space, analogous to lots of marbles sitting in lots of dents on the cosmic table. Moreover, as space expands rapidly, the number of marbles increases. Each of these marbles is jostled by the random processes inherent in quantum mechanics, and some of the marbles will begin rolling across the table and onto the floor. Each marble starts a new Big Bang, essentially a new universe. Thus, the original, rapidly expanding universe spawns a multitude of new universes, in a never-ending process.

STRING THEORY

String theory, too, predicts the possibility of the multiverse. Originally conceived in the late 1960s as a theory of the strong nuclear force but soon enlarged far beyond that ambition, string theory postulates that the smallest constituents of matter are not subatomic particles like the electron but extremely tiny one-dimensional "strings" of energy. These elemental strings can vibrate at different frequencies, like the strings of a violin, and the different modes of vibration correspond to different fundamental particles and forces.

String theories typically require seven dimensions of space in addition to the usual three, which are compacted down to such small sizes that we never experience them, like a three-dimensional garden hose that appears as a one-dimensional line when seen from a great distance. There are, in fact, a vast number of ways that the extra dimensions in string theory can be folded up, and each of the different ways corresponds to a different universe with different physical properties.

It was originally hoped that from a theory of these strings, with very few additional parameters, physicists would be able to explain all the forces and particles of nature—all of reality would be a manifestation of the vibrations of elemental strings. String theory would then be the ultimate realization of the Platonic ideal of a fully explicable cosmos. In the past few years, however, physicists have discovered that ***string theory predicts not a unique universe but a huge number of possible universes with different properties***. It has been estimated that the "string landscape" contains $\mathbf{10^{500}}$ different possible universes. For all practical purposes, that number is infinite.

It is important to point out that neither eternal inflation nor string theory has anywhere near the experimental support of many previous theories in physics, such as special relativity or quantum electrodynamics, mentioned earlier. Eternal inflation or string theory, or both, could turn out to be wrong. However, some of the world's leading physicists have devoted their careers to the study of these two theories.

Back to the ***intelligent fish***. The wizened old fish conjecture that there are many other worlds, some with dry land and some with water. Some of the fish grudgingly accept this explanation. Some feel relieved. Some feel like their lifelong ruminations have been pointless. And some remain deeply concerned. Because there is no way they can prove this conjecture. That same uncertainty disturbs many physicists who are adjusting to the idea of the multiverse. Not only must we accept that basic properties of our universe are accidental and incalculable. In addition, we must believe in the existence of many other universes.

No Conceivable way to Prove their Existence

But we have no conceivable way of observing these other universes and cannot prove their existence. It is science fiction. Thus, to explain what we see in the world and in our mental deductions, we must illogically believe in what we cannot prove.

Sound familiar? Theologians are accustomed to taking some beliefs on faith. Scientists are not. All we can do is hope that the same theories that predict the multiverse also produce many other predictions that we can test here in our own universe. But the other universes themselves will almost certainly remain a science fiction conjecture. Science in the absence in the absence of God is a Degenerate Science!

THE INTELLIGENT DESIGNER

A Designer is Unscientific—Even if All the Evidence Supports One!

Most important, it should be made clear in the classrooms and lecture theaters that science, including evolution, has not disproved God's existence or an Intelligent Designer, because it cannot be allowed to consider it (presumably). Even if all the data point to an intelligent designer, such a hypothesis is excluded from science because it is not naturalistic. Of course, the scientist, as an individual, is free to embrace a reality that transcends naturalism.'

Speedy Species Surprise

The rapid appearance today, of new varieties of fish, lizards, and more defies evolutionary expectations … but fits perfectly with the Bible.

Researchers in Trinidad relocated guppies (*Poecilia reticulata*) from a waterfall pool teeming with predators to previously guppy-free pools above the falls where there was only one known possible predator (of small guppies only, therefore large guppies would be safe), Figure 24.21.

Fig.24.21: *Guppies (Poecilia reticulata) – Curtsy Wikimedia commons: L Church*

The descendants of the transplanted guppies adjusted to their new circumstances by growing bigger, maturing later, and having fewer and bigger offspring.

The speed of these changes bewildered evolutionists, because their standard millions-of-years view is that the guppies would require long periods of time to adapt.

One evolutionist said, 'The guppies adapted to their new environment in a mere four years—a rate of change some 10,000 to 10 million times faster than the average rates determined from the fossil record.'

Leggy Lizards

And it's not just guppies. In the Bahamas, small numbers of anole lizards (*Anolis sagrei*) were transplanted from an island with tall trees to nearby islands where there were previously no lizards and only smaller bushy vegetation.

Body form rapidly changed in succeeding generations. In particular, the relative length of hindlimbs was greatly decreased—thought to be an adaptation for life amongst the twigs of the scrubby vegetation in the lizards' new habitat. (Lizards that live on tree trunks have longer legs than those that live on twigs—an apparent trade-off between the agility necessary for twig-to-twig jumping and the speed that longer limbs provide on the broad surface of tree trunks.)

Nevertheless, again it was the *speed* of adaptation, many thousands of times higher than (their interpretation of) the 'fossil record' that surprised evolutionists.

Daisy Diaspora

On small islands off British Columbia, the seeds of wind-dispersed weedy plants in the daisy family (Asteraceae) are rapidly losing their ability to 'fly.' Specifically, the embryo part of the seed is becoming fatter while the parachute-like 'pappus' that keeps each seed aloft is becoming smaller. These changes are advantageous because they reduce dispersal—otherwise, on such tiny islands, lightweight windblown seeds would be lost in the ocean (which is why they have left fewer descendants). Note that these changes involve the *loss* of the capacity for long-range airborne dispersal.

FLIES, FISH AND FINCHES

Other examples of rapid adaptation, even to the extent of producing 'new species'—speciation—abound. (If a population arises from another which cannot interbreed anymore with its parent population, it is generally defined as a new species.)

Creation magazine recently reported how evolutionists described as 'alarming' the rate of change in the wingspan of European fruit flies introduced accidentally to America. Similarly, rapid changes have been reported recently for *Drosophila* fruit flies and sockeye salmon—within just nine and thirteen generations respectively.

In the case of Darwin's famous finches, it had been estimated that from one million to five million years would have been necessary for today's Galapagos Island species to radiate from their parent populations. But actual observations of rapid finch adaptation have forced evolutionists to scale that back to a timeframe of just a few centuries.

MOSQUITOES AND MICE

Not long ago, evolutionists were astonished to find that bird-biting mosquitoes, which moved into the London Underground train network (and are now biting humans and rats instead), have already become a separate species. And now a study of house mice in Madeira (thought to have been introduced to the island following 15th century Portuguese settlement) has found that 'several reproductively isolated chromosomal races' (in effect, new 'species') have appeared in less than 500 years.

In all of these instances, the speedy changes have nothing to do with the production of any new genes by mutation (the imagined mechanism of molecules-to-man evolution), but result mostly from selection of genes that already exist. Here we have real, observed evidence that (downhill) adaptive formation of new forms and species from the one created kind can take place rapidly. It doesn't need millions of years.

Shouldn't evolutionists rejoice, and creationists despair, at all this observed change? Hardly. Informed creationists have long stressed that natural selection can easily cause major variation in short time periods, by acting on the created genetic information already present. But this does not support the idea of evolution in the molecules-to-man sense, because no new information has been added.

Fig.24.22: Anole lizards have been observed to change rapidly under the right conditions ... observable evidence in favor of biblical natural history. Wikimedia commons: Paul Hirst

Selection by itself gets rid of information, and of all observed mutations which have some effect on survival or function, so far even the rare 'beneficial' ones are also losses of information, Figure 24.22. The late-maturing, larger guppies resulted simply from a re-shuffling of existing genetic material. Such variation can even be sufficient to prevent two groups from interbreeding with each other anymore, thus forming new 'species' by definition, without involving any new information, Figure 24.

THE BIBLICAL ACCOUNT

The biblical account of history not only accommodates such rapid changes in body form, but actually *requires* that it would have happened much faster than evolutionists would expect. As the animals left the Ark, multiplying to fill the Earth and all those empty ecological niches, natural selection could easily have caused an original 'dog kind' (e.g.,) on the Ark to 'split' into wolves, coyotes, dingoes, etc. Because there are historical records showing some of these subtypes in existence only a few hundred years after the Flood, this means that there had to have been some *very rapid* (non-evolutionary) speciation. Accordingly, it is encouragingly supportive of biblical history when some such rapid changes are seen still occurring today. And this is being repeatedly confirmed.

But since evolutionists mistakenly interpret all such adaptation/speciation as 'evolution happening,' they are left stunned when it happens *much* faster than their traditional interpretations of the fossil record would allow. (This is, of course, easy to understand when it is realized that the standard idea about the fossil record—that it is a 'tape-recording' of millions of years—is in fact a misinterpretation. The record reflects the way in which a global Flood and some of its after-effects buried a world of plants and animals, in a time sequence which did not involve millions of years.)

Because such fast changes challenge traditional evolutionary ideas, the findings are often disputed, but with little success. Rapid 'evolution' (a misnomer, as we have seen) is welcomed by some fossil experts who support the idea of 'punctuated equilibrium'. This is the notion that the evolutionary history of life is one of mostly no change, 'punctuated' by short, sharp bursts of evolution (which, conveniently, happen too briefly to be recorded in the fossils). However, not only is this still a minority view among evolutionists, it begs the question of why, if fast change is everywhere, has not a *vastly greater* number of new species been generated over 'geologic time'? I.e., the observed changes are still too fast for comfort.

Not only is this rapid change not adding information, even some evolutionists point out that evolution in the molecules-to-man sense was not observed in any of these studies. The finches are still finches, the mosquitoes stay mosquitoes, and the mice remain mice. One evolutionary geneticist, referring to the guppy data, said, 'As far as I know, these are still guppies.'

Setting the Record Straight

If we start with the Word of the Omniscient, the One who knows all, the evidence of today's world makes a great deal of sense. Creatures were to reproduce 'after their kind', so mice come from mice, lizards from lizards, daisies from daisies. Evolution has never occurred, nor does it occur today. But organisms have a wonderful 'built-in' genetic capacity for rapid change in response to environmental pressures—most easily observed today in isolated island environments.

Such examples of rapid adaptation give us an insight into how the Earth's many vacant ecological niches were recolonized after the Flood—a global event in real history. This event buried the 'world that then was' (*2 Peter 3:6*). Because this was already a fallen world, the fossils record death, suffering and disease. Because it was a created world, the fossil record consists of the remains of some creatures that no longer exist, and some that still do, but no sequence of one type changing by stages into a totally different type, whether slowly *or* quickly.

APPENDIX

i. Evolutionists have invented a unit called the 'darwin' for measuring the speed of change in the form (body size, leg length, etc.) of a species. In the case of the *Anolis sagrei* lizards, the rate of change ranged up to 2,117 darwins—whereas evolutionists had only 'measured' rates of 0.1 to 1.0 darwins over the 'millions of years in the fossil record'. For the guppies in Trinidad, the rates were even higher: from 3,700 to 45,000 darwins. Artificial selection experiments on laboratory mice show rates of up to 200,000 darwins. (Ref. 2 and Ref. 4.).

ii. Many mutations are 'neutral' in this sense, i.e. just meaningless gobbledygook, and are often described as 'transparent to selection pressures'. For instance, the non-critical portion of the amino acid chain of some proteins can vary without apparently affecting the way the protein works.

iii. The major variation being observed seems to be from small, isolated populations starting with a mere subset of the total genetic information (thus variability). Natural selection acts on this, as does genetic drift, i.e., the statistical tendency for some forms of genes to be lost within small populations by pure chance.

iv. The 'long-age creation' ministry of Dr Hugh Ross, *Reasons to Believe*, not long ago devoted a radio program to their apparent denial of observed speciation, repeatedly accusing CMI of believing in 'evolution'. They completely failed to understand the point we have been making for years about *information*. Evolution is allegedly capable of having 'created' a huge amount of new information by natural processes. Thus, one may have a 'new species' by some man-made definition, but if nothing new has been added to the pool of biological information (but rather that information has been decreased), it boggles the mind to see how that could support 'real' evolution. The real issue is of course that rapid speciation further highlights the non-necessity of their local-flood compromise of the Genesis record.

v. Stephen Jay Gould was probably the best-known exponent of this view. However, the majority of evolutionists hold to the classical neo-Darwinian view of 'mostly slow-and-gradual' evolution. A leading member of this camp reportedly said that Gould 'should not be publicly criticized because he is at least on our side against the creationists'; see *Creation* **21**(4):9, 1999.

vi. This philosophy or worldview, promoted under various names (philosophical materialism, atheism or secular humanism), says that nature (or matter) is all there is and everything can and must be explained by time plus chance plus the laws of nature working on matter. This worldview includes not only the way the world operates, but how it came into being. These materialists either believe that matter is eternal (and merely changes form) or that the initial simple matter somehow came into existence by chance.

vii. For example, Phillip Johnson recently wrote, 'To avoid endless confusion and distraction and to keep attention focused on the most important point, I have firmly put aside all questions of Biblical interpretation and religious authority, in order to concentrate my energies on one theme. My theme is that, in Fr. Seraphim's words, "evolution is not 'scientific fact' at all, but philosophy." The philosophy in question is naturalism.' See his introduction to Fr. Seraphim Rose, *Genesis, Creation and Early Man*, St. Herman of Alaska Brotherhood, Platina, California, 50, 2000.

viii. A fully documented analysis of the scriptural geologists and their opposition to old-earth geology may be found in my Ph.D. thesis: T.J. Mortenson, British scriptural geologists in the first half of the nineteenth century, Coventry University, Coventry, U.K., 1996. This is available from the British Library Thesis Service, <www.bl.uk/services/document/brittheses.html>, either on microfilm for loan or on paper for purchase. New Leaf Press published a revised version in Spring 2004 under the title, *The Great Turning Point: The Church's Catastrophic Mistake on Geology—Before Darwin*

ix. It was the Scottish editor and publisher of Cuvier's English editions, Robert Jameson, who made the clear connection between Cuvier's last catastrophe and Noah's Flood, no doubt to make it more compatible with British thinking at the time. The Oxford geologist, William Buckland, made this idea even more popular. See Martin Rudwick, *The Meaning of Fossils*, University of Chicago Press, Chicago, pp. 133–35, 1985.

REFERENCES

1. Reznick, D.N., Shaw, F.H., Rodd, F.H. and Shaw, R.G., Evaluation of the rate of evolution in natural populations of guppies (*Poecilia reticulata*), *Science* 275(5308):1934–1937, 1997.
2. Morell, V., Predator-free guppies take an evolutionary leap forward, *Science* 275(5308):1880, 1997.
3. Losos, J.B., Warheit, K.I. and Schoener, T.W., Adaptive differentiation following experimental island colonization in *Anolis* lizards, *Nature* 387(6628):70–73, 1997.
4. Case, T.J., Natural selection out on a limb, *Nature* 387(6628):15–16, 1997.
5. Morell, V., Catching lizards in the act of adapting, *Science* 276(5313):682–683, 1997.
6. In the insect world, too, the loss of genetic information for flight can be advantageous, as shown by the success of wingless beetles on windswept islands. See Wieland, C., Beetle bloopers, *Creation* 19(3):30, 1997.
7. Fruit flies spread wings, *Creation* 22(4):5, 2000.
8. Walker, M., Flying out of control—alien species can evolve at an alarming rate, *New Scientist* 165(2222):15, 2000.
9. Huey, R.B. *et al.*, Rapid evolution of a geographic cline in size in an introduced fly, *Science* 287(5451):308–309, 2000.
10. Marchant, J., Darwin strikes back—one modern idea about evolution turns out to be wrong, *New Scientist* 168(2262):11, 2000.
11. Wieland, C., Darwin's finches: evidence supporting rapid post-Flood 'adaptation', *Creation* 14(3):22–23, 1992.
12. Britton-Davidian, J. *et al.*, Rapid chromosomal evolution in island mice, *Nature* 403(6766):158, 2000.
13. Galileo, *Letter to the Grand Duchess Christina* (1615), translated and reprinted in Stillman Drake, *Discoveries and Opinions of Galileo*, Doubleday, New York, p. 186, 1957, reprinted in D. C. Goodman, ed., *Science and*

Religious Belief 1600-1900: A Selection of Primary Sources, The Open University Press, Milton Keynes, U.K., p. 34, 1973.

14. Much has been written about this complex Galileo affair. Helpful analyses can be found in Thomas Schirrmacher, The Galileo Affair: history or heroic hagiography? *TJ* 14(1):91–100, 2000, and in William R. Shea, Galileo and the Church, in *God and Nature*, eds. David C. Lindberg and Ronald L. Numbers, University of California Press, Berkeley, California, pp. 114–35, 1986.
15. Francis Bacon, *The Works of Francis Bacon*, London, 2:480–88, 1819.
16. Francis Bacon, translated by Andrew Johnson from the 1620 original *Novum Organum,* London, p. 43, 1859 (Book I, part LXV). See also Francis Bacon, *Advancement of Learning,* Oxford, p. 46, 1906 (Book I, part VI.16).
17. Georges Comte de Buffon, *Les Ã©poques de la nature,* Paris, 1778. According to de Buffon's unpublished manuscript, he actually believed that the sedimentary rocks probably took at least three million years to form. But Buffon's fear of contemporary reaction to this great date led him to put 75,000 years in the published book. See 'Buffon, Georges-Louis LeClerc, Comte de,' in Charles C. Gillispie, ed., *Dictionary of Scientific Biography* [hereafter *DSB*], 16 vols, Scribner's, New York, p. 579, 1970-1990.
18. Pierre Laplace, *Exposition du systÃ¨me du monde*, 2 vols., Cercle Social, Paris, 1796.
19. John H. Brooke, *Science and Religion*, Cambridge University Press, Cambridge, p. 243, 1991.
20. Leroy E. Page, Diluvialism and Its Critics in Great Britain in the Early Nineteenth Century, in *Toward a History of Geology*, ed. Cecil J. Schneer, MIT, Cambridge, Mass., p. 257, 1969.
21. Dennis R. Dean, James Hutton on Religion and Geology: the Unpublished Preface to His *Theory of the Earth* (1788), *Annals of Science* 32:187–93, 1975.
22. William Smith, *Strata Identified by Organized Fossils*, London, 1816, and *Stratigraphical System of Organized Fossils,* London, 1817.
23. Smith's own writings suggest this, as do comments by geologist John Phillips, Smith's nephew and geology student. See John Phillips, *Memoirs of William Smith*, London, p. 25, 1844.
24. Georges Cuvier, *Theory of the Earth*, Blackwood, Edinburgh, 1813. This was the first English translation of the French original, 'Discours PrÃ©liminaire' in *Recherches sur les ossemens fossils de quadrupÃ¨des*, Paris, 1812.
25. Colin A. Russell, *Cross-currents: Interactions Between Science & Faith*, InterVarsity, Leicester, p. 136, 1985.
26. William Hanna, *Memoirs of the Life and Writings of Thomas Chalmers*, Edinburgh, pp. 1.80–81, 1849-52; Thomas Chalmers, Remarks on Curvier' *Theory of the Earth*, in *The Christian Instructor,* 1814, reprinted in *The Works of Thomas Chalmers*, Glasgow 12:347–72, 1836–42.
27. George S. Faber, *Treatise on the Genius and Object of the Patriarchal, the Levitical, and the Christian Dispensations,* London, 1:chap. 3, 1823.
28. Hugh Miller, *The Two Records: Mosaic and the Geological*, London, 1854, and *Testimony of the Rocks* (reprint of 1957 ed.), W.P. Nimmo, Hay & Mitchell, Edinburgh, pp. 107–74, 1897.
29. John Fleming, The Geological Deluge as Interpreted by Baron Cuvier and Buckland Inconsistent with Moses and Nature, *Edinburgh Philosophical Journal* 14(1826):205–39.
30. John Pye Smith, *Relation between the Holy Scriptures and some parts of Geological Science*, Jackson & Walford, London, 1839.
31. See the detailed analysis of commentaries before and during this period in my thesis (footnote 7 above) 53–67, also at British scriptural geologists in the first half of the nineteenth century—part 1: Historical Setting

32. Henning G. Reventlow, *The Authority of the Bible and the Rise of the Modern World*, trans. John Bowden, SCM, London, p. 412, 1984.
33. James A. Secord, *Controversy in Victorian Geology: The Cambrian-Silurian Dispute*, Princeton Univ. Press, Princeton, N.J., p. 6, 1986.
34. Colin A. Russell, The Conflict Metaphor and Its Social Origins, *Science and Christian Belief* **1**(1):25, 1989.
35. Martin J.S. Rudwick, *The Great Devonian Controversy: The Shaping of Scientific Knowledge among Gentlemanly Specialists, University of Chicago Press,* Chicago, pp. 431–32 1985.
36. Charles Lyell, Review of Scrope's *Memoir on the Geology of Central France, Quarterly Review* 36(72):480, 1827.
37. Thomas S. Kuhn, *The Structure of Scientific Revolutions*, University of Chicago Press, Chicago, p. 76, 1970.
38. Henry Cole, *Popular Geology, J. Hatchard,* London, p. 31, 1834; George Young, *Scriptural Geology*, Simpkin, Marshall and Co., London, p. 74, 1838.
39. Quoted in John H. Brooke, 'The Natural Theology of the Geologists: Some Theological Strata,' *Images of the Earth*, eds. L.J. Jordanova and Roy S. Porter, British Society for the History of Science, Monograph 1, p. 45, 1979.
40. Quoted in Roy Porter, Charles Lyell and the Principles of the History of Geology, *The British Journal for the History of Science* 9(2)32:93, July 1976.
41. James Hutton, 'Theory of the Earth,' *Transactions of the Royal Society of Edinburgh*, 1785, quoted in A. Holmes, *Principles of Physical Geology*, Thomas Nelson and Sons Ltd., U.K., pp. 43–44, 1965.
42. John K. Reed, Demythologizing Uniformitarian History, *Creation Research Society Quarterly* (*CRSQ*) 35(3):156-65, December 1998, and *idem,* Historiography and Natural History, *CRSQ* 37(3):160-75, December 2000.
43. Leading creationist researchers on this subject believe there is unequivocal evidence for only one Ice Age and that it was triggered by climatic, atmospheric, geological and oceanic factors existing at the end of the 371-day Flood at the time of Noah. See for example, Michael Oard, *An Ice Age Caused by the Genesis Flood*, Institute for Creation Research, El Cajon, California, 1990, and Larry Vardiman, *Ice Cores and the Age of the Earth*, Institute for Creation Research, El Cajon, California, 1996. For a less technical treatment, see Don Batten, ed., *The Answers Book*, Master Books, Green Forest, Arkansas, pp. 199–210, 1990.
44. For a recent scholarly comparison of the way early nineteenth-century old-earth and young-earth proponents dealt with this issue of evil in the creation, see Thane Hutcherson Ury, 'The Evolving Face of God as Creator: Earth Nineteenth-Century Traditionalist and Accommodationist Theodical Responses in British Religious Thought to Paleonatural Evil in the Fossil Record,' Ph.D. dissertation, Andrews University, 2001.
45. John M. Robson, 'The Fiat and Finger of God: The Bridgewater Treatises,' in *Victorian Faith in Crisis*, eds. Richard J. Helmstadter and Bernard Lightman, MacMillan, Basingston, U.K., pp. 111–13, 1990.
46. Bertrand Russell, 'Why I Am Not A Christian,' lecture to the National Secular Society, South London Branch, at Battersea Town Hall, 6 March 1927<www.users.drew.edu/~jlenz/whynot.html>.
47. David Hull, The God of the Galápagos, *Nature* 352:485–86, 8 August 1991.
48. Phillip Johnson, 'Afterword: How to Sink a Battleship,' in *Mere Creation: Science, Faith and Intelligent Design*, ed. William Dembski, InterVarsity, Downers Grove, Illinois, pp. 448–49, 1998.

49. Paul Nelson and Mark John Reynolds, 'Young-Earth Creationism: Conclusion,' in *Three Views of Creation and Evolution*, eds. J. P. Moreland and John Mark Reynolds, Zondervan, Grand Rapids, p. 100, 1999.
50. See Phillip Johnson, *The Wedge of Truth: Splitting the Foundations of Naturalism*, InterVarsity, Downers Grove, Illinois, p. 151, 2000.
51. P. Hastie, 'Designer genes: Phillip E. Johnson talks to Peter Hastie,' *Australian Presbyterian* 531:4–8, October 2001; see Johanson's reply to 'In your opinion, what are the secondary issues in the creation-evolution debate?' <members.iinet.net.au/~sejones/pjaustpr.html>.
52. Norman Geisler, Beware of Philosophy: A Warning to Biblical Scholars, *JETS* 42(1):3–19, March 1999.
53. Norman L. Geisler, *Encyclopedia of Christian Apologetics*, Baker, Grand Rapids, p. 272, 1999.
54. D.G.A. Whitten and J.R.V. Brooks, *The Penguin Dictionary of Geology*, Penguin Books, London, p. 74, 1972. In a classic example of evolutionary inconsistency, this same dictionary's definitional entry for *uniformitarianism* contradicts what it says about uniformitarianism in this definition of *catastrophism*!
55. Derek Ager, *The Nature of the Stratigraphical Record*, Macmillan, London, pp. 46–47, 1981.
56. A creationist critique of homology, *Creation Research Society Quarterly* 19(3):166–75, 1982 and 20(2):122, 1983.
57. *Developmental Studies and Speciation in Cichlid Fish*. Ph.D. thesis, Department of Zoology and Comparative Physiology, Birmingham University, United Kingdom, 1972, Diss S2 B72.
58. The genetic integrity of the "kinds" (baramins)—a working hypothesis, *Creation Research Society Quarterly* 19(1):13–18, 1982.
59. D.M.S. Watson, Adaptation, *Nature* 124:233, 1929.
60. Boyce Rensberger, *How the World Works* (NY: William Morrow 1986), p. 17–18.
61. Richard Lewontin, Billions and Billions of Demons, *The New York Review*, 9 January 1997, p. 31.
62. C. Wieland, Science: the rules of the game, *Creation* 11(1):47–50, December 1988–February 1989, creation.com/rules.
63. R.E. Dickerson, *J. Molecular Evolution* 34:277, 1992; *Perspectives on Science and the Christian Faith* **44**:137–138, 1992.
64. D. Batten, A Who's Who of evolutionists, *Creation* 20(1):32, December 1997–February 1998, How Religiously Neutral Are the Anti-Creationist Organizations?, creation.com/who.
65. C. Wieland, Darwin's Real Message: Have You Missed It? *Creation* 14(4):16–19, September–November 1992, creation.com/ realmessage.
66. R. Dawkins, *The Blind Watchmaker: Why the Evidence of Evolution Reveals a Universe without Design*, (NY: W.W. Norton, 1986), p. 6.
67. J. Dunphy, A Religion for a New Age, *The Humanist*, Jan.–Feb. 1983, 23, 26 (emphases added), cited by Wendell R. Bird, *Origin of the Species Revisited*, vol. 2, p. 257.
68. Symposium titled The New Anti-Evolutionism (during the 1993 annual meeting of the American Association for the Advancement of Science). See C. Wieland, The Religious Nature of Evolution, *Journal of Creation* 8(1):3–4.
69. W. Gitt, W. Gitt, *Did God Use Evolution?* (Bielefeld, Germany: CLV, 1993); Theistic evolution questions creation.com/theistic.
70. For refutations of Dawkins' books, see: J.D. Sarfati, Review of *Climbing Mt Improbable*, *Journal of Creation* 12(1):29–34, 1998, creation.com/dawkins; J.D. Sarfati, Misotheist's Misology: Dawkins attacks Behe but digs himself into logical potholes, creation.com/dawkbehe, 13 July 2007; P. Bell, Review of The God Delusion, creation.com/delusion, *Journal of Creation* 21(2):28–34, 2007.

71. E.J. Larson and L. Witham, Leading Scientists Still Reject God, *Nature* 394(6691):313, 23 July 1998. The sole criterion for being classified as a 'leading' or 'greater' scientist was membership of the NAS.
72. S. Jaki, *Science and Creation* (Edinburgh and London: Scottish Academic Press, 1974).
73. L. Eiseley: *Darwin's Century: Evolution and the Men who Discovered It* (Anchor, NY: Doubleday, 1961). Return to text.
74. A. Lamont, *21 Great Scientists Who Believed the Bible* (Australia: Creation Science Foundation, 1995), p. 120–131; H.M. Morris, *Men of Science Men of God* (Green Forest, AR: Master Books, 1982).
75. J. Mattson and Merrill Simon, *The Pioneers of NMR in Magnetic Resonance in Medicine: The Story of MRI* (Jericho, NY: Bar-Ilan University Press, 1996), chapter 8. See also J.D. Sarfati, Dr Damadian's vital contribution to MRI: Nobel prize controversy returns, 21–22 October 2006, creation.com/damadian.
76. Standing Firm [Interview of Raymond Jones with Don Batten and Carl Wieland], *Creation* **21**(1):20–22, December 1998–February 1999.
77. Prize-winning Professor Rejects Evolution: Brian Stone Speaks to Don Batten and Carl Wieland, *Creation* **20**(4):52–53, September–November 1998.
78. Sidney W. Fox, *The Emergence of Life: Darwinian Evolution from the Inside* (NY: Basic Books, 1988), p. 46. Fox is a leading chemical evolutionist who believes life evolved from 'proteinoid microspheres.'
79. C.S. Lewis, *God in the Dock* (Grand Rapids, MI: Wm. B. Eerdmans Publishing Co., 1970), p. 52–53.
80. J.D. Sarfati, If God Created the Universe, Then Who Created God? *Journal of Creation* 12(1)20–22, 1998.

THE ENTITY OF THE CREATOR

THE ATTRIBUTES OF GOD

What are God's attributes?

When we contemplate about the attributes of God, we are attempting to find answers to vital questions regarding who is God, what is God like, and what kind of God is he?

An attribute of God is a vast true entity about himself. We as human beings, no matter how great intellect we may possess, we certainly have limited intellectual capacity, that makes it impossible to fully comprehend "Who God Is!"

God is making himself known in a variety of ways, and through what He permits revealing through His Word and His creation. Only then we can begin to wrap our minds around our awesome Creator and God.

God is unlike anything or anyone we could ever know or imagine. He is the absolute, one of a kind, infinitely unique and without comparison. Even describing Him with mere human words falls short of capturing who He really is. Our limited human minds and words are incapable to hardly describe His limitless attributes, in particularly His incomprehensible Holiness, and infinite Wisdom.

Nonetheless, we humbly attempt to embody His attributes in the very depth of our hearts and the upper most of our minds to adequately perceive Him. God possess attributes that no created mind on earth or even in heaven could ever fully fathom them. It is like having a glimpse of a single golden ray of light emanating from the overwhelming sun.

Nevertheless, He gave us His eternal Word as a means to understand Him. Thus, the compiled attributes below are considered "incommunicable," as qualities possessed by Him alone. While other attributes are also considered "communicable," as qualities encompassing both God and man--- though only He possess them perfectly.

I. GOD IS INFINITE – HE IS SELF-EXISTING, WITHOUT ORIGIN

"And he is before all things, and in him all things hold together." - Colossians 1:17, *He is before all things, and in him all things hold together,* Figure 25.1. "Great is our Lord, and abundant in power; his understanding is beyond measure" – Psalm 147:5, *Great is our Lord, and of great power: his understanding is infinite.* The fact that God is self-existent -- that he was created by nothing and has always existed forever -- is perhaps one of the hardest attributes of God for the believer to understand. In our limitedness, grasping the nature of our limit-less God is like holding onto water as it rages down a river. Indeed, this is the most confusing, head-spinning attribute of God's infinity:

Fig.25.1: Infinity of God – Curtsy Ketubah Store Infinity of GoD

'To admit that there is One who lies beyond us, who exists outside of all our categories, who will not be dismissed with a name, who will not appear before the bar of our reason, nor submit to our curious inquiries: this requires a great deal of humility, more than most of us possess, so we save face by thinking God down to our level, or at least down to where we can manage Him."

Dr. Adrian Rogers writes about the self-existence of God: "The name Jehovah is used some 6,800 times in the Bible. It is the personal covenant name of Israel's God. In the King James Version of the Bible, it's translated Lord God. Not only does it speak of God's strength, but also it speaks of the sovereignty of God and the goodness of God. The root of this name means "self-existing," one who never came into being, and one who always will be. When Moses asked God, "Who shall I tell Pharaoh has sent me?" God said, "**I AM THAT I AM**." Jehovah or Yahweh (YHWH) is the most intensely sacred name to Jewish scribes and many will not even pronounce the name. When possible, they use another name."

II. GOD IS IMMUTABLE – HE NEVER CHANGES

"I the Lord do not change. So, you, the descendants of Jacob, are not destroyed." Malachi 3:6. *"I the LORD do not change. So, you, the descendants of Jacob, are not destroyed.*

God does not change. Who he is never changes, Figure 25.2. His attributes are the same from before the beginning of time into eternity. His character never changes – he never gets "better" or "worse." His plans do not change. His promises do not change.

This ought to be a source of incredible joy for believers. Sam Storms writes this about the good news of God's unchanging nature: "What all this means, very simply, is that God is dependable! Our trust in him is therefore a confident trust, for we know that he will not, indeed cannot, change, Figure 25.2. His purposes are unfailing, his promises unassailable. It is because the God who promised us eternal life is immutable that we may rest assured that nothing, not trouble or hardship or persecution or famine or nakedness or danger or sword shall separate us from the love of Christ. It is because Jesus Christ is the same yesterday, today, and forever that neither angels nor demons, neither the present nor the future, not even powers, height, depth, nor anything else in all creation, will be able to separate us from the love of God that is in Christ Jesus our Lord (Rom. 8:35-39)!" *Who shall separate us from the love of Christ? Shall trouble or hardship or persecution or famine or nakedness or danger or sword? As it is written: "For your sake we face death all day long; we are considered as sheep to be slaughtered." No, in all these things we are more than conquerors through him who loved us.*

Fig.25.2: The Immutable God – Curtsy **Carey Helmink**

For I am convinced that neither death nor life, neither angels nor demons, neither the present nor the future, nor any powers, neither height nor depth, nor anything else in all creation, will be able to separate us from the love of God that is in Christ Jesus our Lord.

III. GOD IS SELF-SUFFICIENT – HE HAS NO NEEDS

"For as the Father has life in himself, so he has granted the Son also to have life in himself." – John 5:26, *For as the Father has life in himself, so he has granted the Son also to have life in himself.*

As limited humans, we have incredible needs, which left unfulfilled, result in death. God, however, has never once been in need of anything. As Tim Temple writes, "God is perfectly complete within his own being, Figure 25.3." Scott Swain writes the self-sufficiency of God means he "possesses infinite riches of being, wisdom, goodness, and power in and of himself (Gen 17:1; John 5:26; Eph 3:16), *For as the Father has life in himself, so he has granted the Son also to have life in himself.* Because he possesses these unfathomable riches in the perfect knowledge and love of the Father, Son, and Holy Spirit (Matt 11.25-27; John 17:24-26), God is the "blessed"\" or "happy" God (1 Tim 1.11; 6:15)."

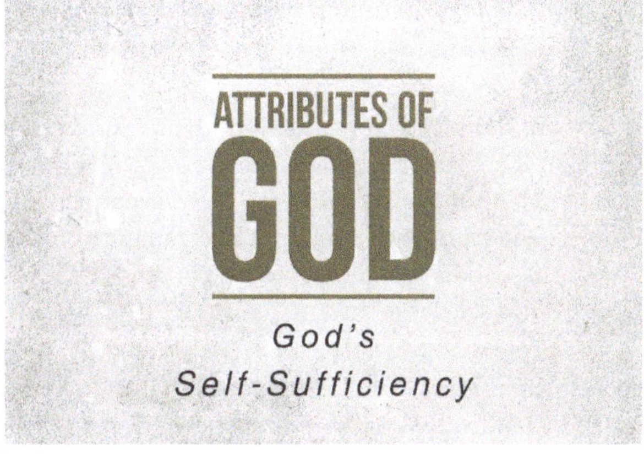

Fig.25.3: God is Self Sufficient

Because God is self-sufficient, we can go to him to satisfy all our needs. We never have to worry about "drying up" his never-ending well of goodness, peace, mercy and grace. "Now to him who is able to do immeasurably more than all we ask or imagine, according to his power that is at work within us…" (Ephesians 3:20), *Now to him who is able to do immeasurably more than all we ask or imagine, according to his power that is at work within us.*

IV. GOD IS OMNIPOTENT – HE IS ALL POWERFUL

"By the word of the LORD (YHWH) the heavens were made, their starry host by the breath of his mouth." – Psalm 33:6, *By the word of the LORD the heavens were made, their starry host by the breath of his mouth*

Fig.25.4: God is Omnipotent

"Can you fathom the mysteries of God? Can you probe the limits of the Almighty? They are higher than the heavens above—what can you do? They are deeper than the depths below—what can you know? Their measure is longer than the earth and wider than the sea. If he comes along and confines you in prison and convenes a court, who can oppose him? Surely, he recognizes deceivers; and when he sees evil, does he not take note?" – Job 11:7-11, *"Can you fathom the mysteries of God? Can you probe the limits of the Almighty? They are higher than the heavens above—what can you do? They are deeper than the depths below—what can you know?*

Their measure is longer than the earth and wider than the sea. "If he comes along and confines you in prison and convenes a court, who can oppose him? Surely, he recognizes deceivers; and when he sees evil, does he not take note?

Omnipotent means to have unlimited power (omni = all; potent = powerful), Figure 25.4. God is able and powerful to do anything he wills without any effort on his part.

It's important to grasp that anything He wills cannot be contradictory or contrary to his nature. It is impossible for God to lie.

In his devotional Forward, Ron Moore expressed his thoughts: "God's attribute of omnipotence means that God is able to do all that He desires to do. When He plans anything, it will come to be. If He purposes something, it will happen. Nothing can prevent His plan. When His hand is stretched out to do something, no one can turn it back. Omnipotence comes from two Latin words. Omni means "all," and potent means «powerful.» God›s decisions are always in perfect harmony with His character, and He has all the power to do whatever He decides to do."

"Scripture is clear that God is strong and mighty (Psalm 24:8). Nothing is too hard for Him to accomplish (Genesis 18:14; Jeremiah 32:17, 27; Luke 1:37). Often God is called "Almighty," describing Him as the One who possesses all power and authority (2 Corinthians 6:18; Revelation 1:8). In fact, Paul says that God is "able to do immeasurably more than all we ask or imagine" (Ephesians 3:20)."

"Although such power might seem frightful, remember that God is good. He can do anything according to His infinite ability, but will do only those things that are consistent with Himself. That's why He can't lie, tolerate sin, or save impenitent sinners." – John MacArthur

V. GOD IS OMNISCIENT – HE IS ALL-KNOWING

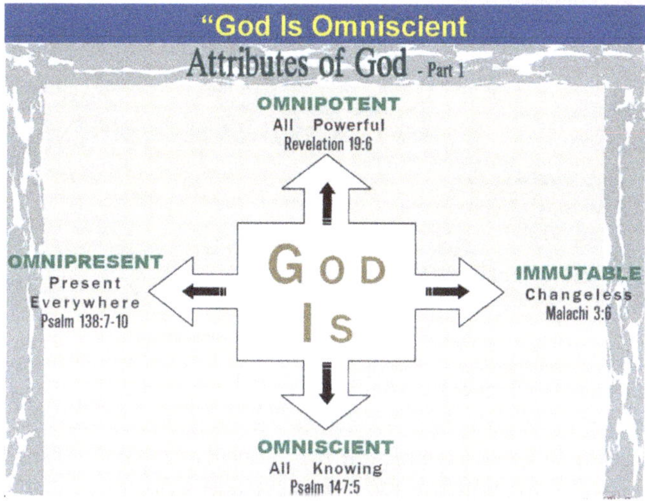

Fig.25.5: God is Omniscient

"Remember the former things, those of long ago; I am God, and there is no other; I am God, and there is none like me. I make known the end from the beginning, from ancient times, what is still to come. I say: My purpose will stand, and I will do all that I please," Figure 25.5 - Isaiah 46:9-10

God is omniscient, which means he knows everything. Debbie McDaniel writes this about the omniscience of God, "He can be everywhere, at the same time. And He never sleeps or slumbers, He's aware every moment of every day, exactly what we're up against. He knows our way, and is with us always. There's no place on this earth we can go that He doesn't see and know of."

Tozer writes this about God's omniscience: "God perfectly knows Himself and, being the source and author of all things, it follows that He knows all that can be known. And this He knows instantly and with a fullness of perfection that includes every possible item of knowledge concerning everything that exists or could have existed anywhere in the universe at any time in the past or that may exist in the centuries or ages yet unborn."

Because God is all-knowing, we can trust that he knows everything we're going through today and everything we will go through tomorrow. When we meditate on this truth, especially in light of his other attributes of goodness and love, it makes it easier to trust him with all we have going on in our lives, from the very serious to the silly and mundane.

VI. GOD IS OMNIPRESENT – HE IS ALWAYS EVERYWHERE

"Where can I go from Your Spirit? Or where can I flee from Your presence? If I ascend to heaven, You are there; If I make my bed in Sheol, behold, You are there. If I take the wings of the dawn, If I dwell in the remotest part of the sea, even there Your hand will lead me, And Your right hand will lay hold of me." Psalm 139:7-10 "'Am I a God at hand,' declares the Lord, 'and not a God afar off? Can a man hide himself in secret places so that I cannot see him?' declares the Lord. 'Do I not fill heaven and earth?' declares the Lord" - Jeremiah 23:23-24.

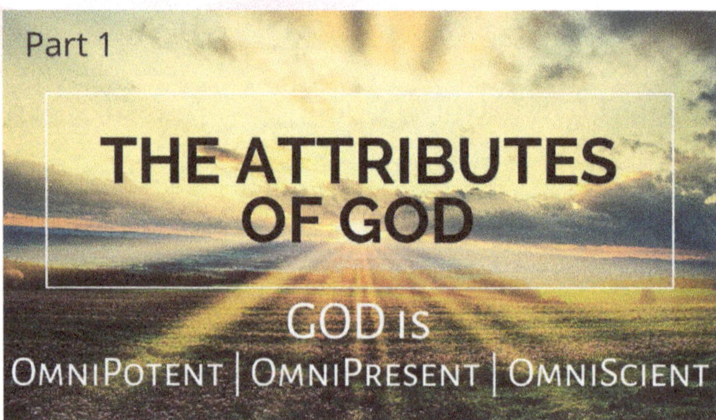

Fig.25.6: God is Omnipresent

To be omnipresent is to be in all places, at all times. Yet, it is important to understand that for God "to be" in a place is not the same way we are in a place. "God's being is altogether different from physical matter." "He exists on a plane wholly distinguishable from the one readily available to the five senses."

Nevertheless, He is with us, the fullness of his presence is all around us, Figure 25.6. "Where shall I go from your Spirit? Or where shall I flee from your presence? If I ascend to heaven, you are there! If I make my bed in Sheol, you are there! If I take the wings of the morning and dwell in the uttermost parts of the sea, even there your hand shall lead me, and your right hand shall hold me." The psalmist proclaims God's omnipresence in Psalm 137.

This ought to bring deep comfort to Christians who struggle with loneliness and deep sorrow. In a very real way, God is always near us, "closer than our thoughts." "The knowledge that we are never alone calms the troubled sea of our lives and speaks peace to our soul."

VII. GOD IS WISE – HE IS FULL OF PERFECT, UNCHANGING WISDOM

"Oh, the depth of the riches both of the wisdom and knowledge of God! How unsearchable are His judgments and unfathomable His ways!" – Romans 11:33 Wisdom is more than just head knowledge and intelligence. A truly wise person is someone who possesses great knowledge and understands all the facts to distinguish insight to make the best decisions. A wise person uses his heart, soul and mind together with skill and competence. But even the wisest man on earth would never come close to being as wise as God.

Fig.25.7: God if Infinitely Wise

God is infinitely wise, consistently wise, perfectly wise, Figure 25.7. "The Wisdom of God, among other things, is the ability to devise perfect ends and to achieve those ends by the most perfect means. He sees the end from the beginning, so there can be no need to guess or conjecture. His Wisdom sees everything in focus, each in proper relation to all, and is thus able to work toward predestined goals with flawless precision."

Indeed, when we see wisdom like this, we realize just how much our limited, finite wisdom compares with the limitless, infinite wisdom of God. And how comforting and wonderful this is for man to dwell on! The fact that God can never be more-wise means He is always doing the wisest thing in our lives. No plan we could make for our lives could be better than the plan He has already crafted and is carrying out for us. We might not understand his ways today, but we can trust that because God is infinitely wise, He truly is working all things out in the best possible way for each individual trust in Him.

VIII. GOD IS FAITHFUL – HE IS INFINITELY, UNCHANGINGLY TRUE

"Know therefore that the LORD your God is God; He is the faithful God, keeping his covenant of love to a thousand generations of those who love Him and keep his commands." - *Deuteronomy 7:9* "[I]f we are faithless, he remains faithful— for he cannot deny himself." 2 Timothy 2:13. As with all of God's attributes, they are not separate, isolated traits but interconnected parts the fact that He never changes. So, when we read that God remains faithful, for He cannot deny himself, we see his attributes working together. The fact that He is unchanging means he can never not be faithful.

Fig.25.8: Cod is Faithful

A. W. Pink writes this about God's faithfulness: "God is true. His Word of Promise is sure. In all His relations with His people God is faithful, Figure 25.8. He is safely relied upon. No one ever yet really trusted Him in vain. We find this precious truth expressed almost everywhere in the Scriptures, for His people need to know that faithfulness is an essential part of His Divine character. This is the basis of our confidence in Him."

The fact that God is infinitely, unchangingly faithful means that He never forgets anything, never fails to do anything He has set out to do, never changes His mind or takes back a promise. And His faithfulness pours out from his love, so we can trust Paul's word that "in all things God works for the good of those who love him."

Of course, we don't always understand or see how His plan is faithful. In our limited understanding and finite minds, God's faithfulness might look a lot like abandonment. For how could a faithful God allow His children to suffer, to hurt, to die?

Thus, Christians can take comfort in these moments by remembering these attributes of God, for when we go through hard times, we know that God is nevertheless unchangingly faithful, good, always with us and wise. Faithfully trusting what God says, is a great comfort to us. "For now, we see in a mirror dimly, but then face to face. Now I know in part; then I shall know fully, even as I have been fully known." 1 Corinthians 13:12.

IX GOD IS GOOD – HE IS INFINITELY, UNCHANGINGLY KIND AND FULL OF GOOD WILL

"O, taste and see that the Lord is good" – Psalm 34:8. Accordingly, the goodness of God "disposes Him to be kind, cordial, benevolent, and full of good will toward men. He is tenderhearted and of quick sympathy, and His unfailing attitude toward all moral beings is open, frank, and friendly. By His nature He is inclined to bestow blessedness and He takes holy pleasure in the happiness of His people."

Fig.25.9: *God is Good*

Just like his other attributes, God's goodness exists within his immutability, and infinite nature, so that he is unchangingly, always good, Figure 25.9. His mercy flows from his goodness. "In his goodness to us, we see that He has purposed to be good in a special way to his people."

As with God's other perfect attributes, Christians find it easier to affirm the goodness of God when things are going well. When life takes a nosedive, though, that's when we begin to question God's goodness to and for us.

When the Psalmist writes "O, taste and see that the Lord is good," (Psalm 34) He is inviting us not just to believe that God is good but to experience God's goodness. And, interestingly, "the psalmist affirms his experience of God's goodness from a place of suffering.

In Psalm 34:19, he makes the remarkable announcement, "Many are the afflictions of the righteous." Even with a good God, who is sovereign over everything and has the power to do whatever he likes, good people still suffer. His punchline, though, comes in the next phrase: "but Yahweh delivers him out of them all." Evil happens, but "none of those who take refuge in him will be condemned" (34:22).

X. GOD IS JUST – HE IS INFINITELY, UNCHANGEABLY RIGHT AND PERFECT IN ALL HE DOES

"The Rock! His work is perfect, For all His ways are just; A God of faithfulness and without injustice, Righteous and upright is He." – Duet 32:4.

Fig.25.10: *God is Just*

What does it mean that God is just? It means more than he is simply fair. It means he always does what is right and good toward all men. Likewise, although this is hard for many to accept, his sentencing of evil, unrepentant sinners to hell is also right and good.

A natural question that arises from this is, how then can a just God justify the unjust? We find the answer through the Christian doctrine of justification and redemption. "Through the work of Christ in atonement, justice is not violated but satisfied when God spares a sinner." His mercy does not forbid him to exercise his justice, nor does his justice forbid him to exercise his mercy. He is both fully merciful and fully just, Figure 25.10.

In light of God's other attributes of goodness, mercy, love and grace, there are some who might, in error, say that God is too kind to punish the ungodly. But to believe this means we dull the reality of his infinite, unchanging justice. God will have justice for sin, either from Christ's atoning death or, for those who will not accept it, eternal wrath in hell.

"Let's assume that all men are guilty of sin in the sight of God. From the mass of humanity, God sovereignly decides to give mercy to some of them. What do the rest get? They get justice. The saved get mercy and the unsaved get justice. Nobody gets injustice"

XI. GOD IS MERCIFUL – HE IS INFINITELY, UNCHANGEABLY COMPASSIONATE AND KIND

Romans 9:15-16 "I will have mercy on whom I have mercy, and I will have compassion on whom I have compassion." So, then it does not depend on the man who wills or the man who runs, but on God who has mercy, Figure 25.11."

Fig.25.11: *God is Merciful*

As stated above, God's mercy is inseparable from his justness. He is infinitely, unchangeably, unfailingly merciful – forgiving, lovingly kind toward us. He is inexhaustibly, actively compassionate. His mercy is also undeserved by us. Spurgeon writes that, "It is undeserved mercy, as indeed all true mercy must be, for deserved mercy is only a misnomer for justice. There was no right on the sinner's part, to the saving mercy of the Highest God. Had the rebel been doomed at once to eternal fire — he would have justly merited the doom; and if delivered from wrath, sovereign love alone has found a cause, for there was none in the sinner himself. "

Without the mercy of God, we would have no hope of heaven. Because of our disobedient hearts, we deserve death. "For all have sinned and fall short glory of God," and, "the wages of sin is death." But because of mercy, we don't receive what we deserve. Instead, because of the mercy of God, we obtain life through faith in Christ.

Regarding the mercy of God. "Judgment is God's justice confronting moral inequity. However, mercy is the goodness of God confronting human suffering and guilt. Were there no guilt in the world, no pain and no tears, God would yet be infinitely merciful; but His mercy might well remain hidden in His heart, unknown to the created universe. No voice would be raised to celebrate the mercy of which none felt the need. It is human misery and sin that call forth the divine mercy."

XII. GOD IS GRACIOUS – GOD IS INFINITELY INCLINED TO SPARE THE GUILTY

"The LORD is gracious and merciful; Slow to anger and great in lovingkindness." – Psalm 145:8. If mercy is not getting what we do deserve (damnation), grace is getting what we don't deserve (eternal life). "As mercy is God's goodness confronting human misery and guilt," Accordingly, "grace" is His goodness directed toward human debt and demerit. It is by his grace that God imputes merit where none previously existed and declares no debt to be where one had been before."

He is GRACIOUS
The Lord is gracious and righteous: our God is compassionate Psalm 1165

Fig.25.12: *God is Gracious*

Because grace is a part of who God is and not just an action He bestows, it means we can trust that grace is eternal. His grace is something we do not earn or lose ("For it is by grace you have been saved, through faith – and this is not from yourselves, it is the gift of God…" Eph. 2:8). His grace is also sovereign. "I will be gracious to whom I will be gracious" (Exodus 33:19).

When talking about the grace of God, theologians will often differentiate between God's common grace and his saving grace, Figure 25.12.

Christianity Today writer Patrick Mabilog writes this about the difference. "His common grace is a gift to all of mankind. It is the reason that everyone – Christian or non-Christian - enjoys the blessings of life, provision and abundance. Matthew 5:45 tells us, 'For he makes his sun rise on the evil and on the good, and sends rain on the just and on the unjust.'"

While all of humanity benefits from **common grace**, only those who profess believe and put their faith in Christ receive **saving grace**. This is what results in our sanctification and our glorification of God, that we might live for him and enjoy him for all eternity.

XIII. GOD IS LOVING – GOD INFINITELY, UNCHANGINGLY LOVES US

"Beloved, let us love one another, for love is from God, and whoever loves has been born of God and knows God. Anyone who does not love does not know God, because God is love." - 1 John 4:7-8. Love, Figure 25.13. The word "staggers before its task of even describing the reality." As with all attributes, we can only begin to comprehend God's love in light of his other attributes. The love of God is eternal, sovereign, unchanging, and infinite.

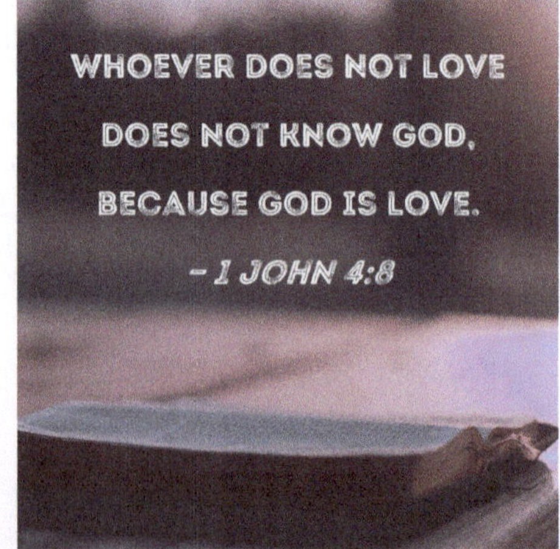

Fig.25.13: *God is Love*

"It is a strange and beautiful eccentricity of the free God," that He has allowed His heart to be emotionally identified with men. Self-sufficient as He is, He wants our love and will not be satisfied till He gets it. Free as He is, He has let His heart be bound to us forever. God's love is active, drawing us to himself. His love is personal. He doesn't love humanity in some vague sense, He loves humans. He loves you and me. And his love for us knows no beginning and no end.

XIV. GOD IS HOLY – HE IS INFINITELY, UNCHANGINGLY PERFECT

"Holy, Holy, Holy, is the Lord Almighty" – Revelation 4:8. The word holy means sacred, set apart, revered, or Devine. And yet none of those words is adequate to describe the awesome holiness of our God, Figure 25.14.

Fig.25.14: *God is Holy*

"Of all the attributes of God, holiness is the one that most uniquely describes Him and in reality is a summation of all His other attributes. The word holiness refers to His separateness, His otherness, the fact that He is unlike any other being. It indicates His complete and infinite perfection. Holiness is the attribute of God that binds all the others together."

That God is Holy means He is endlessly, always perfect. And his standard for us is perfection as well. "Therefore, you are to be perfect, as your Heavenly Father is perfect," Jesus says in Matthew 5:48. That's why we need Christ. Without Christ taking the place for us and dying for our sins, we would all fall short of God's holy standard. This about what God's holiness demands:

"Since God's first concern for His universe is its moral health, that is, its holiness, whatever is contrary to this is necessarily under His eternal displeasure. To preserve His creation God must destroy whatever would destroy it. When He arises to put down iniquity and save the world from irreparable moral collapse, He is said to be angry. Every wrathful judgment in the history of the world has been a holy act of preservation. The holiness of God, the wrath of God, and the health of the creation are inseparably united. God's wrath is His utter intolerance of whatever degrades and destroys."

Thankfully, the Christian will never have to experience God's holy wrath poured out. Through Christ's death and resurrection, the penalty for our sins was paid and we were imputed (credited) with Christ's righteousness. Now, when God looks on us, he sees Christ's perfect holiness. Hallelujah! It is only in this that we can hope to stand in the presence of the blindingly pure, perfect, Holy One of Israel.

XV. GOD IS GLORIOUS – HE IS INFINITELY BEAUTIFUL AND GREAT

"His radiance is like the sunlight; He has rays flashing from His hand and there is the hiding of His power." - Habakkuk 3:4. "The glory of God is the infinite beauty and the greatness of God's manifold perfections. The infinite beauty—Let us focus on the manifestation of His character, His worth and His attributes — all of His perfections and greatness are beautiful as they are seen, and there are many of them. They are presented as "manifold."

Regarding God's Glory: "When we think of the glory of the Lord, the image of brilliant light often comes to our minds. That is certainly appropriate, as the scripture often describes the glory of God in terms of a light that shines brighter than anything that we experience on earth."

Fig.25.15: *God is Glorious*

The glory of God is of course, inseparable from His other attributes, Figure 25.15. Therefore, God is eternally, infinitely, unchangingly glorious. His radiance and beauty emanate from all that his is and all that He does. Isaiah 43:7 says that man was created by God for his glory. So, our whole existence and purpose is to glorify Him, as we are created in his image and do the good work He has prepared for us to do. Inevitably, man will try to find glory in other things, or to try and make himself an object of glory. And when those things fail to bring us satisfaction, we must decide to humble ourselves and turn our gaze back to the only one who is undoubtedly worthy of all glory.

THE DEVINE GOODNESS OF GOD

When considering the divine goodness of God, we find His goodness is one of the most frequently mentioned divine attributes in Scripture, but it is also one that is all too often misunderstood. Therefore, it is vital for us to remember that divine goodness, like God's other attributes, cannot be considered in isolation. God will never exercise His goodness in any way that would cause Him to set aside another of His attributes, Figure 25.16.

Fig.25.16: *God is Divine*

As a divine attribute, goodness is first a description of God's essential character. It means that the Lord is not evil, that He does not love sin and, indeed, cannot even be tempted with evil (v. 13). In this way, it is synonymous with some aspects of what we typically call divine holiness, which refers both to God's being set apart from everything else and to His moral character. Divine goodness is also closely connected to divine justice. Goodness abhors evil, so punishing evil is intrinsic to what it means for God to be good and just (*Ex. 34:6–7*). The Lord forbids human judges from perverting justice (*23:2, 6*), and that is not surprising because all His ways are just (*Deut. 32:4*). Consequently, divine wrath in the service of divine justice is one way in which God manifests His goodness to His creation.

Our Creator shows goodness in other ways as well. First, the Lord reveals His goodness in His benevolence to His creation. God's benevolence is the kindness the Lord bestows on all people, and includes such things as His giving rain to both the just and the unjust (*Matt. 5:45b*). God has a specific love only for believers, and by this love He works out all things for the good of His people (*John 1:12–13; Rom. 8:28*). His benevolence, however, is a more generalized display of goodness that is not the love that leads to salvation.

God's special love for His people also manifests His goodness. This love is a holy love, which means that our sins are punished, but they are punished in Christ, who is the propitiation for our sins (*Rom. 3:21–26*). In saving us, God does not set aside His love for what is good and His hatred for what is evil, but He judges us in Christ so as to save us without compromising His justice. In His holy love, God also disciplines us for our good and His glory (*Heb. 12:5–11*).

Finally, God's mercy flows from His goodness. The Lord would still be good even if He never showed mercy, for mercy is not obligated (*Rom. 9:14–24*). Yet in His mercy to us, we see that He has purposed to be good in a special way to His people.

CORAM DEO – IN THE PRESENCE OF GOD

God's love is not a wishy-washy love that overlooks evil. Even in His love, God manifests His justice. He does not love sinners without dealing with their sin, and if we are in Christ, our sin has been dealt with in our Savior's atonement. Let us proclaim all aspects of God's goodness and call people to repent so that they will receive God's goodness and His mercy.

God's Goodness and Man

O, taste and see that the Lord is good! Blessed is the man who takes refuge in him! (*Psalm 34:8*). The goodness of God is one of those things we affirm happily when things are going well, and find rather tricky when they aren't, Figure 25.17.

Fig.25.17: God's Goodness for Man – The Crucifixion

A friend of mine both of my children are quite severely autistic. Until they were two, they had no real idea anything was wrong, but somewhere around two and a half, they began losing skills rapidly: words, songs, physical skills, eye contact, social awareness and so on.

Not many children have autism, and of those that do, not many have this particular type ("regressive autism"). The parents have had the devastating experience of seeing autism in their two children, a son (age 5) and a daughter (age 3), with all the traumatic upheaval — medical, educational, marital, financial, social, and above all emotional — that comes with it.

It's easy to affirm the goodness of God in the abstract. It's easy to affirm it when things are going well. However, when your children are going backwards on a daily basis, it becomes much harder. When the worship leader goes for a bit of call-and-response — "God is good, all the time, and all the time . . . " — it can by a real physical challenge to squeeze out the words you know come next: " . . . God is good!" Singing becomes a fight between the truths you know and the emotions you feel. Pastoral ministry involves you exhorting others to hold onto things you're struggling to hold onto yourself.

UNDERSTANDING HIS GOODNESS

Over time, the only way through this is theological: you have to wrestle with what it actually *means* for the psalmist (no stranger to suffering himself!) to say that Yahweh is good.

That tiny statement, you see, could mean one of at least three things.

- It could mean that goodness is a *property* of God.
- It could be the description of an *experience* of him. Or
- it could be somehow a *definition* of God.

In my view, in the context of Psalm 34, it is probably all three.

In one sense, to say "God is good" is to give one of his attributes. If I say "milk is white," I am describing what milk is like by saying something which happens to be true about it. It doesn't define it — milk is white, but then so are window frames and teeth and fridges — but it explains that it is white as opposed to brown or pink or blue, and so if you see something colored, it might be a lot of things, but it isn't milk.

Saying God is good, in one sense, is like this. The psalmist is identifying a *property* of God, so we know that (among other things) he is good, which means he does good things and doesn't do bad things or mediocre things or things which don't quite work out.

Now imagine, instead of saying "milk is white," I said, "milk is a whitish liquid containing proteins, fats, lactose, and various vitamins and minerals produced by the mammary glands of all mature female mammals after they have given birth." Suddenly my statement has become larger. This is not just a description, but a *definition*: anywhere you find this type of substance, you will, by definition, have milk, and vice versa.

In the same way, God is good by definition; anywhere you find goodness you will, by definition, have God, and vice versa. You can't have God without goodness, and you can't have goodness without God.

This is the teaching of the whole of Scripture. Everything God made was very good *(Genesis 1:31)*. No one is good except God (*Mark 10:18*). All things work together for good for those who love God (*Romans 8:28*). Those who seek Yahweh lack no good thing (*Psalm 34:10*). And so on.

DEEPER THAN THE DEFINITION

Personally, even in the darkest moments, I've had little problem affirming these two senses of the phrase "God is good." I can divorce what I believe from what I feel, and say things like this in a theoretical, abstract sort of way. But the really tough thing about Psalm 34, when you're suffering, is the experiential dimension. O *taste*, and *see*, that Yahweh is good. That can stick in the throat sometimes.

It's like saying "milk is nice." Instead of an objective statement about milk, I'm now saying something subjective, an opinion based on my *experience*. It's something which others can only verify by trying themselves.

This is the clearest meaning of the cry, "***O, taste and see that Yahweh is good***!" The psalmist has got a list of examples of God's goodness to him — like

- deliverance (verse 4),
- provision (verse 10),
- being heard (verse 15), and so on — and he is urging us as readers to experience this goodness ourselves.

Knowing God is good but never experiencing His goodness is as useless as knowing the definition of milk and never drinking it.

Yet the psalmist affirms his experience of God's goodness from a place of suffering. In verse 19, he makes the remarkable announcement, "***Many are the afflictions of the righteous***." Even with a good God, who is sovereign over everything and has the power to do whatever he likes, good people still suffer. His punchline, though, comes in the next phrase: "***but Yahweh delivers him out of them all***." Evil happens, but "none of those who take refuge in him will be condemned" (34:22).

It could not be any other way. God is shown to be good from our *experience*, and he has the *property* of goodness, but he is also good by *definition*. He has never been faced with a catch-22 situation, forced to choose between the lesser of two evils, or flummoxed into a decision that was anything less than completely good. Therefore — and this bit is both the hardest, and the most powerful, when we're suffering — if God has done something, *it is good*.

We may well not understand why God has done it, of course. Job didn't either. But we can be confident, based on Scripture and on our experience, that as sure as milk is white, Yahweh is good. *Taste and see!*

Yahweh (YHWH) God the Creator

Fig.25.18: God's Character and Attributes Based on His Creation

Most people think that the topic of creation is limited to the first couple of chapters of Genesis and perhaps a few isolated texts in the rest of Scripture. A previous article established the importance of the doctrine of Creation in the New Testament, but creation is also important throughout the Old Testament. No less than 52 passages in the Old Testament give us important insights into God's character and attributes based on His creation, Figure 25.18.

Some Christians may feel that studying the Old Testament isn't relevant to us as Christians, because the Old Testament was written to Jews. Nonetheless, while Jews were the primary audience for the Old Testament, all of Scripture is *for* Christians to use.

God is the same now as He was during the time the Old Testament was written, so these passages establish a theology of Yahweh (YHWH), as well as His relationship with His creation and humanity in particular.

Let's take a look at how these attributes of God's character are reflected through His creation. This is only a brief overview of the various ways the Old Testament uses creation, but it has profound implications. Not only did the Old Testament writers believe in the Genesis account of creation, but they used it to clearly teach us many important aspects about God's nature.

Furthermore, recognizing these uses brings out some of the continuity between the Old Testament and the New Testament, because both use Genesis in the same way. Jesus in the New Testament refers to the Bible as the totally authoritative Word of God—e.g.m His repeated use of "it is written" and "have you not read?".

YAHWEH - (YHWH): THE CREATOR

The Genesis account of Creation calls God *Elohim*, a more generic name, but Genesis 2:4 introduces the covenant name, Yahweh. Here, this is combined with Elohim—YHWH-Elohim—to show that the covenant God of Israel is none other than the creator of the universe. Other places in the Old Testament use God's covenant name, Yahweh, Figure 25.19, when they describe His creative work; for instance:

Fig.25.19: YHWH – The Creator

Yahweh who made the earth, Yahweh who formed it to establish it—Yahweh is his name (*Jeremiah 33:2*).

This is a primary identifier of the true God. When Jonah wants to identify himself and his God, he tells the men on the ship, "I am a Hebrew, and I fear Yahweh, the God of heaven, who made the sea and the dry land" (*Jonah 1:9*). And this knowledge was not limited to the Hebrews: Hiram the king of Tyre in Solomon's day recognized that Israel's God was the Creator: "Blessed is Yahweh God of Israel, who made heaven and earth" (*2 Chronicles 2:12*).

When Scripture wants to emphasize the authority of Yahweh's words, the prophets often appeal to God's creative power:

Thus says Yahweh, who gives the sun for light by day and the fixed order of the moon and stars for light by night, who stirs up the sea so that its waves roar—Yahweh of hosts is his name (*Jeremiah 31:35*; see also *Zechariah 12:1*).

As God created the world, He has the authority and the power to make sure that His word is fulfilled—so we should listen to Him!

YAHWEH: THE ONLY GOD

Yahweh is repeatedly asserted to have created the heavens and earth alone, demonstrating that He is God alone:

Thus says Yahweh, your Redeemer, who formed you from the womb: 'I am Yahweh, who made all things, who alone stretched out the heavens, who spread out the earth by myself' (*Isaiah 44:24*, see also *Isaiah 45:18, Job 9:8–9*).

God's creative activity is a major basis for praising Yahweh:

Praise Yahweh! Praise Yahweh from the heavens; Praise Him in the heights ... Let them praise the name of Yahweh, for He commanded and they were created (*Psalm 148:1–5*).

You are Yahweh, you alone. You have made heaven, the heaven of heavens, with all their host, the earth and all that is on it, the seas and all that is in them; and you preserve all of them; and the host of heaven worships you (*Nehemiah 9:6*).

Scripture is clear that God's creative activity, which brought everything into existence, and His sustaining power, without which none of us would continue to exist, are both major reasons to worship God.

YAHWEH: THE TRANSCENDENT GOD

God could not have created heaven and earth if He were contained inside or limited to it. So, it follows that God preceded and is outside of the created heavens and earth, Figure 25.20.

Fig.25.20: God, Infinite, Personal, and Transcendent

Of old you founded the earth, and the heavens are the work of your hands. Even they will perish, but you endure; and all of them will wear out like a garment; like clothing you will change them and they will be changed (*Psalm 102:25–26*).

Even though God had a temple in Jerusalem, it was clear that unlike temples for false gods which were filled with lifeless idols, it was not a place for Israel's God to actually live:

But will God indeed dwell on the earth? Behold heaven and the highest heaven cannot contain you; how much less this house that I have built! (*1 Kings 8:27*; see also *2 Chronicles 2:5–6; 6:18*).

God is also Beyond Human Understanding:

Just as you do not know the path of the wind and how bones are formed in the womb of the pregnant woman, so you do not know the activity of God who makes all things (*Ecclesiastes 11:5*).

Because God is so far above the earth and its inhabitants, it is awe-inspiring that He concerns Himself with human beings:

When I look at your heavens, the work of your fingers,
The moon and the stars, which you have set in place,
what is man that you are mindful of him,
and the son of man that you care for him? (*Psalm 8:3–4*).
Yahweh: the ruler and judge of the earth

It follows logically that since Yahweh created the heavens and earth, they belong to Him:

The earth is Yahweh's and the fullness thereof, the world and those who dwell therein, for he has founded it upon the seas, and established it upon the rivers (*Psalm 24:1–2*; see also *89:11–12; Deuteronomy 10:14*).

When we offer things to God, whether sacrifices in the Old Testament covenant, or time, talents or monetary offerings today, we are only giving back to God what already belongs to Him. David prays:

But who am I, and what is my people, that we should be able thus to offer willingly? For all things come from you, and of your own we have given you (*1 Chronicles 29:14*).

Because God owns the world, He also rules it:

For Yahweh is a great God and a great King above all gods in whose hand are the depths of the earth; the peaks of the mountains are his also. The sea is his, for it was he who made it, and his hands formed the dry land (*Psalm 95:3–5*).

Because Yahweh is our Creator, Scripture indicates that we lack the standing to bring a case even accusing God of being unjust, because we belong to Him, and anything He could ever give us or deprive us of belongs to Him; whether existence or health or material goods, it is His.

Woe to him who strives with him who formed him, a pot among earthen pots! Does the clay say to him who forms it; 'What are you making?' or 'Your work has no handles'? (*Isaiah 45:9)*.

For example, if anyone had a reason to bring a complaint against God, it would seem (humanly speaking) that no one could have a better case than Job. Job faithfully served God, but lost his riches, children, and health in rapid succession. At first, he had a righteous response: "Yahweh gave, and Yahweh has taken away; blessed be the name of Yahweh" (*Job 1:21*). But then when his friends argued that Job must have committed some sin that caused his calamity, he said, "Oh, that I knew where I might find him, that I might come even to his seat! I would lay my case before him and fill my mouth with arguments. I would know what he would answer me and understand what he would say to me. Would he contend with me in the greatness of his power? No; he would pay attention to me. There an upright man could argue with him, and I would be acquitted forever by my judge." (Job 23:2–7).

Job gets his wish, but instead God challenges Job:
Where were you when I laid the foundation of the earth?
Tell me if you have understanding.
Who determined its measurements—surely you know!
Or who stretched the line upon it?
On what were its bases sunk,
Or who laid its cornerstone,
when the morning stars sang together
and all the sons of God shouted for joy? (*Job 38:2–7*)

God's challenge to Job reveals that Job's knowledge and wisdom are so far below his Creator's that Job simply doesn't have the standing or the perspective to even bring an accusation before God. In fact, the book of Job tells us what was going on 'behind the scenes' which caused God to allow this intense, but ultimately temporary, suffering in Job's life.

YAHWEH: THE ALL-POWERFUL GOD

God's creation of the universe is the first and greatest demonstration of his power. As Jeremiah said:

Ah, Lord Yahweh! It is you who have made the heavens and earth by your great power and by your outstretched arm! Nothing is too hard for you (*Jeremiah 32:17*).

He is able to do anything:

Whatever Yahweh pleases, He does, in heaven and earth, in the seas and all deeps (*Psalm 135:6*).

This leads to several conclusions in Scripture.

- First, when He reveals Himself, the enemies of God are terrified. Rahab tells the Israelite spies, "And as soon as we heard it, our hearts melted, and there was no spirit left in any man because of you, for Yahweh your God, he is God in the heavens above and on the earth beneath" (*Joshua 2:11*).
- Second, those who trust in God can utterly rely on Him:

How blessed is he whose help in is in the God of Jacob, whose hope is in Yahweh his God, who made heaven and earth, the sea and all that is in them (*Psalm 146:5–6*, see also *121:2; 124:8*).

In Isaiah, God says,

I, I am he who comforts you;
who are you that you are afraid of man who dies,
of the son of man who is made like grass,
and have forgotten your Maker;
who stretched out the heavens,
and laid the foundations of the earth? (*Isaiah 51:12–13*)

One powerful example is when Sennacherib's armies were threatening Judah. The Rabshakeh (a vizier for the king) sent this message to Hezekiah:

Do not let your God in whom you trust deceive you by promising that Jerusalem will not be given into the hand of the king of Assyria. Behold, you have heard what the kings of Assyria have done to all lands, devoting them to destruction. And shall you be delivered? Have the gods of the nations delivered them, the nations that my fathers destroyed, Gozan, Haran, Rezeph, and the people of Eden who were in Telassar? Where is the king of Hamath, the king of Arpad, the king of the city of Sepharvaim, the king of Hena, or the king of Ivvah? (*2 Kings 19:10–13*).

Hezekiah, however, trusted in the Lord: he prayed:

O Yahweh, the God of Israel, enthroned above the cherubim, you are the God, you alone, of all the kingdoms of the earth; you have made heaven and earth. … Truly, O Yahweh, the kings of Assyria have laid waste the nations and their lands and have cast their gods into the fire, for they were not gods, but the work of men's hands, wood and stone. Therefore, they were destroyed (*2 Kings 19:15, 17–18*, see also *Isaiah 37:16–20*).

The reason Hezekiah trusted that his God, unlike the gods of other nations, would not fail is because Yahweh alone is the Creator.

Yahweh: **Superior to Idols and False gods**

God often claims the right to exclusive worship:

I am Yahweh; that is my name; my glory I give to no other, nor my praise to carved idols (*Isaiah 42:8*), Figure 25.21.

The Old Testament is full of anti-idol polemics, invoking Yahweh's creatorship in contrast to the idols: *1 Chronicles 16:26* says "all the gods of the peoples are idols, but Yahweh made the heavens" (see also *Psalm 96:4–5*). *Isaiah 40–43* has some quite developed arguments against idols:

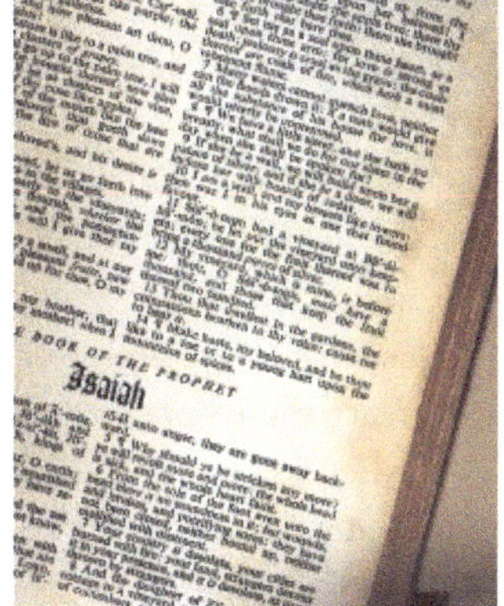

Fig.25.21: The Book of Isiah

To whom then will you liken God,
or what likeness compare with him?
An idol! A craftsman casts it,
and a goldsmith overlays it with gold
and casts for it silver chains.
He who is too impoverished for an offering
chooses wood that will not rot;
he seeks out a skillful craftsman
to set up an idol that will not move.
Do you not know? Do you not hear?
Has it not been told you from the beginning?
Have you not understood from the foundations of the earth?
It is he who sits above the circle of the earth,
and its inhabitants are like grasshoppers;
who stretches out the heavens like a curtain,
and spreads them like a tent to dwell in;
who brings princes to nothing.
and makes the rulers of the earth as emptiness (*40:18–23*).

God also promises the destruction of all idols that did not create, unlike Yahweh:

Thus, you shall say to them: 'The gods who did not make the heavens and earth shall perish from the earth and from under the heavens' (*Jeremiah 10:11*).

Only the Creator deserves to be called God, because everything else is only created and finite.

YAHWEH: OUR WISE AND LOVING CREATOR

The above attributes would be terrifying if we were not also assured of God's love. Scripture indicates that creation is also a demonstration of God's love:

To him who alone does wonders,
for his steadfast love endures forever;
to him who by understanding made the heavens,
for his steadfast love endures forever;
to him who spread out the earth above the waters,
for his steadfast love endures forever;
to him who made the great lights.
for his steadfast love endures forever;
the sun to rule over the day,for his steadfast love endures forever;

the moon and stars to rule over the night,
for his steadfast love endures forever (*Psalm 136:4–9*).

Scripture also indicates creation is a great indication of His wisdom: "It is he who made the earth by his power, who established the world by his wisdom, and by his understanding stretched out the heavens" (*Jeremiah 10:16*). In Proverbs, the divine attribute of wisdom is personified as God's companion at creation:

Yahweh possessed me at the beginning of His way,
Before His works of old.
From everlasting I was established,
From the beginning, from the earliest times of the earth.
When there were no depths I was brought forth,
When there were no springs abounding with water.
Before the mountains were settled,
Before the hills I was brought forth;
While he had not yet made the earth and the fields,
Nor the first dust of the world.
When He established the heavens, I was there,
When He inscribed a circle on the face of the deep,
When He made firm the skies above,
When the springs of the deep became fixed,
When He set for the sea its boundary
So that the water would not transgress His command,
When He marked out the foundations of the earth;
Then I was beside him, as a master workman;
And I was daily His delight,
Rejoicing always before Him.
Rejoicing in the world, His earth.
And having my delight in the sons of men (*Proverbs 8:22–31*).

JESUS: OUR CREATOR AND SAVIOR

All of the above demonstrates that Scripture uses God's creation of the heavens and earth as the basis for asserting His attributes. But of course, we also believe that the entire Old Testament points forward to salvation in Christ. *Psalm 33:6* says:

By the word of Yahweh, the heavens were made, and by the breath of His mouth all their host.

John asserts that this Word is God the Son, who became incarnate as Jesus Christ:

In the beginning was the Word, and the Word was with God, and the Word was God. He was in the beginning with God. All things were made through him, and without him nothing was made that was made (*John 1:1–3*).

Paul concurs:

For by him all things were created, in heaven and on earth, visible and invisible, whether thrones or dominions or rulers or authorities—all things were created through him and for him. And he is before all things, and in him all things hold together (*Colossians 1:16–18*).

Because Jesus is the agent of God's creation, *everything creation says about Yahweh also applies to Jesus.* So, it is important for us to understand the creation basis for asserting Yahweh's attributes, because they're also Jesus' attributes, Figure 25.22.

Fig.25.22: Jesus Washes the Feet of His Disciples

Because creation is so intertwined with how the rest of Scripture sees Yahweh and Jesus, and because their creative work is used as the basis for so many doctrines, it's problematic when various compromising creation theories attempt to reinterpret Genesis to fit with secular ideas about origins that leave no room for a supernatural Creator. Not only is it unnecessary, but it removes the foundation for these statements about the nature of God and His relationship with creation.

JESUS THE CREATOR GOD

The Bible affirms in several places that Jesus Christ is the Creator God, as stated that 'All things were made by him ' (John 1:1,3), In the beginning was the Word, and the Word was with God, and the Word was God. [the Word, in Greek ὁ λógos, = Jesus Christ], and 'For by him [Jesus Christ] were all things created' (Colossians 1:16), For by him all things were created, in heaven and on earth, visible and invisible, whether thrones or dominions or rulers or authorities—all things were created through him and for him.

Because this is true, we should expect to see some parallelism between what happened at creation and the works of Jesus during his ministry on earth.

What do we find?

CONSIDER THE EVIDENCE

Some of the essential and distinctive elements of creation, as revealed in Genesis chapter l, as well as elsewhere in the Bible, are: Creation involved the act of God in bringing into being immediately and instantaneously matter which did not previously exist, without the use of pre-existing materials in the creation of the heavens and the earth, as recorded in Genesis 1:1, In the beginning, God created the heavens and the earth. Creation also involved the shaping, combining, or transforming of existing materials, as when God created Adam from the dust of the ground (Genesis 2:7), and Eve from Adam's rib (Genesis 2:21–22). So, the LORD God caused a deep sleep to fall upon the man, and while he slept took one of his ribs and closed up its place with flesh. And the rib that the LORD God had taken from the man he made into a woman and brought her to the man.

- Creation involved the imparting of life to otherwise lifeless matter.
- The mechanism of creation, or the means whereby the above aspects were accomplished, was by the ***Word*** of the Lord, that is, God said (= God willed it to happen) … and it happened.
- The purpose or motive of God in creating was to display His glory, to make known His power, His wisdom, His will, and His holy name, and that He might receive glory from His created beings.

We should not expect to find exact parallels between the miracles of Jesus and what happened at Creation, as Jesus did not come to re-create the universe, but 'to seek and to save that which was lost,' and 'to give his life a ransom for many.' With this in mind, let us compare these four aspects of creation with the works of Jesus.

CREATION OUT OF NOTHING AND/OR FROM EXISTING MATERIALS

Several of Jesus' miracles involved the aspect of creation. The major aspect of Jesus' creation is focused on the fact of the miracles and the effects they produced (John emphasizes the teaching that Jesus drew from them), rather than on any analyses of the *modus operandi*.

Jesus' first *miraculous sign to His disciples* involved the creation of wine (His first *miracle* recorded in the Gospels is actually the creation of the universe (John 1:3), as mentioned above). At a wedding breakfast, Jesus instructed the waiters to fill six stone water-pots with water, and then to take them to the master of ceremonies of the wedding banquet.

When they arrived, the water had been turned into wine, that is, there had been the instantaneous creation of the carbon atoms and chemical molecules that made up the grape sugar, carbon dioxide, coloring matter, etc., of the wine.

Other examples are the two times when Jesus fed a multitude: on the first occasion more than 5,000 people from five loaves and two fish, and on the second occasion more than 4,000 people from seven loaves and a few little fish.

Here there were bread and fish to begin with on both occasions. Jesus either caused these original items to multiply, or He may have dispensed all the original food and then created new loaves and fishes until everyone was fed. Either way, Jesus created sufficient extra bread and fish, not only to feed many thousands of people, but also to provide 12 basketfuls of leftovers on the first occasion and seven basketfuls of leftovers on the second.

This involved not just the creation of the appropriate carbohydrate, protein and other molecules, but their immediate arrangement into the complex forms and structures needed to make baked bread and fish (albeit dead and cooked).

Some of Jesus' miracles of healing, for example, of lepers, the blind, and paralytics, involved the instant repair of tissues, nerves, muscles, etc., and the instantaneous growth or regrowth of healthy cells. The net result was the creation of healthy functioning parts of the body to replace diseased, non-functioning or atrophied parts.

THE GIVING OF LIFE

Jesus gave life to the dead on three occasions: to a widow's son, Figure 25.23, to Jairus' daughter, and to his friend Lazarus.

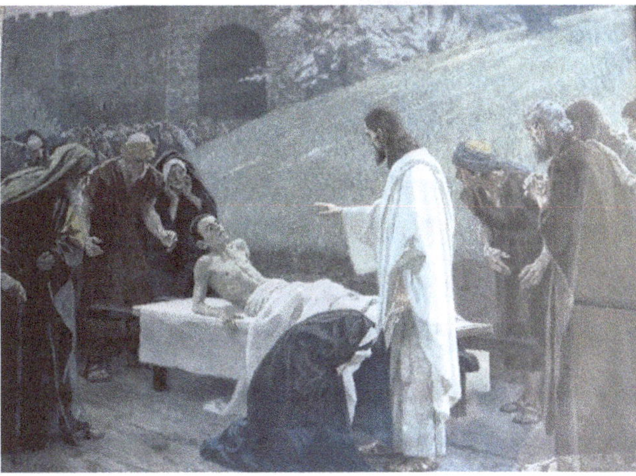

Fig.25.23: Jesus Raised the Widow's Son from the Dead

In the case of Lazarus, the body had been in the grave for four days, and Martha's words are recorded for us: '...by this time there is a bad odor, for he has been there four days.'

This shows that the process of decomposition whereby a dead body eventually becomes dust had already begun. So here we have a parallel with what happened on the sixth day of creation when God formed Adam from the dust of the ground and breathed into his nostrils the breath of life, and Adam became a living being. Jesus called Lazarus back to life, Figure 25.24, and the molecules of matter that were in the process of becoming dust became, again, a living human being.

Fig.25.24: Jesus Called Lazarus Back to Life

In the case of the widow's son and of Jairus' daughter, death was more recent, that is, probably on the same day that Jesus gave life to their dead bodies. The principle still applies.

THE METHODOLOGY JESUS IMPLEMENTED

Jesus appeared to employ a variety of means in performing His miracles. These included touching lepers, the blind, and the deaf; the use of saliva to heal a deaf mute and a blindman; the use of clay (with instructions to wash) to heal a blind man; and the word of command to heal, to raise the dead, and to exorcise demons.

THE WILL TO ACCOMPLISH MIRACLES

However, what happened in these and in all of Jesus' miracles was that Jesus willed the event to happen and it did. This is nowhere better illustrated than in the healing of the nobleman's son. Jesus was at Cana in Galilee and a certain royal official asked Him to travel to Capernaum to heal his son who was close to death. The Apostle John records what happened, as follows:

'So, He came again to Cana in Galilee, where He had made the water wine. And at Capernaum there was an official whose son was ill.'

'When this man heard that Jesus had come from Judea to Galilee, he went to Him and asked Him to come down and heal his son, for he was at the point of death.'

'So, Jesus said to him, "Unless you see signs and wonders you will not believe."'

'The official said to Him, "Sir, come down before my child dies."'

'Jesus said to him, "Go; your son will live." The man believed the word that Jesus spoke to him and went on his way.'

'As he was going down, his servants met him and told him that his son was recovering.'

'So, he asked them the hour when he began to get better, and they said to him, "Yesterday at the seventh hour the fever left him."'

'The father knew that was the hour when Jesus had said to him, "Your son will live." And he himself believed, and all his household.' (*John 4:46–53*).

Capernaum was about 27 kilometers (17 miles) from Cana as the crow flies, which means there was no way that the sick son, or anyone else in Capernaum, could have heard Jesus or been influenced by His physical presence in Cana.

Jesus Willed the sick boy to recover, at a distance of 27 kilometers, and he did so. Similarly, **Jesus Willed** the water to become wine, as it was being taken into the wedding feast in Cana, and it did so. **Jesus Willed** the bread and fish to form and they did, Figure 25.25, and He willed the 10 lepers to become well after they had left Him and were on their way to the priests, and they were healed.

Fig.25.25: "Jesus took the loaves, gave thanks, and distributed them to those who were reclining, and also as much of the fish as they wanted." —John 6: 11

It is interesting that a Gentile centurion recognized this authority of Jesus. The centurion had sent servants to request Jesus to come and heal his servant, as Luke records:

'And Jesus went with them. When He was not far from the house, the centurion sent friends, saying to Him, "Lord, do not trouble yourself, for I am not worthy to have you come under my roof.'

'Therefore, I did not presume to come to you. But say the word, and let my servant be healed.'

'For I too am a man set under authority, with soldiers under me: and I say to one, 'Go,' and he goes; and to another, 'Come,' and he comes; and to my servant, 'Do this,' and he does it."'

'When Jesus heard these things, he marveled at him, and turning to the crowd that followed him, said, "I tell you, not even in Israel have I found such faith."'

'And when those who had been sent returned to the house, they found the servant well.' (*Luke 7:6–10*)

The centurion recognized that the voice of Jesus could not be heard by his sick servant, but the result, brought about by the exercise of Jesus' authority, would be no less effective because of this.

JESUS' GLORY SEEN IN HIS MIRACLES

After narrating Jesus' first miraculous sign to His disciples—the turning of water into wine—the Apostle John says, He 'manifested forth his glory; and his disciples believed on him.'

When Jesus heard that Lazarus was sick He said, 'This sickness is not unto death, but for the glory of God, that the Son of God might be glorified'.

And then, after Lazarus had died and before Jesus raised him to life, He said to Martha, 'Said I not unto thee, that, if thou wouldest believe, thou shouldest see the glory of God?'

John calls Jesus' miracles' signs and in his Gospel John shows which way the signs point: 'these are written, that ye might believe that Jesus is the Christ, the Son of God; and that believing ye might have life through his name.'

Jesus Christ is the Creator God. Not only does Scripture affirm it, but during His earthly life and ministry He did the very things we would expect the Creator God to do. He did them in the way that we would expect the Creator God to do them—by His word of authority and the exercise of His will. And the doing of them displayed His glory. (John 14:9-10) "Anyone who has seen me has seen the Father." "I am in the Father, and that the Father is in me." (John 10:30) "I and the Father are one."

This is a source of praise and inspiration for those who believe the Word of God, and at the same time it is a reproof of the doctrine of theistic evolution. The thought that Jesus might have used evolutionary chance random processes to heal the sick or give life to the dead is as unsustainable as the idea that He used such processes to create and give life to all things 'in the beginning'.

Jesus taught with a unique sense of divine authority. He claimed to have authority to forgive sins. He said that following Him determined one's eternal destiny. He prophesied on His own authority. He claimed to be Lord of the Sabbath (*Mark 2:27*). His favorite title for himself ("the Son of Man") also revealed that He believed Himself to be the heir of God's eternal kingdom (having taken this term from *Daniel 7:13–14*)!

Jesus left a distinct impression that *in His very person* Israel's God was at last returning to Zion to establish His kingdom. And this was the impression He clearly left on His earliest followers, e.g., *James 2:1, John 1:1, Romans 10:9–13, and Mark 6:45–53.*

But why believe such outlandish claims? Jesus said, numerous times, that He would be killed, and then raised from the dead (e.g., *Matthew 12:38–40, Mark 8:31*). This is how Jesus believed God would vindicate him. The resurrection verifies His claims, if true. Jesus was supposedly raised never to die again, which is a miracle only the controller of the cosmos, i.e., God, could pull off.

But are miracles even possible? A miracle is nothing more than a historical event with a supernatural cause. And we have no problem inferring unseen entities like quarks or ancient artisans to explain observational evidence.

Miracles are only problematic if we limit our pool of possible causes to natural causes *before* looking at the evidence. But why so constrain ourselves, especially if a supernatural cause is the best, or even only, explanation of the evidence?

So what evidence is there for Jesus' resurrection?

- First, Jesus died. You can't have a resurrection if the person didn't die! Jesus' death by crucifixion is better attested than pretty much any historical event in antiquity.

It's everywhere in the New Testament and the Apostolic Fathers. And it's mentioned by Josephus, Tacitus, and Pliny the Younger, just to name a few non-Christian sources that mention it.

- Second, we have a church tradition for the resurrection that dates to about three years after Jesus' death: *1 Corinthians 15:3–8*. It testifies of multiple appearances of Jesus to individuals and groups in a variety of settings. Paul wrote it about 25 years after Jesus' death, and said of it: "I delivered to you as of first importance what I also received". He must have received this message from the apostles he equates his message with in verse 11. And he first met these apostles about 3 years after his conversion (*Galatians 1:19–20*), which was about 3 or so years after Jesus' death.
- Third, Jesus' tomb was empty. All the Gospels (explicitly) and 1 Corinthians 15 (implicitly) testify to this. Even the church's enemies presuppose this in their alternative explanation, that the disciples stole Jesus' body *(Matthew 28:11–15)*. Even if Matthew was putting words in his enemies' mouths, it makes no sense to have them admit an empty tomb if the tomb wasn't actually empty.
- Fourth, the first eyewitnesses to both the empty tomb and the resurrection were women, and the church didn't hide this. Women's testimony was worth nothing in the 1st century, so the early Christians had every reason to leave this detail out. It makes even less sense to start a 'plausible' lie with women as your primary witnesses! As such, it makes no sense to consider the Gospel resurrection narratives as legends, since they all start with women witnesses.
- Fifth, the disciples really believed that Jesus appeared to them risen from the dead. They were not expecting it; their leader was dead. A crucified messiah was a contradiction in terms. And there was nothing in Jewish tradition to suggest that one man would rise immortal from the dead in the middle of history. And if the claim was a lie, the apostles had to have known it. And yet most suffered for the claim, and many even died for it. Why would multiple disciples die for a lie they made up?
- Finally, Paul and Jesus' brother James believed, despite being skeptical before Jesus appeared to them. Jesus' brothers didn't believe in Him before his death (*John 7:5*), but James became a leader in the Jerusalem church, and was eventually martyred for his Christian faith because "he appeared to James" (*1 Corinthians 15:7*). Paul tried to destroy the church, but after having what he believed was an encounter with the risen Jesus, became its greatest missionary, and also ended up martyred. Why would *skeptics* lie for the church, let alone *die* for its central message?

No explanation can account for all these facts, as well as the context that Jesus' claims to be God incarnate, other than God raising Jesus from the dead. And this would entail that the God who raised Jesus from the dead exists.

We can summarize this argument as follows:

1. There are several solid facts concerning the fate of Jesus: His death by crucifixion, the empty tomb, His appearances after having died, and the disciples' genuine belief in His resurrection.
2. These facts are best explained by the thesis 'God raised Jesus from the dead'.
3. If God raised Jesus from the dead, then the God revealed by Jesus exists.
4. Therefore, the God revealed by Jesus exists.[8]

MUST GOD BE GOOD?

God is *by definition* worthy of worship. Can such the perfect being be indifferent?

Of course not. For a being truly worthy of worship, there can't be anything greater than or equal to it. It is ***singularly supreme***, and necessarily so. And such a being will be good in the best way possible. That is *essentially* good: the very source and ground of all goodness.

Obviously, if something is the very ground of goodness, it cannot fail to be good, and thus can't be evil or indifferent. Otherwise, it wouldn't be the standard of goodness.

However, are we justified in believing such a God exists? Here, we propose two arguments, which are useful to illustrate our points: t

- he ontological argument and
- the moral argument.

THE ONTOLOGICAL ARGUMENT

It says that since a worship-worthy perfect being can exist, it must, since part of what it means for something to be worthy of worship is that it necessarily exists. Thus, if such a being can exist, it must, and thus it does.

The key to this argument, however, is whether such a being is possible. If it isn't, then it can't exist. Nevertheless, we believe in most cases, while we may be willing to grant the need for some sort of evidence to declare that something *does* exist, most people will generally admit that, as long as there are no inconsistencies or absurdities entailed in the existence of some being, that such a being *can* exist.

In other words, we're generally biased towards needing a reason to think that something *does* exist, but the opposite is the case for *possible* existence; we're biased towards granting possible existence to anything for which there are no obvious inconsistencies or absurdities. Thus, all else being equal, we think that most people would grant the premise that a ***singularly supreme*** being could exist. But if so, then the ontological argument shows that they are committed to the existence of such a being by force of logic.

Still, a general *a priori* assumption in favor of possible existence absent any inconsistencies or absurdities may seem to many a rather flimsy foundation on which to build a case for God. Is there anything that might 'buttress' our thinking on this matter?

There are some arguments for the possible existence of a supreme being, but they tend to be rather conceptual, convoluted, and arcane. They're not easily accessible, in our experience. So, this is where we would turn to more traditional arguments for God to buttress the presumption that He is possible. Of special interest here is the moral argument for God:

MORAL ARGUMENT

1. If God does not exist, objective morals don't exist.
2. Objective morals exist. Therefore, God exists

We believe we can take premise "b" as a given, here: we know torturing babies just for fun really is wrong, regardless of what anyone thinks. It's *objectively* wrong; thus, objective morals exist.

The key issue is premise "a": if God doesn't exist, objective morals don't exist. In other words, if atheism is true, there is no such thing as objective morals. The only way for this to be false is if objective morals exist in an atheistic world. But many atheists themselves reject this. Dawkins said it really well: "The universe that we observe has precisely the properties we should expect if there is, at bottom, no design, no purpose, no evil, no good, nothing but pitiless indifference."

But positively, what makes God a fitting foundation for morality? He is worthy of worship; the greatest conceivable being. More particularly, He is the ultimate source and standard of goodness. We *need* such a standard for objective morality to exist. Otherwise, moral states are left untethered to reality.

And that standard must itself be perfectly *good*. After all, which is primal, good or evil? Goodness can exist *of itself*. It doesn't need evil to exist. Evil, however, is merely perverted goodness. As C.S. Lewis explains:

"... wickedness, when you examine it, turns out to be the pursuit of some good in the wrong way. You can be good for the mere sake of goodness: you cannot be bad for the mere sake of badness. ..."

To be bad, he [an inherently Evil Power] must exist and have intelligence and will. But existence, intelligence and will are in themselves good. Therefore, he must be getting them from the Good Power: even to be bad he must borrow or steal from his opponent. And now do you begin to see why Christianity has always said that the devil is a fallen angel? That is not a mere story for the children. It is a real recognition of the fact that evil is a parasite, not an original thing. The powers which enable evil to carry on are the powers given it by goodness. All the things which enable a bad man to be effectively bad are in themselves good things—resolution, cleverness, good looks, existence itself.[2]

And so, we come full circle. God being essentially good is necessary for objective morality to exist. Objective morality exists. Therefore, God is essentially good.

THE EXISTENCE OF GOD

Why believe God exists?

Skeptics often demand that theists need to *conclusively prove* that God is there before either of us can believe He is there. But just because I may not be able to convince a skeptic that God exists doesn't mean I cannot know God exists. God can reveal Himself to people in numerous ways, some of which don't involve arguments. For instance, the Spirit Himself testifies to Christians' spirits that we are God's children (*Romans 8:16*). And God can also *withdraw* knowledge of Himself (*Romans 1:18–32*). We don't have to be a master debater for God to reveal himself to us sufficiently to know that He exists.

In fact, there is no such thing as a 'conclusive proof', if by this one means an argument that compels universal acceptance. No argument can *make* people believe its conclusion. Humans are not logic-chopping robots; they come with biases, experiences, and tastes that affect the way they view arguments.

For instance, consider the contrast between C.S. Lewis, a former atheist who later embraced idealistic philosophy, and Antony Flew, a hard-nosed evidentialist philosopher. Lewis was convinced by the moral argument for God, which shows God to be the moral ideal: "it is more important that Heaven should exist than that any of us should reach it." Flew rejected Lewis' moral argument for God, although late in life he was convinced of deism by the argument from design. He adopted a 'convince me with hard evidence' stand (a position he apparently never abandoned).

But if there are no conclusive proofs for God (in the above sense), what use are arguments for God? If an argument is sound and solid, it acts as a sign pointing to God. But signs only convey limited information, and people looking at a sign need to properly read and respond to the sign.

A person who ignores a stop sign, or misreads a speed limit sign, will act accordingly. Their response may even have disastrous consequences. But that's hardly the fault of the sign! In the same way, good arguments for God don't have to tell us everything about God. Nor can we make people read and respond to them properly.

All they offer is a public case commending belief in God as reasonable. And all we can do is faithfully portray the signs. We plant and water, but only God can give the increase (1 Cor. 3:7).

As such, when reading these arguments below, it must be understood that they are offered in that very spirit. They are not conclusive proofs, but signs pointing to God, showing at the very least that belief in Him is reasonable, if not rationally obligatory.

Can we be Good Without God?

The Moral Argument

Moral values and duties impress themselves upon us every day. For instance, practically everyone knows that torturing babies just for fun is objectively bad, and compassion for the helpless is objectively good. And we readily recognize those who disagree as abnormal (e.g., sociopaths). But why?

What makes the world a moral world?

The best explanation is God. God is the ultimate standard of goodness, and all morality is measured by His character, and meted out to us by His commands. Nothing else, whether evolution, or finite persons, or even moral facts themselves, provide a sufficient ground for moral values, duties, and accountability. We can formalize this argument like this:

1. If God does not exist, objective morals do not exist.
2. Objective morals exist.
3. Therefore, God exists.

Kalām Cosmological Argument:

In the beginning ...

In the beginning ... what? *Genesis 1:1* says "In the beginning God created the heavens and the earth", but why think that's true? First, nothing comes into being without a cause, a basic principle of science and rationality. Everything we see that started to be has some sort of cause.

But we also know that the universe itself had a beginning. The laws of thermodynamics powerfully imply that the universe had a beginning. And an infinite regress of secondary causes can't even exist, because it can be shown mathematically that this would lead to absurdities!

However, that means the universe itself had a cause. But what could cause the universe? The universe is all of space-time-matter reality, so the cause can't be bound by those things. And it must be powerful to cause the universe! The simplest solution is an eternal, non-material, uncaused cause.

But how to get a temporal effect from an eternal cause? That cause must have *freely chosen* to create, so it must be a *personal* cause. So, the simplest cause for the universe is a single, powerful, personal, eternal, immaterial, uncaused cause—it sounds a lot like God! We can formalize the argument like this:

1. Everything that has a beginning has a cause.
2. The universe had a beginning.
3. Therefore, the universe had a cause.

For more on this argument, please see:

THE REASON FOR EXISTENCE

Why is there something rather than nothing? Underlying the question is a principle: everything has a reason why it exists. What sort of reasons could they be? It turns out there are really only two basic sorts. A thing might be caused by something else, or it might be such that it must exist by its own nature.

But could the universe be exempt from needing a reason for why it exists? If so, why should it be exempted? Size doesn't matter. If we replaced the universe with a universe-sized book, the book would still need a reason for why it is there. Therefore, so does the universe. And if we exempt the universe, there's no reason not to exempt other things as well, such as things in the universe.

However, no sane person believes that ducks, stars, and chairs could exist for no reason. And imagine what would happen to science if we adopted this principle. Science is all about finding explanations for things. So, if some things have no explanation, how could we know whether science applies to them or not? We wouldn't be able to do science at all! So much for atheism being the friend of science.

THE EXISTENCE OF THE UNIVERSE

So then, why does the universe exist? It clearly doesn't have to exist. It's comprised of things that don't have to exist! Any number of universes, or none at all, could have existed instead. And yet it does exist. So, if there's a reason why the universe exists, what would it be? God made it, Figure 25.26.

Fig.25.26: Where did the universe come from? Curtsy - Getty

However, we stated: '*if* the universe has a reason why it exists, that reason is God'. Atheists will often respond: 'But if atheism is true, the universe doesn't have a reason why it exists.' That's precisely the point!

If they say 'if atheism is true, there's no reason why the universe exists,' it's also true that 'if the universe has a reason why it exists, atheism is false.' And why would atheism be false if there's a reason why the universe exists? Simple: God is the reason, if it has one.

But if everything has a reason why it exists, and the reason why the universe exists is that God made it, then God exists. The ultimate answer to why there is something rather than nothing is God. We can summarize this as follows:

1. Everything has a reason why it exists—either by the necessity of its own nature, or because it was caused by something else.
2. If the universe has a reason why it exists, it is that God caused it to exist.

3. The universe exists.
4. Therefore, God caused the universe to exist.
5. Therefore, God exists.

DESIGN ARGUMENT

Usefulness of Math in Science

Why is mathematics such a useful tool in science? It's as if nature is written in the language of math. It seems like a massive fluke! At least, apart from God.

But is it really likely that the usefulness of math in science is just dumb luck apart from God's existence? After all, it's very hard to imagine a self-consistent world without basic mathematics applying (e.g., 2+3=5). But it's not basic math that make the link between math and science look like a coincidence. Rather, it's complex mathematical ideas like imaginary numbers, tensor calculus, and Hilbert space. Many of these ideas have no physical existence (such as Hilbert space).

But they are vital for describing how nature works. Even if the physical world must be mathematical, that doesn't explain why the particular complex math we use works in describing the physical world.

Could mathematical structures cause the physical world? Only if they could cause things to be. But does e.g., the object "5" *cause* it to be true that my hand has 5 digits? That doesn't even make sense! The number "5" *describes* how many digits are on my left hand; it doesn't *cause* that fact to be true. If they exist at all, mathematical objects don't cause anything.

But instead, might the universe actually *be* a mathematical structure? Physicist 'Max Tegmark' believes just this. But he also provides us with powerful reason to reject his view:

"This crazy-sounding belief of mine … makes us self-aware parts of a giant mathematical object. … [T] his ultimately demotes familiar notions such as randomness, complexity and even change to the status of illusions …"

Tegmark's belief sounds crazy because it *is* crazy. It forces us to regard almost everything basic to human experience as an illusion! Any argument for this will always have premises less convincing than our conviction that our experiences are real. Better to believe a much less ambitious and counter-intuitive idea: that the universe was structured so mathematical concepts would be useful in describing it.

At least that doesn't mean we live in a mathematical Matrix! In essence, even if mathematical objects are real, they can't explain by themselves why nature is written in the language of math's.

But what if mathematical structures don't exist? Apart from God, the problem is even worse. The usefulness of math in science is something we *discovered*. We didn't invent it! Nature really is written in the language of math regardless of us. But if mathematical objects can't explain it, and we can't explain it, and it's not a coincidence, then why is nature written in the language of math? A transcendent mind. In other words, *God*. We can summarize the argument like this:

1. If God does not exist, the applicability of math to the physical world is just a coincidence.
2. The applicability of math to the physical world is not just a coincidence.
3. Therefore, God exists.

This is just one of many different design arguments. Some focus on the fine tuning of the universe for life, or the origin of life, or the origin of biological structures, or the origin of consciousness. See Created or evolved? for an overview. On creation.com, you will find many examples of design in nature that point to the God of the Bible.

ONTOLOGICAL ARGUMENT: GOD IS UNIQUELY SUPREME

Ever since Anselm of Canterbury first put forward his version in the 11th century, ontological arguments have been the source of much discussion and debate. They rest on two basic ideas. The first comes from Anselm himself—God is 'that than which nothing greater can be conceived.' In other words, *by definition* there's nothing conceivably equal to or greater than God. The second is that if it's possible that God exists, He must exist. God can't just *happen* to exist, because part of what 'God' means is a necessarily-existent being. So, either God *must* exist, or He *can't* exist.

So, what do these arguments look like? Here is an example:

1. 'That than which nothing equal to or greater can be conceived' (i.e., God) possibly exists. (Premise)
2. Suppose that God can fail to exist. (Supposition)
3. A being that cannot fail to exist is greater than a being that can fail to exist. (Premise)
4. If 'that than which nothing equal to or greater can be conceived' can fail to exist, then it is not 'that than which nothing equal to or greater can be conceived' (from (3)).
5. But this is a contradiction.
6. Therefore, 'that than which nothing equal to or greater can be conceived', i.e., God, cannot fail to exist.

What does all this mean? First, posit that God possibly exists. Then, suppose that God doesn't exist. But what does it mean for there to be nothing conceivably greater than or equal to God? Think about how things can exist. A thing either *can't* fail to exist (i.e., it exists *necessarily*), or it *can* fail to exist (i.e., it exists *contingently*).

Which is better? Necessary existence, right? Now apply this to God. If there's nothing conceivably greater than or equal to God, can He fail to exist? Clearly not. Why? If He could, then God could not be 'that than which nothing greater or equal to can be conceived.' But that's just nonsense! There's nothing conceivably greater than or equal to God *by definition*. Therefore, supposing that God doesn't exist makes no sense (provided that God possibly exists). Since by definition there's nothing conceivably greater than or equal to God, if He exists, He has to exist in the greatest way possible. That is necessary existence. But if God necessarily exists, then He *actually* exists.

Many regard this as just a word trick. But it isn't. The statement 'There is nothing conceivably greater than or equal to God' does entail that 'God cannot fail to exist,' But the two statements don't *mean* the same thing. People often think 'God' is a coherent concept without realizing it implies that God must exist. These arguments can help us see this. The conclusion is implicit in the premises, but that's true for every valid deductive argument.

Critics also often say that similar arguments could be made for any so-called necessary being. In other words, they say that if something *could* exist necessarily, it *does* exist necessarily. And that's a valid way to argue. The trick is showing that other so-called necessary beings are possible. But at most it would only show that God is not the only necessary being around. But that's not even relevant to this version! This version uses the *greatness* of necessary existence (compared to contingent existence) to show that God must exist. If there are other things that must exist, God must still be greater than them.

Or the critic might say that 'that than which nothing greater or equal to can be conceived' is incoherent, like the idea of a '***Married Bachelor***.' This is probably the 'best' objection of the lot, because it's the easiest to imagine. But notice what it forces the atheist to say: not simply that God *doesn't* exist, but that the very idea 'God exists' is *incoherent*. Saying that is easy enough. But how to *defend* such a claim? That is anything but easy!

And there are reasons to think God is possible. For instance, the other arguments listed here give us reason to think that God at least possibly exists. Atheists have also tried, and failed, to find a clear-cut incoherence in the idea of God for millennia. Even the very idea of a being nothing can possibly be greater than or equal to plausibly entails its own possibility.

Anselm thought he had found in the ontological argument a conclusive proof for God. But it's not. It's hard to understand. And it's not immediately obvious that the statement 'God does not exist' is necessarily false. But it can help undercut doubts about God for those inclined to thinking that God possibly exists. And even for the atheist, it can help show them the intellectual cost of rejecting the possibility of God.

OBJECTIVE MORALS

How can we believe in objective morals if people disagree on moral questions?

Is the Moral Argument a poor argument for theism? The only evidence of a moral law existing would be our sense of it. What if our conscience conflicts with the Bible? My conscience tells me that it would be evil to kill my dog, but the Bible says it wouldn't be. Not everyone has the same moral feelings. Does this significantly weaken the Moral Argument?

We clearly don't think it's a poor argument for God. However, that doesn't mean it will convince everyone. No argument for God (or against God) will do that.

But do conflicting moral intuitions undermine our warrant for believing in objective morals? The example you give is: 'is it permissible to kill my pet?' You even raise what you see as a conflict between your moral feelings and what the Bible says: "My conscience tells me that it would be evil to kill my dog, but the Bible says it wouldn't be."

First, the Bible doesn't say that you can kill your dog for just any reason. "Whoever is righteous has regard for the life of his beast" (Proverbs 12:10). While we have authority over animals (Genesis 1:28), and in a fallen world that authority has been extended to authority over their lives to a certain extent (e.g., for food, or in sacrifice), we still need a good reason to kill our animals, according to the Bible. And while the Bible certainly gives us principles to help discern the matter, and in some cases, we do need to kill our animals (e.g., if they become a danger to human life), it will in many cases come down to a question of personal conscience.

And on that count, the Bible says that violating personal conscience is bad: "whatever does not proceed from faith is sin" (Romans 14:23). But that doesn't mean you should expect everyone to agree with you, and because the Bible's advice is so generic, there's plenty of room for reasonable disagreement among Christians. So, while the Bible doesn't mandate your views, it doesn't condemn them either.

Second, if the mere existence of conflicting moral feelings undermines our warrant for believing in objective morals based on our moral experience, that seems to imply we can only have reasonable warrant for accepting objective morality if *everyone* agrees on *every* moral question. But disagreements could arise for many different reasons. Perhaps we're limited. Perhaps some of these questions are difficult. Perhaps some of us have bought into falsehoods that distort our moral views. Perhaps we're sinners with moral compasses that are somewhat distorted naturally.

All these reasons plausibly explain why disagreements might exist, despite the reality of objective morals and us having a generally reliable grasp of basic moral questions.

Besides, universal agreement doesn't mean much. For instance, say that the Nazis killed or brainwashed everyone who thought the Holocaust was evil. Would that have made the Holocaust good? Would it mean that there's no truth to the matter? Neither. The Holocaust is evil, regardless of what anyone thinks.

Moreover, the question 'is it permissible to kill my pet?' is not the best question to gauge the reality of objective morals by. The reality of objective morals doesn't mean that all moral questions are easy to answer. Rather, we use clear-cut examples like 'it's wrong to torture a baby just for fun' to show that, despite the difficulty of many moral questions, there are some situations where merely a moment's reflection will make it clear to most that we can't escape the reality of morality. Indeed, the 'is it right to torture X just for fun' is a really handy one, because we can substitute 'babies' for whatever the person we're dealing with really loves, like 'your mum', or 'your dog', or 'your son', or whatever. Making it personal heightens the person's disgust at the idea, and usually forces people to admit that they can't escape the reality of morality.

Transfer this sort of thinking to another subject. For instance, Muslims and Christians typically disagree on whether Jesus died by crucifixion. Does this disagreement undermine our ability to know *any* historical truth? Does it call into question the very notion of historical truth? Surely not! Even if we couldn't know who was right in this instance (e.g., for lack of evidence), it still wouldn't undermine our ability to know *any* historical truth. And it certainly wouldn't be reason to think that there's no historical truth to be known. But that's exactly parallel to what you're suggesting regarding morals: just because our moral experience might fail us in some instances does not mean that it's untrustworthy in all other cases.

You will always find diehard moral skeptics. They will reject objective morality, no matter how much you try to convince them otherwise. But that doesn't make you wrong, and it's not cause for you to doubt. Rather, it just makes them crazy, or it means they have an atheistic axe to grind (the latter is usually more likely).

And press them for reasons why they reject objective morals! What justification for their views do they have? We've seen that the mere existence of conflicting moral feelings doesn't work.

What about evolution? Evolution is irrelevant.

- First, evolutionary accounts of the origin of our moral beliefs are little more than a pack of conflicting just-so stories.
- Second, even if it could explain the origin of our moral beliefs, that doesn't mean they're not objective.

To say otherwise is to commit the genetic fallacy. For instance, even if evolution were true, God could've planned the process to produce in us generally reliable moral faculties. (Though, of course, God did not use evolution, and the Bible contradicts such an idea).

Evolution by itself doesn't undermine our moral faculties. It's only when evolution is combined with naturalism that it does that.

But then naturalism, which is atheistic, becomes the only operative reason for rejecting objective morality. In other words, the only reason for rejecting objective morality turns out to be a rejection of God. But that just assumes what the skeptic needs to prove.

The skeptic really doesn't have much hope in justifying a rejection of objective morals. Which is why you should press them to try. And remind them that if they buckle under the pressure, they have to deal with their prior acceptance of the other premise of the moral argument (i.e., If God doesn't exist, objective morals don't exist.'

The diehard skeptic will likely just hop between denying each premise, ultimately doing so only because they can't accept the conclusion of the argument. As frustrating as that might seem, it's actually a really good position to leave them in. It's hard to live with contradiction once it's been pointed out to us, and it may be the catalyst God will use to move them closer to Himself.

WITHOUT GOD

God and Morality

Atheists and Christians often debate such questions. In this case, the politician's answer is true: it really does depend upon what you mean by 'God' and 'without'.

In fact, atheists not only can, but *must* be (at least to some extent) good without believing in God—even if they hate God with every inch of their being. If they are really made in the image of God as the Bible teaches (*Genesis 1:27*), then that fact must have some results. They, like all of us, are fallen (as explained in *Genesis 3*), but even so must still have an in-built sense of the reality and the importance of right and wrong.

The very fact that atheists routinely argue that this or that is moral or immoral, and that such matters are important, bespeaks that fact. Unless that were so then the Bible would start to look suspect. When society comes across someone who really does seem to have mostly wiped out the ideas of right and wrong from their mind, we label them as insane and lock them in padded cells.

We don't just say, 'If that's what you like, then we'll respect your choice.'

If atheists were generally able to throw off all the shackles of morality and live their lives consistently with atheism, we'd be worried. If they could *consistently* live out such ideas as, 'We're just here to pass on our selfish genes,' 'Survival of the fittest' or 'Life is ultimately all without meaning or purpose.' it would put a serious question mark over the record given to us in Genesis. It would be evidence that maybe they weren't creatures made by God after all, and that atheism might actually be true.

The fact, though, that most atheists find themselves unable to live out such ideas is reassuring; instead, they find it necessary to live as if morality were real, hunting around for far-fetched arguments to justify this.

There are of course some atheists who have been more consistent, at least in their theoretical thinking. Believing that man is nothing more than a cosmic fluke, they realize that this means that ultimately morality is just something that was put together in the human mind, a product of evolution, which has no more real authority over our behavior than any other activity of the human mind. It's got no more compulsion ('you *ought* to do this') than anything else thrown up by our brain cells—such as, for example, immorality! One person thinks that we should not hurt our neighbor; the cannibal, though, thinks that it's OK to eat him. And both of those ideas are nothing more than the result of chemicals fizzing around in our heads. Neither has any real *authority*—they are ultimately just personal preferences.

Fig.25.27: Frederich Nietzsche - Curtsy - Photo wikipedia

FREDERICH NIETZSCHE

Frederich Nietzsche, Figure 25.27, the famous atheist who said, 'God is dead', saw that this was where the logic led. He wrote that *'our moral judgments and evaluations are only images and fantasies based on a physiological process unknown to us.'*

The last century saw what happened when evolutionary philosophy was put into practice: Nazi genocide and euthanasia, millions butchered by Stalin, Mao and Pol Pot. Even today, 'ethicists' like Peter Singer support infanticide, and evolutionary enviro-fanatics propose population extermination.

Two evolutionists wrote a book claiming that rape was a device for men to perpetuate their genes—one of the authors tied himself in knots trying to explain why rape was still wrong under his own philosophy. Fortunately, most atheists don't carry their atheism to its logical conclusion like these horrific examples.

In any case, any atheist philosophizing nicely about such systems of amorality would, in real life, be quickly brought back to his senses by a punch on the nose. He would quickly get back his old feelings about the reality of right and wrong, and start talking like a theist again, telling his assailant that what was done to him was 'wrong'—no arguments! Don't try telling him that his attacker's brain is wired differently, the result of genes mutating in a different direction so that for him punching the atheist was right—he won't accept it!

The fact is, though, that when atheists are concerned about good, or are being good, none of that is 'being good without God.' It's the opposite—being good *with* God, because God really exists and they are made in His image.

To actually talk about being 'good without God,' we would need to take a journey into a different universe: the mental universe of atheism. Because the image of God is impressed upon our nature as created beings, we've all assumed some ideas about right and wrong. But what kind of idea of morality is logically consistent with atheism, the idea of a universe in which we're just highly-evolved pond goo?

Fig.25.28: Richard Dawkins Curtsy - Photo wikipedia

British atheologian Richard Dawkins, Figure 25.28, says:

'Atheists and humanists tend to define good and bad deeds in terms of the welfare and suffering of others. Murder, torture, and cruelty are bad because they cause people to suffer.'

Defining good and bad in terms of welfare and suffering sounds reasonable—pretty close to the Christian commandment to love our neighbor. Hurting them is bad, helping them is good.

The problems do not come with the second half of Dawkins' first sentence, but the first. 'Atheists and humanists tend to define good and bad ...'

In fact, there's no reason to read anything that comes after that point. Whether we *choose* to define good and bad in terms of helping society or in terms of crushing it with an iron fist makes no difference from here. If good and bad are merely what atheists, humanists or anyone else *chooses to define them as*, then good and bad are merely a product of the human brain. They have no binding moral authority over us, any more than any other mere construction of the human brain has. They might make us happy, but happiness is not the same as righteousness—even a serial killer might feel that he gains 'happiness' from his crimes. They exist only within our cerebral chemistry, and nowhere outside of it. Like opinions on the best England football XI, or on the finest vintage of South African wine, morality is no more than one of the moveable's and ever-moving feasts of human thought.

With no external or transcendent source of values, Richard Dawkins' opinion on what is good or bad has no more authority over me or objective basis that should guide me than my preference for classical music over grunge. Both have precisely the same foundation—the ever-evolving activities of the human brain. It's just a matter of however I happen to like or want things to be! Indeed, Dawkins has himself recognized that ultimately evolution 'leads to a moral vacuum ... in which [people's] best impulses have no basis in nature'.

He scoffs at the idea of righteous indignation and retribution against child murderers and other vile criminals, claiming that it is as irrational as "Basil Fawlty" (*In on of Faulty Tower –Comedy Episode*) beating his car.

To be real, though, morality must be a matter of authority: You 'ought' or 'ought not' to do this or that. Its very essence depends upon transcendence. That is, something that is 'bigger' than you, and tells you what to do. It cannot be something that is just a part of you or humanity in general: it must be 'outside' of humanity, something over and above us.

Dawkins' morality is not morality at all, but personal preference. He *prefers* to not cause suffering; rapists *prefer* to maximize their own gratification. In atheism, there's no ultimate authority we can appeal to in order to determine whose thoughts are 'better.' Both are just human brain activity, without any ultimate reference point by which to evaluate them.

Morality is precisely real because God is real. As our Creator, He is the transcendent authority—the law-giver who gets to tell us what we 'ought' or 'ought not' to do. It is because we are made by Him and are like Him that we know we cannot really treat morality as just an invention. It is because existence is more than just molecules that right and wrong are important.

It is because we are made in the image of God the Creator that morality really is bigger than we are. Which means that God ultimately defines what is right and what is wrong.

At this point, the atheist is in dire straits. The first of God's laws is to love Him with our whole being (*Mark 12:30*). Can we therefore be *truly* good without honoring God?

No, because by refusing to honor our Maker we break the first and greatest of all the commandments.

Atheists need to face up to logic—ultimately, either nothing is immoral (because there is no God, and thus no such thing as morality) or atheism is itself immoral. There are no coherent alternatives.

GOD MORAL OBLIGATION!

I dislike the term 'moral obligation' when applied to God because God has done so much *more* than He would ever be obligated to. His grace is so overabundant that to speak in terms of 'obligation' when we understand His grace is just sort of repugnant.

It also gives the implication that we have the right to demand something of God, and that He might be less than completely righteous, both of which are wrongheaded.

It further puts us as 'judge' over God as to whether He's fulfilled that 'obligation.' When Job tried to 'sue' God, God appeared, but didn't answer Job, and Job realized his unworthiness to challenge God after God challenged him. Paul, in Romans 9, is very blunt: we have no more right to challenge our Creator than clay pots have to challenge their potter "why have you made me thus?" God is not in any way limited by His creation or intrinsically obligated to it.

Also, it is very important to understand the framework of biblical history. God created a 'very good' world, and existence itself is a gift. He created the pinnacle of His creation, human beings (Adam and Eve) with everything they needed in a perfect environment, in a perfect relationship with Himself. So *certainly,* at this point, there's no question about whether God fulfills any obligations (to the extent He had any to begin with) He might conceivably have.

But then of course sin enters the picture. All the bad stuff that's happened in the 6,000 years since is attributable to that sin. We can't accuse God of *any* of the 'bad things', because it's our ancestor Adam's fault that there is sin and death and suffering in the world.

God doesn't have any moral obligation to mitigate the effects of sin; in fact, He has a 'moral obligation' to blot out sin (and before you start cheering for that solution, remember that *we're sinners*!). If there were no sin, there would be no suffering.

But we know that He is merciful, and He continues to care providentially for His creation (the theological term for this is 'common grace'—Jesus said, "For [God] makes his sun rise on the evil and on the good, and sends rain on the just and on the unjust" *Matthew 5:45*). Even in this sin-cursed world, we have a lot of good things, and God is the author of those things, so we should be grateful to Him.

Furthermore, God wasn't content to just leave us in our sin, or to destroy us so that His universe would be sinless again. Instead, He went completely beyond *any* obligation to us. God Himself, the Second Person of the Trinity, took on human nature in Jesus Christ, who was absolutely sinless, to die to pay the price for our sin.

His Resurrection serves as a type of promise that God will eventually restore the whole creation, and raise all those who love Him to live with Him forever in a perfect New Heavens and Earth.

Therefore, in other words, I think we're not knowledgeable enough about what goes on in the heavenly realm to even evaluate God's actions beyond what's revealed in Scripture, but then even what we do know reveals that not only does He fulfill every obligation perfectly; He goes unimaginably further than that in providing salvation through Christ.

CREATION IN ISAIAH

Fig.25.29: *Isaiah – Curtsy Photo: Trounce/Wikimedia Commons*

The truth of creation is a theme running through the whole of the Holy Scriptures. In Genesis we have the basic statements of God creating the heavens and the earth in the beginning, and some details of His work in the six days. The Psalms draw attention to the wonder, beauty and care shown in the creation, which show the power, glory and wisdom of the Lord. It was John Ray, the seventeenth century Puritan biologist, who put the text of *Psalm 104:24* on the title page of the great book, *The Ornithology of Francis Willoughby*, that he edited for his patron and friend in 1678: "O Lord, how manifold are thy works! in wisdom hast thou made them all: the earth is full of thy riches."

While three particular Hebrew words are used frequently elsewhere in the Old Testament, they occur together in *Isaiah 45:18* and are translated "created," "formed" and "made." It is interesting that these words, *bara* (to prepare, form, fashion, create), *yatsar* (to form, fashion, frame, constitute), and *asah* (to do, make), are much more frequently used with reference to the creation in one particular book than in any of the others in the Old Testament: Isaiah.

The Old Testament prophets, with reference to Israel and the nations, portray God as the supreme ruler, king of the whole world. He is shown to be able and active. Like earthly monarchs, but of course superior to them, He is described as being capable of building or destroying, producing ruin or prosperity.

Emphasizing this authority and power, references are made to God as the Eternal and the Creator. This makes Him not only vastly superior to any conceivable world ruler, whatever his powers and resources, but in a different category and unique. This approach is particularly apparent in Isaiah. It is a truth implied in the earlier chapters of the prophecy, but the statements are clearer and more frequent in the later passages of the book.

God's power is indicated in that He will in judgment "shake terribly the earth," in contrast with "man whose breath is in his nostrils" (*Isaiah 2:19-22*). This theme is continued when it is stated that He will assemble the outcasts of Israel, using geographical changes, for He "shall utterly destroy the tongue of the Egyptian sea: and with his mighty wind shall shake his hand over the river, and shall smite it in the seven streams, and make men go over dry-shod" (*Isaiah 11:12,15-16*), Figure 25.29.

HEAVENS SHAKEN

Further in the judgment of Babylon, described in chapter 13, God states He will stop the light of the moon and stars, and "I will shake the heavens, and the earth shall remove out of her place." Again, in chapter 24, we have the statement that "the Lord maketh the earth empty [*baqaq*—to make void], and maketh it waste, and turneth it upside down." That "the moon shall be confounded and the sun ashamed when the Lord of hosts shall reign in Mount Zion and in Jerusalem" is recorded in the same chapter. This is similar to the state described in *Revelation 21:23*. Control of the Creation in Judgment is further illustrated in the prophecy against Ariel in chapter 29: "thou shalt be visited of the Lord of hosts, with storm and tempest, and the flame of devouring fire."

The greatness, power and wisdom of the Lord is described in the well-known gems of *Isaiah 40*: "who hath measured the waters in the hollow of his hand, and meted out heaven with a span, and comprehended the dust of the earth in a measure, and weighed the mountains in scales, and the hills in a balance?" To Him "the nations are as a drop in the bucket, and are counted as the small dust of the balance: behold he taketh up the isles as a very little thing."

Also "He sitteth upon the circle of the earth, and the inhabitants thereof are as grasshoppers: that stretcheth out [*natah*—to stretch out] the heavens as a curtain, and spreadeth them out as a tent to dwell in." His people are encouraged by the questions and statements: "Hast thou not known? hast thou not heard, that the everlasting God, the Lord, the Creator [*bara*] of the ends of the earth fainteth not, neither is weary? there is no searching of his understanding."

In chapter 41:18-20, God's promised creation of fertility for Israel is mentioned; and in chapter 42:5, God is described as "He that created the heavens, and stretched [*natah*] them out; he that giveth breath [*nishma*] unto the people upon it, and spirit [*ruach*] to them that walk therein." Jacob as a nation is also not to fear as the Lord has created (*bara*), formed (*yatsar*), made (*asah*) and redeemed him and will protect him (chapter 43:1,2,7,15]. This is repeated in chapter 44:21 and 24.

CREATOR REVEALED

King Cyrus is told (chapter 45:7,12). "I form the light and create [*bara*] darkness"; "I have made [*asah*] the earth, and created [*bara*] man upon it; I, even my hands, have stretched out [*natah*] the heavens, and all their host have I commanded." Also that Israel shall be saved with an everlasting salvation: "For thus saith the Lord that created [*bara*] the heavens; God himself that formed the earth and made [*asah*] it not in vain [*tohu*]—without form, as in *Genesis 1:2*], he formed [*yatsar*] it to be inhabited" (chapter 45:17-18). Thus, an eternal, purposeful and benevolent Creator is revealed to this heathen Persian king.

Israel is reminded of her God being the Eternal and the Creator in chapter 48:12-13: "I am the first and I am also the last. Mine hand hath also laid [*yasad*] the foundation of the earth, and my right hand hath spanned the heavens; when I call unto them, they stand together." In chapter 51:12-13, the nation is comforted and reminded not to forget "The Lord thy Maker [*asah*], that hath stretched forth [*natah*] the heavens, and laid [*yasad*] the foundations of the earth." There is more comfort in chapter 54:5, where her Maker is described as her husband.

Numerous other references are made by the prophet, to God acting as Creator and Controller of the natural world to fulfil His purposes. The climax in both prophecy and powerful activity is reached on a happy note in the last chapters, as the Lord states, "For behold I create [*bara*] new heavens and a new earth; and the former shall not be remembered, nor come to mind. But be ye glad and rejoice forever in that which I create [*bara*] for behold I create [*bara*] Jerusalem a rejoicing, and her people a joy" (chapter 65:17-18).

He also promises: "For as the new heavens and the new earth, which I will make [*asah*] shall remain before me ... so shall your seed and your name remain" (chapter 66:22).

MIGHTY CREATION

No Evolution Process in Isaiah's Creation

Throughout this book of Isaiah there is no suggestion of the use of an evolutionary process, but God's action in Creation being mighty and immediate. Thus, in picturesque, but accurate and not extravagant language, the inspired prophet records for his day and ours the importance of the doctrine of Creation.

God, who is more than a superior ruler or earth mover, brings prophecy to pass, just as He brought forth the heavens and the earth, and He controls them. This is comfort for His people and a warning to man who is possessed of a rebellious nature and is proud of his material achievements.

MAN'S DOMINION

Man is God's vice-regent in the earth since the day Adam was given dominion over every living thing (*Genesis 1:28* and *Psalm 8:6*). He was put in 'the garden of Eden to dress it and keep it' (*Genesis 2:15*). A number of other scriptures point to his being responsible to treat the Creation properly.

The avoidance of disturbing the mother bird (*Deuteronomy 22:6-7*) is one example. The care of the domestic beast is commended (*Proverbs 12:10*). The need for the observation of the sabbath for man, beast and the land are frequently enjoined. This last provides for the refreshment and renewal of these principal parts of the Creation. Such considerations should be of value to modern ecologists.

Of course, the New Testament references to Creation are also numerous. Here we have the further revelation of the Personality who created all things, the Lord Jesus Christ, who was in the beginning (*John 1:3*) and who is the Source, Sustainer and Savior of the physical world (*Colossians 1:16-17*) as well as of the spiritual.

The created universe is for His pleasure (*Revelation 4:11*). Man, His creature, is expected to recognize the Creator through the observation of his created world (*Romans 1:20-1*). Here surely is justification for scientific study and research!

MAN'S MORAL RUIN AND JUDGMENT

It was Sir Francis Bacon in the seventeenth century who stated that such activity was "for the glory of God and the relief of man's estate." However, failure to accept this doctrine of Creation through pride and unbelief shows man's foolishness which leads to moral ruin and judgment. This is the ultimate result of evolutionary teaching. But the acceptance of the Divine teaching brings present blessing and the promise of participation in the new heavens and new earth, wherein dwelleth righteousness (*2 Peter 3:13*).

GOD CREATED MAN TO BE AN ETERNAL COMPANION FOR JESUS

The idea of redeemed mankind being an eternal companion to the Lord Jesus Christ is more a summary of certain parts of the Bible, or perhaps a conclusion based on these parts, rather than something for which we can produce a proof text, so consider the following:

John 17:24 (Jesus' prayer for His disciples before His crucifixion):

Father, I desire that they also whom You gave Me may be with Me where I am, that they may behold My glory which You have given Me; for You loved Me before the foundation of the world.

Philippians 3:20–21:

For our citizenship is in heaven, from which we also eagerly wait for the Savior, the Lord Jesus Christ, who will transform our lowly body that it may be conformed to His glorious body,

Ephesians 5:31–32, where Paul writes:

For this reason, a man shall leave his father and mother and be joined to his wife, and the two shall become one flesh. This is a great mystery, but I speak concerning Christ and the church.

In *Revelation 21* we read of a wedding between the bridegroom, called 'the Lamb' in *Rev. 21:9*, 'Come I will show you the Bride, the wife of the Lamb' and His Bride. The Bride is said to be 'the New Jerusalem,' whose citizens are God's people (v. 3), and those who enter the city are those 'whose names are written in the Lamb's book of life' (v. 27).

Putting all of this together, I think it is a fair and accurate summary or conclusion for us to say that God created man to be an eternal companion for His Son Jesus Christ. By 'man' of course I mean 'redeemed mankind,' not 'sinful mankind.'

Why would an infinite all-powerful God need a companion?

We did not mean to imply that God *needed* a companion. God does not need anything. What we see in the Bible is God choosing to do various things, and one of these is to create the human race, some of whom receive God's gift of salvation and dwell with Him forever.

Was He lonely?

The Bible does not suggest that He was. There was love within the Trinity prior to creation (*John 17:24*). C.S. Lewis wrote: 'God did not make us because He was bored, lonely, or had run out of things to do. He created us to be the objects of His love! Sometimes our actions make us unlovely, but we are never unloved. And because God loves us—we have value. And nobody can take that value away.

God's love revealed at Calvary fastens itself onto flawed creatures like us, and for reasons none of us can ever quite figure out, makes us precious and valued beyond calculation. This is love beyond reason. And this is the love with which God loves us.'

If He needed a companion then surely, He'd want something better than humanity.

John Newton wrote in his famous hymn: 'Amazing grace that saved a wretch like me … '. There is nothing in me or in any Christian that can possibly be of eternal value to a holy God, except for one thing: When God created mankind originally, He made man 'in the image of God,' which image was marred because of sin, beginning with the Fall (*Romans 5:12*). However, 'if anyone is in Christ, he is a new creature (*2 Corinthians 5:17*).

We have Christ's righteousness imputed to us. 'Christ has become for us wisdom, righteousness, holiness and redemption' (*1 Corinthians 1:30*). Preachers sometimes express this by saying that we are clothed in the robes of Christ's righteousness, so that when God looks at us, He sees the righteousness of Christ. A strong case can be made for this from reference to Christ's parable about the wedding garment (*Matthew 22:1–14*) and the robes of white in *Revelation 7:9*).

The idea of God creating a companion for His son Jesus Christ sounds awfully close to a heresy that fails to identify Father and Son as being essentially the same—perhaps some form of Arianism [denial of the complete deity of Jesus—Ed.]?'

No. The Bible is a progressive revelation, or we can say that there is a progress of doctrine throughout the Bible, or that we learn more about God the further we progress through the Bible. In the Old Testament God is seen essentially in His unity—He usually appears as one person, as in the appearances to Moses.

In the New Testament the three persons of the Trinity are clearly seen and differentiated, especially so during the life of Christ on Earth, e.g., at His baptism (*Luke 3:21–22*). In Christian theology the term 'Trinity' means that there are three eternal distinctions in the one divine essence, known respectively as Father, Son, and Holy Spirit. In *Revelation 21:9 & 10* we read of ' … the bride, the wife of the Lamb … coming down out of heaven from God.' Just as in a human wedding a father gives the bride to another person, so in the heavenly wedding the same thought applies—God the Father is depicted giving the bride to His own Son. God the Father and God the Son are seen as distinct persons having distinct roles.

What are Christians looking forward to in eternity future, if not the prospect of being with Christ, the One who loved us enough to die in our place, in an unending relationship of love, worship and sharing in His glory (*Romans 8:17*)?

In fact, we have all been invited to a wedding—not just to be spectators, but to make up the bridal party! There is an invitation in the Bible to you and to me to be part of that immense throng of believers who form 'the bride of Christ.' It reads, 'The Spirit and the bride say, "Come!"' (*Revelation 22:17*).

But a wedding invitation requires a response, an RSVP. Those who will be attending the wedding, i.e., being united with Jesus Christ in Heaven, and escaping the Judgment in the life to come, are those who respond in this life in repentance and faith.

There's one other thing we would like to say … . We were sure, we were among the redeemed.

Suppose that you and I were both to die at the same time and together we front up to 'the Pearly Gates,' where St Peter takes one look at us both and says, 'Tell me why I should let you into God's Heaven.'

My reply might be, 'The Book which I believe to be the Word of God tells me that the penalty for all my sin has been paid by Christ through His death on the Cross. This I believe. God's Word further tells me that because I have received Christ, I am a child of God, a joint-heir with Christ, and a citizen of Heaven. Step aside, Peter. On the authority of the Word of God, neither you nor angels nor devils have the right to bar me entrance.' (Or words to that effect!)

Now I don't really believe that Peter is the gatekeeper of Heaven, but suppose he was, or suppose that the Lord Himself should ask us both the same question, how would you answer?

THE CREATION OF THE UNIVERSE

Fig.25.30: *The Creation of the Universe – Curtsy - coffeekai—Getty Images*

The cosmic home that we live in is astonishingly beautiful and vast. Just think of the pictures of exquisite planets, galaxies and nebulae that NASA has beamed home for us. Pause for another moment to reflect on the fact that the observable universe is 93 billion light years across. The reality, though, is that there is a vast realm of cosmic space-time beyond the observable universe which we will never be able to peer into, even with the most powerful telescopes that we could ever build, Figure 25.30.

A BRIEF HISTORY OF COSMOLOGY

For thousands of years, some of the greatest minds in the world believed that the universe has always been here. This included the Greek philosopher Aristotle, but goes back even further to Babylonian and Hindu myths. If the universe has always been here, it seems fairly easy and straightforward to conclude that the universe did not need to have a cause. Problem solved. But is it really that easy? Doesn't your curiosity invite you to dig deeper?

In 1714, it was this curiosity that led German mathematician Gottfried Leibniz to mull over the fundamental question: Why is there something rather than nothing? Surely it is possible that, at some point, the universe did not exist. But if that is the case, then where did it come from and why is it here?

Let's wind the clock forward 110 years to 1824, when French engineer Nicolas Carnot was learning how heat engines work. He discovered one of the most fundamental properties of nature, the second law of thermodynamics. This law tells us that things in nature, when they are left alone, slowly wind down to an equilibrium where all of the energy has been spread out evenly. So, if there are things that still have more highly concentrated, useable energy, like the sun, we have not yet reached a state of equilibrium. This provides powerful scientific support for the reality that the universe has not always been here.

A century later in 1917, German-born cosmologist Albert Einstein was working on his general theory of relativity. He was unnerved and then irritated to find that his theory pointed to a universe that could change its size and shape over time. Within a decade, two fellow scientists, Alexander Friedmann and Georges Lemaitre, had used Einstein's work to show that the universe has indeed changed over time and, in fact, had a beginning. This became known as the Big Bang theory. In spite of persistent unresolved problems with the Big Bang theory, which means that it may ultimately be discarded, cosmologists have consistently been confronted with the reality that the universe is gradually expanding and, therefore, had a beginning, no matter what other models of the universe they explore.

The scientific confirmation that the universe had a beginning really brings to the forefront the question of its origin. We can no longer ignore this question simply by assuming that it has always been here. So, where did the universe come from?

A UNIVERSE FROM NOTHING?

There are scientists who believe that the universe and the very concept of space-time did, indeed, come from nothing. This idea is quite mind-boggling; could an entire universe—quarks, quasars and quokkas—come into existence from literally nothing? No matter how implausible this may seem, there are some very high-profile scientists who claim that this is what actually happened. For example, Lawrence Krauss wrote a book which had the title, *A Universe from Nothing: Why There is Something Rather than Nothing*. The famous cosmologist Stephen Hawking made the remark in his book *The Grand Design*, "Because there is a law like gravity, the universe can and will create itself from nothing."

The proposal that the universe came from nothing defies not only logic, but also physics. If things like universes can just appear out of nothing, why don't we see less grandiose things like Volvos, gorillas and jumbo jets just appearing out of nothing before our eyes? While it may sound rather exciting if these sorts of things could suddenly just appear, we just don't experience that happening. It would also be rather confusing, inconvenient and even dangerous if they did. It is much better to recognize the metaphysical truth that things do not appear out of literally nothing.

If you read Krauss and Hawking's books really carefully, what you will suddenly realize is that they are not really talking about the universe coming from literally nothing. Their "nothing" is actually something, what physicists call a quantum vacuum. Even though we ordinarily think of a vacuum as empty space, a quantum vacuum is filled with a fluctuating sea of energy. This means that a quantum vacuum is, as physicists need to remind us, "by no means a simple empty space". So, Krauss and Hawking haven't really answered the question of how the universe could come from literally nothing—they have simply dodged the fundamental question by making a quantum vacuum sound as if it is nothing.

BUBBLE UNIVERSES

Fig.25.31: Bubble Universes - Multiverse hypotheses – Curtsy: cosmology.com

One of the things that can happen in a quantum vacuum is that particles, like small bubbles, suddenly appear and then disappear. If particles can pop into existence like this, what about little bubble universes that keep growing and expanding? This idea opened up a whole new vista of possibilities for cosmologists, Figure 25.31. Perhaps there is a much bigger universe that has been here for all eternity, within which our whole universe is just a quantum bubble that just popped into existence billions of years ago. The idea of multiple universes—a multiverse— is not just a theory in science fiction, but one which is also debated by scientists.

The problem with this idea is that these bubble universes could pop into existence anywhere, and just like the bubbles your kids blow, these bubble universes would start to bump into each other and maybe merge together in larger and larger clumps. If the bigger universe was infinitely old, then all of these bubble universes would have merged together by now, and we should be observing an infinitely old universe.

But, again, that is not what we observe. Even if our universe did suddenly appear as a bubble in a quantum vacuum—which is actually very speculative since we have never observed universes beginning as bubbles—the fact is that the proposed quantum vacuum itself would have had a beginning. Then the follow-up question has to be, where did the quantum vacuum come from?

STRING THEORY

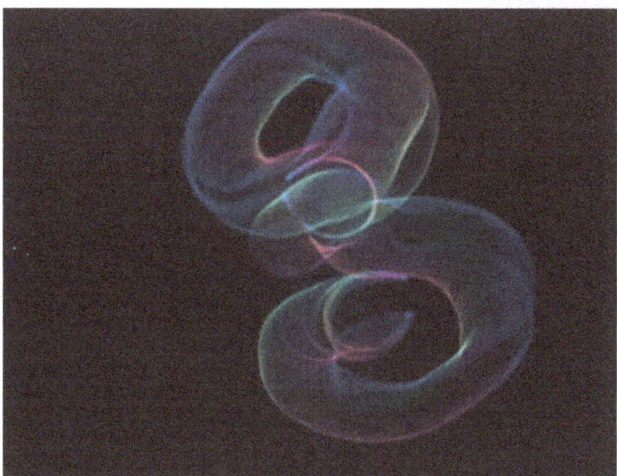

Fig.25.32: String Theory Universes is not Science- But Science Fiction – Curtsy: Image flickr user Trailfan,

In order to try to pull all of reality together into one single Theory of Everything, scientists have proposed that, at its most fundamental level, our world is made up of tiny vibrating strings, Figure 25,32. String theory suggests that particles are more like tiny little balls of thread. Scientists are using the concept of strings to try to unify all of the fundamental physical forces of matter, gravity and electromagnetism, Figure 25.33.

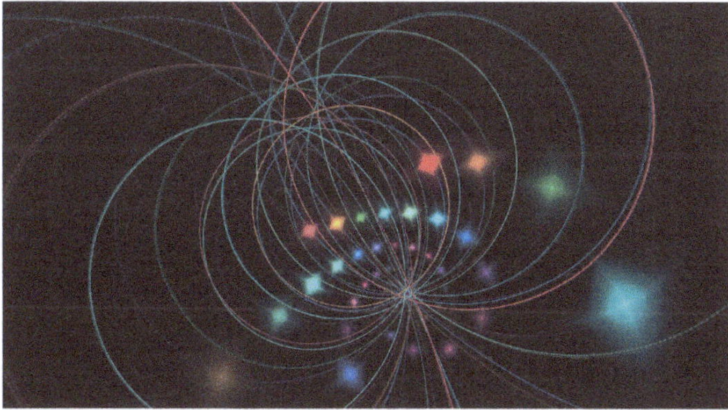

Fig.25.33: Unified Field Theory: Einstein was Unsuccessful - Einstein's failure in achieving a Unified Field Theory didn't stop the others. Nonetheless, reached the same fate - Curtsy: Natali Art collections/Shutterstock

String theory has been used to explain two aspects of the universe. First, scientists believe that string theory could explain why the universe is very finely tuned so that stable structures such as stars, galaxies and even intelligent life can exist. The other way string theory has been used to describe our universe is to suggest that our universe exists within an even higher dimensional space. Our universe can then bump into other "things" in the higher-dimensional space. These interactions may be the causation of contractions and expansions in our own universe, one of which was the proposed Big Bang itself.

While these ideas coming from string theory may sound rather exciting, especially the possibility of arriving at a grand "Theory of Everything," the reality is that such a theory still would not solve the fundamental question. At a cosmology conference at the University of California, Davis, Stephen Hawking—while presenting a talk entitled "Cosmology from the Top Down"— admitted as much: "Even when we understand the ultimate theory, it won't tell us much about how the universe began." Also, where did the strings come from, if they even exist at all, and why do they have the qualities that they apparently have?

THE SUPERNATURAL REASON

Accordingly. if these cosmological explanations are not airtight, what other alternatives are there? There is another reason for the existence of the universe, and that is that existence of God, an almighty being who created it. This is the conclusion reached by Gottfried Liebniz, who originally voiced the question, why is there something rather than nothing? It's also a conclusion of multiple religions, including Islam, Christianity and what some subsets of Hinduism believe.

If an all-powerful, all-knowledgeable Creator exists, then this supreme being would have the capacity and the knowledge to be the creator of the universe that we live in. His creative power would explain why there is something rather than nothing. His knowledge and intentionality would also explain why our universe has been so finely tuned so that human beings could live in it. Since it would not be possible to have infinite chain of causes that depend on other causes, God must be the ultimate ***Uncaused Cause***, which means that He exists necessarily, outside of space and time, and without beginning or end. Finally, the facts that the universe had a beginning at a particular point in time and also has very carefully selected physical parameters means that the ultimate ***Uncaused Cause*** has intelligence and the power of choice. In other words, God is a Person. When He calls Himself "Yahweh" or "I Am" He is referring to His eternal existence and position as Creator as chronicled in Genesis 1.

After all of the exertions of cosmologists to try to understand and describe the universe, God remains the best and only explanation. The American astronomer Robert Jastrow puts it well: "For the scientist who has lived by his faith in the power of reason, the story ends like a bad dream. He has scaled the mountains of ignorance, he is about to conquer the highest peak; as he pulls himself over the final rock, he is greeted by a band of theologians who have been sitting there for centuries." While atheism or agnostics may claim that the concept of God is uncertain, the idea of a Creator God remains a solid explanation for the existence of the universe.

If God is not only all-powerful and all-knowledgeable but also loves us with an everlasting, self-sacrificing love, as the Bible vividly portrays, then the story will not end like a bad dream. The story will end the way the Bible describes—with a new heaven and a new earth where there is no more pain or crying or death, and where we get to live with this amazing God forever.

APPENDIX

i. Most English translations translate God's covenant name, *YHWH*, as "the LORD", but this article will translate it as Yahweh, to emphasize the presence of God's covenant name in these verses. Yahweh is thought to be pronounced 'YHWH', but this is not certain as there has been a long tradition amongst Jews of not saying aloud the holy name of God, and the Hebrew writing does not indicate the vowels, only the consonants.

ii. Basil Fawlty was the character played by John Cleese in the classic 1970s British TV comedy series *Fawlty Towers*. He was prone to angry, irrational outbursts blaming others for his problems. In one episode he took out his frustrations on his stalled vehicle by beating it with a tree branch.

iii. John does this by showing the occasion, the teaching that Jesus drew from them (for example, 'I am the Bread of Life' after the feeding of the 5,000; 'I am the Resurrection and the Life' after the raising of Lazarus), the increased faith of those who were willing to receive truth, and the increased spiritual blindness of those who rejected Christ's claims.

iv. Miracles per se are not necessarily evidence of deity; rather they are evidence of supernatural power. Others in the Bible, from Pharaoh's magicians (Exodus 7:22) to the false prophet (Revelation 19:20), are said to perform miracles.

v. Possibly to increase the sense of expectancy on the part of those who would relate in a particular way to touch—the blind, the deaf, and lepers.

REFERENCES

1. R.M. Grigg, Creation—how did God do it? *Creation* 13(2):36–38.
2. Dawkins, R., *River out of Eden*, Weidenfeld & Nicholson, London, p. 133, 1995.
3. Lewis, C.S., *Mere Christianity*, Geoffery Bles, London, pp. 36–37, 1952.
4. Lewis, C.S., *Surprised by Joy*, Harper Collins, London, p. 245, 2002.
5. But is there a reason why God exists? Yes, there is: God is naturally necessary—by His very nature He can't fail to exist.
6. The form of this argument comes from Craig, W.L., *On Guard*, David, C. Cook, Colorado Springs, CO, p. 54, 2010. See also Kumar, S. and Sarfati, J., *Christianity for Skeptics*, Creation Book Publishers, Atlanta, GA, pp. 17–19, 2012.
7. Wigner, E., The Unreasonable Effectiveness of Mathematics in the Natural Sciences, in: *Communications in Pure and Applied Mathematics* 13(1), John Wiley & Sons, New York, 1960; www.dartmouth.edu/~matc/MathDrama/reading/Wigner.html.
8. Tegmark, M., Is the Universe Made of Math? www.scientificamerican.com, 10 January 2014. http://www.scientificamerican.com/article/is-the-universe-made-of-math-excerpt//.
9. For the original formulation and defense of this argument, see Craig, W.L., God and the 'Unreasonable Effectiveness of Mathematics', *Christian Research Journal* 36:31–35, 2013.
10. See Maydole, R.E., The ontological argument; in: Craig, W.L. (Ed.), *The Blackwell Companion to Natural Theology*, Kindle Locations 15314–15325, John Wiley & Sons Ltd, Chichester, United Kingdom, 2009 (Kindle edition). A helpful exposition of Maydole's modal perfection argument can be found here: Miller, C., Robert Maydole's Modal Perfection Argument, calumsblog.com/apologetics/arguments-for-gods-existence/modal-perfection-argument/, accessed 23 June 2016. The crucial aspect for this point is the first step of Maydole's argument, which argues for the possibility of a supreme being (i.e. a being that nothing can be greater than or equal to).
11. This summary was adapted from Craig, W.L., Does God Exist? reasonablefaith.org, accessed 23 June 2016.
12. Nietzsche's Moral and Political Philosophy, *Stanford Encyclopedia of Philosophy*, 27 July 2007.
13. Dawkins, C.R., Logical Path from Religious Beliefs to Evil Deeds, 2 October 2007.

26

ISAAC NEWTON'S RECOGNITION OF THE CREATOR

SIR ISAAC NEWTON (1642–1727):

A Scientific Genius - Recognition of the Creator

Isaac Newton is well known as one of the greatest scientists who ever lived, Figure 26.1. Less well known is his deep belief in God and his conviction that scientific investigation leads to a greater knowledge of God the Creator of the universe.

Fig.26.1: *Sir Isaac Newton (1642–1727)*

Isaac Newton was born at Woolthorpe, Lincolnshire, England on Christmas Day 1642. On that cold winter night, the sick, premature baby seemed unlikely to live. Gradually, however, he gained strength to survive. But Isaac's first few years were a struggle. His mother had become a widow two months before Isaac was born. Even with the help of her own mother, she had difficulty caring for Isaac in addition to running their farm while the Civil War in England raged around them.

Several years later, his mother married the minister from nearby North Witham, but Isaac remained at Woolthorpe with his grandmother. As he grew, however, he visited his mother frequently. He eagerly read books from his stepfather' well-stocked library, in addition to reading the Bible regularly.

Isaac attended school at King's College in nearby Grantham. Rather than playing outdoor games as a boy, he preferred to make models of such things as windmills and carts. Not only were these in exactly the right proportions, but all of the moving parts actually worked.

Isaac's mother was widowed for the second time when he was 14 years old. Isaac was taken out of school to run the family farm to support his mother and her three younger children. However, Isaac missed his studies greatly and his mother recognized this. When King's College offered to waive tuition fees because of his ability and poor circumstances, Isaac returned and completed his schooling. Teachers and other students were impressed with the boy's knowledge of the Bible.

INTENDED TO BECOME A MINISTER

Isaac then went to *Trinity College at Cambridge University* with the intention of becoming a Church of England minister. Again, life was not easy for him. As he was unable to afford the tuition fees, he worked many hours each day serving meals and doing other jobs for the professors in order to pay his way. Isaac's knowledge of the Bible continued to impress those around him.

At that time the ideas of the ancient Greek scholars still dominated what was taught in science, and recent scientific discoveries were largely ignored. This greatly annoyed Isaac Newton who firmly believed that ideas in science should be tested and only accepted if their usefulness could be demonstrated. He was committed to the experimental method of science.

Isaac graduated in 1665, shortly before an outbreak of Black Death swept through London. All universities were closed while the plague raged. During this time, Isaac returned to his family's farm, now run by his young half-brother. He continued his study and research, working on the binomial theorem, light, telescopes, calculus and theology. After supposedly seeing an apple fall in the garden, he investigated gravity, but was unable to solve the puzzle until some years later. (It should be noted that some authorities question this 'apple' story. They say that the first mention of it came through the antireligious French philosopher and skeptic, Voltaire, who reputedly heard it from Newton's grandniece.)

REVOLUTION IN MATHEMATICS

Newton applied his binomial theorem to infinite series and from there developed calculus, a revolutionary new form of mathematics. For the first time it was possible to accurately calculate the area inside a shape with curved sides, and to calculate the rate of change of one physical quantity with respect to another.

Fig.26.2: Sir Isaac Newton used prisms to show that sunlight was made up of all the colors of the rainbow. This proved that the ancient Greeks ideas about light were wrong.

A similar system of mathematics was developed by German mathematician Gottfried Leibniz. For a long time, there was great confusion, with each being accused of stealing the other's work. It was a distressing time for both. Many years later, it was established that each had developed calculus independently at roughly the same time. Neither was a cheat.

Optics

When Cambridge University reopened in 1667, Isaac Newton returned to do a Master's Degree, while teaching and doing research, Figure 26.2.

Newton used prisms to show that sunlight was made up of all the colors of the rainbow. This proved that the ancient Greeks' ideas about light were wrong. In Newton's time, astronomy was severely hampered because lenses in telescopes broke some of the light into unwanted colors, causing a somewhat unclear view. Although not the first to consider using a curved mirror instead of a lens, Newton was the first to successfully construct a telescope using this principle—a principle still used today in many telescopes, Figure 26.3.

Fig.26.3: Isaac's Optical Principle Still Used Today in Many Telescopes

THE ROYAL SOCIETY

In 1672, Newton became a member of the Royal Society—a group of scientists committed to the experimental method. He presented one of his new telescopes to the Royal Society along with his findings on light. The Royal Society set up a committee led by physicist Robert Hooke to evaluate Newton's findings. Hooke was a scientist employed by the Royal Society to evaluate new inventions. However, Hooke had his own ideas on light and was slow to accept the truth of Newton's findings. This surprised and disappointed Newton, who even considered not circulating his discoveries in the future.

While it is sometimes said that Newton was too sensitive to critical evaluation of his work, he was merely concerned that the time spent justifying past findings was preventing him from making new discoveries.

POLITICAL INTERFERENCE

Isaac Newton lived at a time when politics, religion and education were not separated. King Charles II commanded that everyone who taught at places such as *Trinity College*, where Church of England ministers were trained, must themselves be ordained as Church of England ministers after seven years. This included people such as Newton who taught only mathematics and science, not theology.

Although a devout Christian, Newton was not in full agreement with all the doctrines of the Church of England. Thus, his conscience would not allow him to accept ordination. He was also strongly opposed to political involvement in both religious matters and education. The only way for Newton to keep his job was for the king to make an exception in his case. Others who had previously asked for this had been refused.

So, Newton headed south to London for six weeks to plead his case before the king. During his time in London, he became better acquainted with other scientists in the Royal Society. Those who had known him only through his letters defending his discoveries had mistaken his confidence in his work for arrogance.

His impatience to get on with new work had been mistaken for bad temper. Now the scientists realized what a friendly and considerate person he was and they rallied to his aid. Fortunately, for Newton and for science, the king granted Newton's request to continue at *Trinity College* without being ordained.

FOCUSED ON GRAVITY

In Newton's Day, many people were superstitious or afraid of what they could not understand—such as the appearance of a comet, which was considered a sign of coming disaster. Even scientists generally considered the motion of planets and the motion of bodies on the earth as separate problems.

In contrast, Newton reasoned that since the same God created the heavens as well as the earth, the same laws should apply throughout.

In 1684, Newton again began to consider gravity. He developed his theory of universal gravitation, which used what is known as the inverse square law. ***He developed his three laws of motion (movement) and proved mathematically that the same laws did, in fact, apply both to the heavens and the earth***. His faith had focused his thoughts in the right direction.

When Newton was investigating the movement of the planets, he quite clearly saw the hand of God at work. He wrote:

"This most beautiful system of the sun, planets, and comets, could only proceed from the counsel and dominion of an intelligent Being. ... This Being governs all things, not as the soul of the world, but as Lord over all; and on account of his dominion, he is wont to be called "Lord God" Παντοκράτωρ [Pantokratōr cf. 2 Corinthians 6:18], or "Universal Ruler". ... The Supreme God is a Being eternal, infinite, absolutely perfect.

Opposition to godliness is atheism in profession and idolatry in practice. Atheism is so senseless and odious to mankind that it never had many professors.

When I wrote my treatise about our system, I had my eye upon such principles as might work with considering men for the belief of a deity; and nothing can rejoice me more than to find it useful for that purpose."

Again, Newton encountered difficulties with his old rival Robert Hooke. A number of scientists believed that an inverse square law probably applied, but they had not been able to prove that this would produce the elliptical orbits observed by famous German astronomer Johannes Kepler. Despite Hooke' boasts to the contrary he too failed to be forthcoming with proof.

In contrast, Newton succeeded; but Hooke wanted some of the credit.

The Royal Society did not wish to be seen to take sides. This, together with shortage of finances, made the Royal Society reluctant to publish Newton's landmark book *Principia Mathematical.* Newton's friend, astronomer Edmond Halley, came to his aid and privately financed the publication of Newton's three-part book in 1687. (Halley later used Newton's laws in his work on comets which, like the planets, move in elliptical orbits around the sun.)

ROYAL OPPOSITION

After 1685, Newton again encountered the problem of a monarch who tried to mingle politics, religion and education. The new king, James II, wanted *Trinity College* to award unearned degrees to those whose religious

beliefs agreed with his own. Because they would not do this, Newton and eight other teachers from Trinity College were brought before the High Court on trumped-up charges. Although the charges were rightfully dismissed, the episode had been a great strain on the men.

Isaac Newton's times of hardship and struggle throughout his lifetime did not produce bitterness. Instead, Newton's own words show that this brought him closer to God. 'Trials are medicines which our gracious and wise physician gives because we need them; and the proportions the frequency and weight of them to what the case requires. Let us trust his skill and thank him for the prescription.'

NERVOUS BREAKDOWN

Isaac Newton represented *Cambridge University* as a Member of Parliament in 1689 and 1690. In 1690, his health failed. This illness was probably a nervous breakdown brought on by many years of working long hours and enduring too much stress. Eventually he fully recovered. For the next few years, Newton pursued his other great love—***studying the Bible***. The books he wrote included *Chronology of Ancient Kingdoms* and *Observations Upon the Prophecies of Daniel.*

In 1696, the government appointed Newton to the post of Warden of the Mint. He supervised the replacement of England's old and damaged coins with those which were new and more durable, and even helped break up a counterfeiting ring.

In 1701, Newton began another short term as parliamentarian. Two years later he was elected president of the Royal Society. His re-election to that position every year for the rest of his life showed the high esteem in which he was held by fellow scientists. Now that he had returned to science, Newton published his earlier work on light. His book, *Optiks,* contained both his own findings and suggestions for further research. His country officially recognized his work in 1705 when he became the first person to receive a knighthood for scientific achievement.

Newton died in 1727, at the age of 84. He was buried in Westminster Abbey.

Isaac Newton's contributions to science were many and varied. They covered revolutionary ideas and practical inventions. His work in physics, mathematics and astronomy is of importance even today. His contributions in any one of these fields would have made him famous; collectively, they make him truly outstanding. But Newton remained a modest man who loved his Lord and Savior.

He loved God and believed God's Word—all of it. He wrote, '*I have a fundamental belief in the Bible as the Word of God, written by men who were inspired. I study the Bible daily.*'

(Quotes by Newton are taken from the book by J.H. Tiner, *Isaac Newton—Inventor, Scientist and Teacher,* Mott Media, Milford (Michigan), 1975.)

NEWTON'S BOOK

A Scientific Masterpiece

The year 1987 marked the 300th anniversary of the publication of one of the world's masterpieces in scientific literature.

It was in 1687 that Sir Isaac Newton, English mathematician, physicist and astronomer, published his monumental work *Philosophiae Naturalis Principia Mathematica* (mathematical principles of natural philosophy). In this work, Newton presented his famous three laws of motion:

- force-free motion is uniform;
- accelerated motion is proportional to the impressed force; and
- for every action there is an equal and opposite reaction.

From these, together with the law of universal gravitation, the whole science of matter in motion is derived.

Sir Isaac Newton is regarded by many scholars and historians as the greatest scientist who ever lived. Yet he also firmly believed that the Bible was God's Word. He wrote much on biblical subjects, and even wrote a book defending Archbishop Ussher's chronology of the world (Ussher set the date of Creation as 4004 BC).

Newton believed in a literal six-day creation, and that the worldwide flood of Noah's time accounted for most geological phenomena. He also firmly believed in Christ as his Savior.

In thinking on the publication of Newton's *Principia*, it would be well to reflect on this great scientist's view of Scripture:

"I find more sure marks of authenticity in the Bible than in any profane history whatsoever."

ISAAC NEWTON

Biblical Creation

Isaac Newton has often been cited as a six-day creationist or enrolled favorably in support of biblical creation. Moreover, his chronological writings do seem to allow for a world not much older than 6,000 years. But his less-than-literal treatment of *Genesis 1–11* should make creationists cautious to appropriate Newton to this end.

There is also reason to believe that Newton was possibly the first to advance the day-age theory, and inspired some of the earliest naturalistic reinterpretations of *Genesis 1–11*.

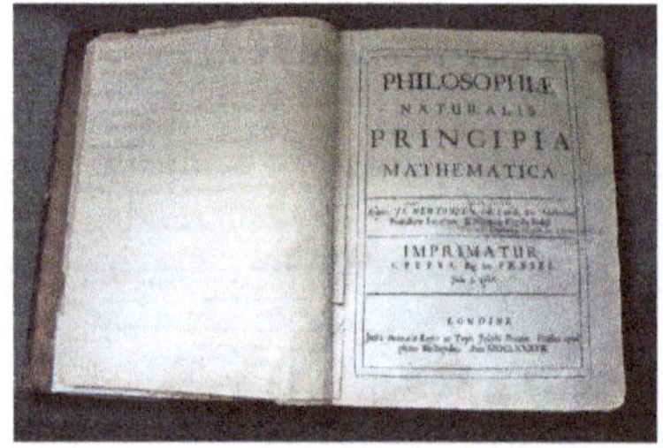

Fig.26.4: *Isaac Newton's Own First Edition copy of Philosophiæ Naturalis Principia Mathematica - Andrew Dunn/CC BY-SA 2.0*

Newton has been called "one of the greatest creative men of genius who ever existed," the "high priest of science," and the "last of the magicians." As early as 1728, one reviewer even said that he was "the greatest man in the world, not only in this age, but in any age, since the world began/" His intellectual impact in the world was such that, "it was not till a century after his death that men freed themselves from his authority sufficiently to do important original work in the subjects of which he had treated."

With a such a legacy it is only natural for creationists to seek to utilize Newton in support of biblical creation. But what were his actual beliefs and how did he treat *Genesis 1–11*? This writing will illustrate that Newton's writings on creation, chronology, and the Christian faith were sometimes enigmatic and less than orthodox.

Newton's *Principia Mathematica* and *Opticks*

Most of Newton's seminal work was completed "between the ages of 21 and 23." In these two years, he "formulated his basic laws of mechanics, his optical observations on the nature of light, the calculus, and the law of universal gravitation." But his greatest literary masterpiece, published several years later in 1687, was *Philosophiæ Naturalis Principia Mathematica* Figure 26.4. This book has been praised as the "most famous scientific work of all time." It is therefore both surprising and significant to realize that Newton wrote more on theology than he ever did on science.

THE RIDDLE OF THE UNIVERSE

Therefore, to what extent did Newton allow his theology to influence his science? Storr, following Keynes, believes that Newton "regarded the riddle of the universe in theological terms." Even if this assessment is justified, Newton apparently disapproved of the only direct reference to God in the first publication of *Principia*: "God placed the planets at different distances from the sun," because in all subsequent editions, he changed it to "the planets were to be placed at different distances from the sun." This alteration was mitigated by the fact that, to all subsequent editions of *Principia*, he also added a short theologically explicit addendum entitled "General Scholium." In the General Scholium, Newton writes:

"This most beautiful system of the sun, planets, and comets, could only proceed from the counsel and dominion of an intelligent and powerful Being .... This Being governs all things, not as the soul of the world, but as Lord over all; and on account of his dominion, he is wont to be called Lord God [emphasis in original]."

This is followed by two more pages of theological philosophizing, after which Newton concludes, "thus much concerning God; to discourse of whom from the appearances of things, *does certainly belong to natural philosophy* [emphasis added]," So, although Newton removed any explicit theologizing from *Principia*, these latter remarks appear to support the idea in principle.

His endorsement of intelligent design can also be found in the second edition of *Opticks*, where Newton wrote, "And tho' every true step made in this philosophy brings us not immediately to the knowledge of the first cause, yet it brings us nearer to it, and on that account is to be highly valued." In the fourth edition, he took this even further, saying:

"... the main business of natural philosophy is to argue from phenomena without feigning hypotheses, and to deduce causes from effects, till we come to the very first cause, which certainly is not mechanical; and not only to unfold the mechanism of the world, but chiefly to resolve these and such like questions."

Newton had no time for atheism which he regards as "so senseless & odious to mankind that it never had many professors."

NEWTON'S SEPARATION OF SCIENCE AND THEOLOGY

Outside of the Royal Society, Newton was also "unambiguously favorable" toward those who used his system in their own apologetics. For example, in 1692, Newton wrote to Richard Bentley (1662–1742): "When I wrote my treatise about our Systeme I had an eye upon such Principles as might work with considering men for the beleife of a Deity & nothing can rejoyce me more then to find it usefull for that purpose."

But in another short paper, *Seven Statements on Religion*, Newton states as his first principle, "religion & philosophy are to be preserved distinct. We are not to introduce divine revelations into philosophy, nor philosophical opinions into religion." Moreover, as Manuel relates, "When Newton was President of the [Royal] Society, the journal-books record, he banned anything remotely touching on religion, even apologetics."

This makes it hard to ascertain, therefore, to what extent Newton allowed theology to infiltrate his science, given that he preferred—if not in principle, at least in practice—to keep the two separate. As Snobelen has observed, there is "more explicit theology in Charles Darwin's *Origin of Species* (1859) than in the first edition of Newton's great work." That said, Snobelen still contends:

"Recent work on early modern science has demonstrated a direct (and positive) relationship between the resurgence of the Hebraic, literal exegesis of the Bible in the Protestant Reformation, and the rise of the empirical method in modern science .... In this, Newton also played a pivotal role. As strange as it may sound, science will forever be in the debt of millenarians and biblical literalists [emphasis added]."

This may be true, but Newton does not fit the classification of a traditional 'biblical literalist' easily.

NEWTON'S HETERODOXY

Whilst it is not difficult to find evidence of Newton's enthusiasm for the Bible, his theological views were often heterodox. As Thomas Hearne (1678–1735) once wrote of him:

"Sir Isaac Newton, tho' a great mathematician, was a man of very little religion, in so much that he is ranked with the heterodox men of the age."

Likewise, Snobelen concedes that although Newton was a "devoted believer," in other ways he was a "damnable heretic." It is hard to dispute the fact that Newton was an Arian. In his treatment on the religion of Newton, Manuel refrains from "pigeonholing" Newton like this, but still concedes that he was anti-trinitarian. This kind of heterodoxy can be seen in 12 propositions he made on the nature of Christ, and in *Irenicum, or Ecclesiastical Polyty tending to Peace*, where he identifies *Jesus as the archangel Michael.* Newton also questioned the reality of a literal devil and evil spirits. Yet, Snobelen maintains that "Newton's demonology was an exegetical option, not a sign of the encroaching Enlightenment." But this is hard to accept, given that, as Snobelen himself admits:

"Newton's denial of evil spirits was well outside the theological mainstream in his own day and for a long time afterward. His position would have been viewed as a runway to infidelity, a capitulation to cold, dark atheism, a disturbing disenchantment of the world or even a delusion inspired by Beelzebub himself. If only his witching-hunting colleagues at the Royal Society had known."

NEWTON'S RADICAL FRIENDS

Newton's secrecy in these matters is telling. Moreover, as King and Popkin have argued, Newton's views were not untouched by the writings of seminal Enlightenment thinkers like Baruch Spinoza (1632–1677) and Richard Simon (1638–1712). As his interpretation of *Genesis 36:31* shows, Newton employed the same exegetical argument used by Simon to dismiss the Mosaic authorship of certain sections of Genesis. He was also good friends of John Locke (1632– 1704), widely regarded as "England's foremost Enlightenment thinker" and "father of liberalism." Newton asked Locke "in the strictest confidence" to help him translate a controversial treatise on *1 John 5:7* and *1 Timothy 3:6* into French and have it published anonymously in the Netherlands. The treatise was entitled *An historical account of two notable corruptions of Scripture* and was addressed to Jean Le Clerc (1657–1736),[39] "Europe's most tenacious protagonist of rationalist Christian theology". Thus, Popkin writes:

"Newton, like Spinoza and Simon, took seriously the problems that had arisen in the collection, editing, and transmission of Scripture that made it difficult if not impossible to find the pure original text. Newton, unlike the fundamentalists of the past century and a half, was not committed to claiming the inerrancy of the biblical text, but was committed to finding its message for mankind [emphasis added]."

Whilst Popkin may be overstating the case, the fact remains that Newton was not untouched by the kind of thinking that fueled the Enlightenment. Put more strongly, Newton's contribution to the sciences and theology did little to hinder its development. As Israel observes:

"Although down to 1750, in Europe as a whole, the struggle for the middle ground remained inconclusive, much of the European mainstream had, by the 1730s and 1740s, firmly espoused the ideas of Locke and Newton which indeed seemed uniquely attuned and suited to the moderate Enlightenment purpose."

So although Newton "devoted close to six decades to a passionate study of the Bible, theology, prophecy, church history and natural theology", his heterodox reading of Scripture left room for a less-than-literal interpretation of Genesis.

NEWTON'S EARLIER TREATMENT OF GENESIS

In 1680, Thomas Burnet (1635–1715) sent Newton a pre-publication copy of *Telluris theoria sacra* (1681) for review. Burnet wanted to establish scientific reasons to "justify the doctrines of the *Universal Deluge*, and of a *Paradisiacal* State, and protect them from the cavils of those that are no well-wishers to sacred history". In the correspondence which ensued, Newton suggests that the "heat of the sun" might explain how the oceans were formed and dry land appeared on the earth. This imaginative re-interpretation of Genesis 1:9 requires the "diurnal revolutions of the Earth" to be "very slow" at the beginning of creation so that "the first 6 revolutions or days might continue time enough for the whole Creation, & ye Sun in that time might convert & shrinke the parts of the Earth about the Æquator". In his reply to Newton, Burnet points out that the sun was only made later in the week:

"… methinkes you forget Moses (whom in another place you will not suffer us to recede from) in this account of the formation of the Earth; for hee makes the seas & dry land to bee diuided & the Earth wholly formd before the Sun or Moon existed. These were made the fourth day according to Moses, & the Earth was finisht the 3d day"

This gives Newton cause to explain, in the letter which follows, that *Genesis 1–2* was written phenomenologically to describe what Moses would have seen, had he been there to witness it himself. The point is not to read Genesis as science or as fiction, says Newton. Instead, Genesis is a "true description" of creation accommodated to the "vulgar" understanding of Moses' first readers. Taken on its own, Newton's overall conclusion is both conservative and orthodox:

"… me thinks one of the tenn commandments given by God in mount Sina, prest by divers of the prophets, observed by our Saviour, his Apostles & first Christians for 300 years & with a day's alteration by all Christians to this day, should not be grounded on a fiction."

But this stance does not preclude him from asserting that sun, moon, and stars were not created on "the fourth day nor in any one day of the creation"; that Moses might not mention their creation at all; that the duration of the first and second days might be "as long as you please"; or that Burnet's theory could allow "a year for each day's work" without misinterpreting the text.

NEWTON'S LATER TREATMENT OF GENESIS

On the whole, it is possible that Newton, at this stage in his career, maintained a more literal approach to Genesis, albeit tenuously. From the mid-to-late 1680s, however, his opinions on Genesis moved in a more radical direction. In a treatise on Revelation, Newton calls ***the story of the fall of man a "parable"; the trees in Eden, "mystical"; and the serpent, "only a symbol of the spirit of delusion***". With regards to the six days of creation he writes:

"And so, the six days of the Creation may signify not only six years but even six thousand years … or any other six long times. For the history of the creation is not in all things litteral. In that Paradise the flaming sword & trees of life & knowledge may be as much figurative descriptions of something we now understand not as the tree of life is in the Paradise to come, & in a parabolical description of the creation a day may be used figuratively as well as other things are especially since there was no light till the end of the first day nor sun till the fourth—to make natural days. The evenings & mornings of Moses respect all parts of the Earth alike so that it was evening all over the Earth in the beginning of each day of Moses & morning all over it in the end of each day: & therefore his evenings & mornings were not natural ones. For had they been natural ones it would have been morning in one part of the Earth when it was evening in another."

For this reason, it is evident that Newton did not advocate a literal six-day creation, nor was he "committed to the literal truth of Holy Scripture." Whether or not it was Burnet's writings that finally convinced him to abandon the literal historicity of the hexameron is not easy to ascertain. What is known, however, is that in Burnet's subsequent cosmological treatise, *Archaeologiae philosophicae sive doctrina antiqua de rerum originibus* (1692), he calls ***the story of Adam and Eve a parable*; *rejects the creation of Eve from Adam's rib; considers a speaking serpent to be utterly nonsensical*; *does not believe that Adam and Eve were capable of sewing their own clothes; denies that the Garden of Eden was guarded by real Cherubim*; *disbelieves that Adam had named all the animals in a single day or that the universe is less than 6,000 years old*.**

NEWTON'S CHRONOLOGY

That said, Newton's chronological writings do treat aspects of *Genesis 1–11* as literal history. He affirms the repopulation of the world from Noah's sons and traces the origin of nations back to Babel. He also asserts, in the conclusion to the first draft of his chronology of ancient kingdoms, that "mankind could not be older then [*sic*] is represented in scripture" which he later revised to "mankind could not be *much* older than is represented in Scripture [emphasis added]." Therefore, it is probable that Newton still believed in an earth not much older than 6,000 years. Some scholars even assert that Newton's chronology depended upon or defended James Ussher's *Annales veteris testamenti, a prima mundi origine deducti* (1650).

But this claim is questionable for the following reasons:

- firstly, Newton's chronology does not begin with Adam, it begins with Noah;
- secondly, the manner in which he places Scripture alongside many other historical sources suggests, as Westfall rightly points out, that "Newton's view of human history did not center on the Bible", but that he "treated the historical books of the Old Testament as human documents to be used in concert with other human documents;"
- thirdly, Arthur Bedford (1668–1745), a contemporary of Isaac Newton, wrote one of the earliest critiques of *The Chronology of Ancient Kingdoms Amended* (1728) in which he demonstrates how Newton's chronology "differs *toto cælo* from all the learned men in the world" including, most notably, James Ussher. In fact, I have also not found a single reference, positive or negative, to Ussher within the Newtonian corpus.
- Finally, in all his chronological writings, Newton never provides an age for the earth or a date for creation. The omission of such an obvious chronological detail is telling.

Newton (2006:191, 376), nevertheless, is helpfully critical of the Egyptian, Persian, and Syrian records, which "out of vanity" have exaggerated the antiquity of their kingdoms by "some thousands of years older than the world". Consequently, his stated objective is to correct these erroneous chronologies by referring to the more reliable records preserved by the Greeks and Hebrews.

But, like Bedford observes: "As to what [Newton] saith, that he hath made it agreeable with sacred history; it is hard to know, whether he was in earnest or in jest." This is because, in Bedford's assessment, Newton's chronology ignores, misquotes, and contradicts the biblical record in several places. His conclusion is forceful: "such poison ought not to go abroad into the world" for it undermines the integrity of the sacred text. For these reasons, we should be hesitant to endorse Newton's chronology uncritically.

NEWTON'S HERMENEUTICAL LEGACY

Burnet was the first to attempt an explanation of *Genesis 1–11* in collaboration with Newton himself and in terms of Newtonian physics. And what was the result? A cosmology that had very little to do with Genesis at all. In Burnet's estimation, "the very letter of the *Hexaemeron* [is] most absolutely contradictory to the nature of things, as well as to all philosophical reasons" and, "people could neither understand nor bear a plain and philosophical explication of it".

What this called for, in practice, was a *scientific* hermeneutic, whereby "philosophy is the interpreter of Scripture in natural things." It was a principle that resonated strongly with Charles Blount (1654–1693), the "chief deist of his age," who eagerly plagiarized sections of Burnet's writings to further his radical agenda in England.

Like Burnet, William Whiston (1667–1752) wrote his own philosophical version of *Genesis 1–11*, *A new theory of the Earth* (1696), which he dedicated to Newton. In it he calls literal six-day creation a "vulgar hypothesis", arguing instead that the days in Genesis should be understood as years. He also argues that *Genesis 1* does not tell us how matter came into being or how the universe was created, being restricted exclusively to the origin of the earth. Far from disapproving of Whiston's theory of the earth, in 1702, Newton appointed him as his successor to the Lucasian chair of mathematics at Cambridge.

Edmund Halley (1656–1742), a close friend and admirer of Newton, also refused to accept that the days in Genesis should be taken as "natural days". He maintained that the Scriptures could not provide a reliable account of the age of the earth. Instead, Halley proposed that the salinity of the oceans could give a better estimate.

In France, Bernard Le Bovier de Fontenelle (1657–1757), the first to publish a biography of Newton, also rejected *Genesis 1–11* as literal history. He used his editorial influence at the French *Académie des sciences* to actively promote geological views that precluded the biblical Flood, whilst at the same time censoring any scientific interpretations that assumed or asserted it. Likewise, ***Voltaire (1694–1778) cites Newton in support of his claim that Genesis was not written by Moses***, and calls the Pentateuch a "***stupid falsehood***" and "absurd fable", "written by fools, commented upon by simpletons, taught by knaves," and filled with "innumerable geographical and chronological errors and contradictions".

Comte de Buffon (1708–1788) also employed Newtonian physics in his rigorously naturalistic reinterpretation of *Genesis 1–11*. This is noteworthy for the simple fact that Buffon, "more than anyone else, was responsible for a new chronology of the earth, that is, for the acceptance of a vast time scale."

In Germany, one of the most influential philosophers of the last three centuries, Immanuel Kant (1724–1804), launched his academic career with a scientific treatise entitled, *General natural history and theory of the heavens, or an essay on the constitution and mechanical origin of the whole universe, treated in accordance with Newtonian principles* (1755). In this controversial book, he constrains God to a first cause whilst trying to explain, on Newtonian grounds, how matter could arrange itself into the present universe over time.

Newton might never have anticipated or desired such a legacy, but his influence directly and indirectly affected how *Genesis 1–11* would be read by future generations.

NEWTON'S DISTORTED TREATMENT OF GENESIS

For the 1680s, Newton's treatment of Genesis was far from orthodox, and possibly the first articulation of the day-age hypothesis in history. If correct, this makes Newton a key figure in the ensuing hermeneutical revolution which shaped how *Genesis 1–11* would be read for the next three centuries. This accords well with Israel's analysis for the onset of the Enlightenment, which he places within the same time period (i.e., from 1650–1680). Thus the timing of Newton's comments on Genesis, dated to the late 1680s, happen to correlate strongly with the inauguration of the Enlightenment period. Prophetically perhaps, in the front matter of the first edition of *Principia*, ***Halley regards Newton as superior to Moses***, "Who opens the treasure chest of hidden truth .... No closer to the gods can any mortal rise." The curious corollary to all this, is that ***Newton accepted such praise*** in print as the foreword to his *magnum opus*.

Did Newton think of himself as Moses' scientific successor?

Perhaps, perhaps not. Either way, Moses did not fare well in the next century. As Manuel has observed, for the Enlightenment to flourish in the 18th century, "certain basic intellectual needs" had to be met, the first being: "***a replacement of Genesis***." Newton did little to hinder such a venture. It is probable that several theologians from the next generation took their lead from him.

For reasons such as these, ***Newton's legacy is a greater hindrance*** than help to biblical creationists. Although his science was inspired by Scripture, his view of Scripture was increasingly shaped by his science.

HOW EINSTEIN HELPED PROVE GOD IS THE CREATOR

Einstein figured out his famous Theory of General Relativity in the early 1900s and put out a paper on it in 1916.

The famed British mathematical physicist Sir Arthur Eddington saw the paper and recognized that there was an opportunity with the 1919 total solar eclipse in Brazil to put the Theory of General Relativity to the test.

WHY THE ECLIPSE WAS CRUCIAL

Why an eclipse? A unique thing about Einstein's theory is it predicted the gravity of the sun would bend the light of stars as that light passed the sun. It would take a total solar eclipse blocking the sun to make it possible for scientists to see that light pass by the sun's limb.

So, Sir Arthur Eddington organized a trip to Brazil." "They observed the sun during the solar eclipse and the stars right next to the limb. And, indeed, they were able to confirm that the starlight was bent by the amount the Theory of General Relativity predicted."

That was an amazing confirmation. The findings was picked up by newspapers around the world and suddenly "Albert Einstein" became a household name. And his Theory of General Relativity became established in the mind of physicists and astronomers."

WHERE THERE'S A BEGINNING, THERE'S SOMEONE TO BEGIN IT

It's that Theory of General Relativity that predicts there's a beginning to the universe. Until Albert Einstein's theory came along, astronomers and physicists thought the universe was infinitely old. The Theory of General Relativity had proved the universe is not old. It's finite in time. It has a beginning, which implies there must be a Beginner who was responsible for bringing the universe into existence.

Those who followed in Einstein's footsteps added more evidence with space/time theorems to back up this hugely important idea. The space and time were created. Space and time had a beginning. Which means there must be an agent beyond space and time that created our universe of matter, energy, space and time.

GOD IN THE SCIENCE

This said to the inquisitive and searching minds of many people that of all the gods of all the religions, the One True God was the God of the Bible.

Non-biblical religions teach that God or gods create within space and time that eternally exists. Nonetheless, with the Biblical God, the universe doesn't exist until space and time exists. The God of the Bible created space and time. It is evident that thanks to the space/time theorems, it is the God of the Bible who created the universe, not the others idols and gods of the world.

Many is the doubter who would like to do away with Einstein's famed Theory and its surrounding theorems and implications, but to no avail.

YOU CAN BE CONFIDENT IN THIS

There are about 30 of these theorems. And basically, the conclusion is indeed are stuck with the implications of those space/time theorems. In fact, today, General Relativity ranks as the most exhaustively tested and best proven principle in all of physics.

The astronomers and scientists are often stating, "There's nothing we can be more confident of than the reliability of the Theory of General Relativity. Which means that likewise we have confidence that there must be a God beyond space and time who created our universe."

ALBERT EINSTEIN'S GOD LETTER

Albert Einstein's so-called God letter first surfaced in 2008, when it fetched four hundred and four thousand dollars in a sale at a British auction house. The letter came back into the news earlier this month, when its owner or owners auctioned it off again, this time at Christie's in New York, and someone paid $2.9 million for it, a pretty good return on investment, and apparently a record in the Einstein-letters market. The former top seller was a copy of a letter to Franklin Roosevelt from 1939, warning that Germany might be developing a nuclear bomb. That one was sold at Christie's for $2.1 million, in 2002.

Although it bears his signature, Einstein didn't actually write the bomb letter. It *was written* by the physicist "Leo Szilard," based on a letter that Einstein had dictated. But, if auction price is at all relative to historical significance, that letter should be way more valuable than the God letter.

IS THERE A GOD?

Do I have free will?

The God letter was cleverly marketed, though. "Not only does the letter contain the words of a great genius who was perhaps feeling the end fast approaching," Christie's said *on its Web site*, "It addresses the philosophical and religious questions that mankind has wrestled with since the dawn of time: Is there a God? Do I have free will?" The press release called it "one of the definitive statements in the Religion vs Science debate." Journalistic interest was stirred up by the question of whether the letter might contradict other comments that Einstein is recorded having made about God.

Fig.26.5: *Einstein - A Night-sky Theology-Sense of Universe - Curtsy - Photograph by Ernst Haas / Getty*

This all made the letter sound a lot more thoughtful than it is. Einstein, Figure 26.5, did have views about God, but he was a physicist, not a moral philosopher, and, along with a tendency to make gnomic utterances—"***God does not play dice with the universe***" is his best-known aperçu on the topic—he seems to have held a standard belief for a scientist of his generation. He regarded organized religion as a superstition, but he believed that, by means of scientific inquiry, a person might gain an insight into the exquisite rationality of the world's structure, and he called this experience "*cosmic religion*."

It was a misleading choice of words. "Cosmic religion" has nothing to do with morality or free will or sin and redemption. It's just a recognition of the way things ultimately are, which is what Einstein meant by "God." The reason that God does not play dice in Einstein's universe is that physical laws are inexorable. And it is precisely by getting that they *are* inexorable that we experience this religious feeling. There are no supernatural entities out there for Einstein, and there is ***no uncaused cause***. The only mystery is why there is something when there could be nothing.

In the God letter, the subject is not the cosmic religion of the scientist. It is the organized religion of the believer, a completely different subject. Einstein wrote the letter, in 1954, to an émigré German writer named Eric Gutkind, whose book "Choose Life: The Biblical Call to Revolt" he had read at the urging of a mutual friend and had disliked so much that he felt compelled to share his opinion of it with the author. A year later, Einstein died. Gutkind died in 1965; it was his heirs who put the letter up for auction, in 2008.

The letter to Gutkind is conspicuously short on metaphysics. It's essentially a complaint about traditional Judaism. Einstein says that he is happy being a Jew, but that he sees nothing special about Jewishness. The word God, he says, is "nothing more than the expression and product of human weakness," and the Hebrew Bible is a collection of "honorable, but still purely primitive legends."

In some news accounts, Einstein is quoted as calling the Biblical stories "nevertheless pretty childish," but that is not what his letter says. That phrase was inserted by a translator, apparently at the time of the first auction. Nor does Einstein call Judaism "the incarnation of the most childish superstitions," also a translation error. The word that he uses is "*primitiven*"—that is, "primitive," meaning pre-scientific. He is saying that, before humans developed science, they had to account for the universe in some way, so they invented supernatural stories. (Such is the nature of our own super-scientific age, however, that if you perform a search for "Einstein childish God," you will get thousands of hits. Einstein will be eternally associated with a characterization he never made.)

Einstein had what might be called a ***night-sky theology***, a sense of the awesomeness of the universe that even atheists and materialists feel when they gaze up at the Milky Way.

Is it too awesome for human minds to know?

A scientist from a generation before Einstein, William James, thought that maybe we can't—maybe our brains are too small. There might indeed be something like God out there; we just can't pick it up with the radar we've got. In James's lovely metaphor, "We may be in the universe as dogs and cats are in our libraries, seeing the books and hearing the conversation, but having no inkling of the meaning of it all."

The best thing in Einstein's letter to Gutkind is not the grouchy dismissal of traditional theology. It's the closing paragraph, where Einstein puts all that aside. "Now that I have expressed our differences in intellectual convictions completely openly," he writes, "it is still clear to me that we are very close to each other in the essentials, that is, in our evaluations of human behavior." He thinks that if he and Gutkind met and talked about "concrete things," they would get along fine. He is saying that it doesn't matter what our religious or our philosophical commitments are. The only thing that matters is how we treat one another. I don't think it took a genius to figure this out, but it's nice that one did.

Did Albert Einstein believe in the existence of a creator of the universe

Albert Einstein said "the more I study science, the more I am amazed by the complexity of the universe and the more I believe in the existence of a creator.

You may find a lot of similar but different quotes by Einstein, such as:

"I want to know how God created this world. I am not interested in this or that phenomenon, in the spectrum of this or that element. I want to know His thoughts; the rest are details.

OVERVIEW

Albert Einstein, was one of the most important man in history. He is perhaps the most famous physicist in history, and his work has contributed to multiple advancements in the sciences. Albert Einstein's is born in March 14th, 1879. He is best known for developing the theory of relativity, as well as his involvement in research that led to the development of the first atomic bombs.

He was an excellent student at math and the sciences from an early age, and it's said that he taught himself algebra and geometry over one of his summer holidays from school.

After graduating from college in 1900, he obtained a job at the patent office in Bern, Switzerland. There, he researched and approved patents on numerous technological advancements of that era, such as the typewriter. He applied his knowledge of technology and physics further by obtaining a PhD from the University of Zurich in 1905 at the age of 26. Due to his great success and the popularity of his published papers among those in the scientific community, he went on to become a lecturer at the University of Bern, the first of many positions he held in higher academics.

Einstein traveled frequently throughout his research and professional career, and visited the United States for the first time with a visit to New York City in 1921. At the outset of World War II, he discovered that his cottage in Belgium was ransacked by Nazi soldiers in the area. He then decided to renounce his German citizenship (though he kept his Swiss citizenship), and officially became a US citizen in 1940.

THE FAMOUS ALBERT EINSTEIN

Albert Einstein Theory of Relativity and

$$E = MC^2$$

One may be asking: "What did Albert Einstein invent?" Though not an inventor specifically, Einstein did make many important contributions to science. The scientific discovery that made Albert Einstein famous was his "theory of relativity."

A basic explanation of this is essentially that huge objects, such as planets, bend space around them as they travel or rotate. This has become an important part of modern astrophysics, and plays a major role in the study of black holes. Within Einstein's theory of relativity, he wrote perhaps the most famous equation of all time: $E=MC^2$.

This is a formula on mass energy equivalence, and is frequently used in depictions of scientific research and intelligence. It would make him one of the most well-known scientists of all time.

PHOTONS

Though there are no specific Albert Einstein inventions is credited as being the first researcher to determine that light is made up of individual particles, which would later come to be referred to as photons. This was a result of his studies into the thermal properties of light. Though this is now thought of as scientific truth, it was met with great skepticism at the time in the science community.

The most well-known critic of the photon theory was Niels Bohr, a fellow physicist. The two scientists would debate frequently throughout their careers, but were still close friends. Bohr eventually conceded to Einstein and recognized the existence of photons in 1925. It is important to note that, though Einstein is the first to identify photons as a concept, they were not given a name until much later.

FAMOUS QUOTES

Technology Quotes:

- "There is a race between mankind and the universe. Mankind is trying to build bigger, better, faster, and more foolproof machines. The universe is trying to build bigger, better, and faster fools. So far, the universe is winning."
- "I fear the day that technology will surpass our human interaction. The world will have a generation of idiots."
- "A new type of thinking is essential if mankind is to survive and move toward higher levels."

About Life Quotes:

- "Stay away from negative people. They have a problem for every solution."
- "It is strange to be known so universally and yet to be so lonely."
- "You have to learn the rules of the game. And then you have to play better than anyone else."
- "If you want to live a happy life, tie it to a goal, not to people or things."
- "A happy man is too satisfied with the present to dwell too much on the future."
- "What is right is not always popular and what is popular is not always right."

About Love Quotes:

- "Women marry men hoping they will change. Men marry women hoping they will not. So, each is inevitably disappointed."
- "We can explain how the universe works, yet we cannot say why people fall in love. Some things remain governed by mysterious, yet intriguing forces."

Quotes about Imagination:

- "Imagination is the highest form of research."
- "Knowledge is a map that guides us while imagination is the territory where we can roam freely and search for answers and opportunities. Imagination knows no restraint and it is the power that puts knowledge to use."

Quotes on Education:

- "Any fool can know. The point is to understand."
- "If you want your children to be intelligent, read them fairy tales. If you want them to be more intelligent, read them more fairy tales."
- "The only source of knowledge is experience."
- "Do not worry about your difficulties in Mathematics. I can assure you mine are still greater."
- "It is a miracle that curiosity survives formal education."
- "Studying, and striving for truth and beauty in general, is a sphere in which we are allowed to be children throughout life."
- "The only thing that you absolutely have to know, is the location of the library."

More Quotes:

- "Weak people revenge. Strong people forgive. Intelligent people ignore."
- "The value of a man should be seen in what he gives and not in what he is able to receive."
- "Black holes are where God divided by zero."
- "If you are out to describe the truth, leave elegance to the tailor."
- "The only thing more dangerous than ignorance is arrogance."

- "Most people say that it is the intellect which makes a great scientist. They are wrong: it is character."
- "Unthinking respect for authority is the greatest enemy of truth."
- "It gives me great pleasure, indeed, to see the stubbornness of an incorrigible nonconformist warmly acclaimed." –

On receiving Lord & Taylor Award, 1953.

- "The most powerful force in the universe is compound interest."
- "The man who regards his own life and that of his fellow creatures as meaningless is not merely unfortunate but almost disqualified for life."
- "It's not that I'm so smart, it's just that I stay with problems longer."
- "The hardest thing to understand in the world is the income tax."
- "We all know that light travels faster than sound. That's why certain people appear bright until you hear them speak."
- "A clever person solves a problem. A wise person avoids it."
- "How unfortunate a state must a community find itself if it cannot produce a more suitable candidate upon whom to confer such a distinction?" – Upon receiving a distinction from the Chicago Decalogue Society
- "Science is a wonderful thing if one does not have to earn one's living at it."
- "Never do anything against conscience, even if the state demands it."
- "Two things are infinite: the universe and human stupidity; and I'm not sure about the universe."
- "Whoever is careless with the truth in small matters cannot be trusted with important matters."
- "The high destiny of the individual is to serve rather than to rule."
- "The only way to escape the corruptible effect of praise is to go on working."
- "I speak to everyone in the same way, whether he is the garbage man or the president of the university."
- "Only one who devotes himself to a cause with his whole strength and soul can be a true master. For this reason, mastery demands all of a person."
- "Joy in looking and comprehending is nature's most beautiful gift."
- "The only reason for time is so that everything doesn't happen at once."
- "One must shy away from questionable undertakings, even when they bear a high-sounding name."
- "Reality is merely an illusion, albeit a very persistent one."
- "My pacifism is not based on any intellectual theory but on a deep antipathy to every form of cruelty and hatred."
- "It is the duty of every citizen according to his best capacities to give validity to his convictions in political affairs."
- "The world as we have created it is a process of our thinking. It cannot be changed without changing our thinking."
- "If you can't explain it to a six-year-old, you don't understand it yourself."
- "Most people stop looking when they find the proverbial needle in the haystack. I would continue looking to see if there were other needles."
- "Anyone who has never made a mistake has never tried anything new."
- "Peace cannot be kept by force. It can only be achieved by understanding."
- "Thinking is hard work; that's why so few do it."
- "Everyone should be respected as an individual, but no one idolized."

FAST FACTS ABOUT ALBERT EINSTEIN

Want to know interesting more about Einstein? Here are the answers to a few frequently asked questions about the life of the famous scientist.

- What was Albert Einstein's Profession?: Scientist, Physicist and Professor
- When was Albert Einstein Born?: March 18th, 1879
- Where was Albert Einstein Born?: Ulm, Württemberg, Germany
- What was Albert Einstein High School and Education?: Aargau Cantonal School, B.A. – Swiss Federal Polytechnic (1900), Ph.D. – University of Zurich (1905).
- When did Albert Einstein Die?: Einstein died April 18th, 1955 in Princeton, NJ.
- How did Albert Einstein Die?: He died from an abdominal aortic aneurysm.
- What is He Best Known For?: Theory of relativity, $E=MC^2$, photons, atomic bomb
- What are Some of His Achievements and Awards?: Nobel Prize in Physics (1921), Matteucci Medal (1921), Copley Medal (1925), Gold Medal of the Royal Astronomical Society (1926).
- What Are Famous Albert Einstein Books?: *Relativity: The Special and the General Theory* (1916); *The World As I See It* (1934); *Ideas and Opinions* (1954)

EINSTEIN – WORLD WAR II

Fleeing Germany

Albert Einstein Stated: *"As long as I have any choice in the matter, I shall live only in a country where civil liberty, tolerance, and equality of all citizens before the law prevail."*

What impact did his German heritage have on his life, and what did Albert Einstein do to escape the Nazis?

Though not a strictly religious man, Einstein was Jewish by heritage. This fact, along with his views on politics and his scientific advancements, made him a threat to the Nazi regime that was gaining control of his native Germany in the early 1930s.

Many of the books of Albert Einstein and his scientific works were burned at the now infamous Nazi book burnings, as intellectualism by Jewish researchers was not accepted by the fascists. After deciding to leave Nazi occupied territory for good in 1933, he came to reside in Princeton, NJ, where he took a position at the school's Institute for Advanced Studies. Up until that time, it was rare for a university to employ Jewish faculty, but the Institute became a haven for other scientists who were fleeing Nazi controlled areas in Europe.

THE NUCLEAR ATOMIC BOMB

Albert Einstein Stated: *"Mankind invented the atomic bomb, but no mouse would ever construct a mousetrap."*

It is a common misconception that Einstein invented the atomic bomb. While he certainly conducted years of research into atomic theory, he was not directly involved with the development of the weapon itself.

His works on relativity, including $E=MC^2$, however, were crucial to the other scientists who worked on the bomb's creation. His greatest contribution the development was of a secondary nature, in that he signed a letter written by other research scientists of the age to then-president Franklin Roosevelt urging that it would be in the interest of national security for the bomb to be researched and built before the German army could do the same.

He would later come to regret this decision, as expressed in the quote "*I made one great mistake in my life, when I signed the letter to Franklin D. Roosevelt recommending that atom bombs be made.*"

PACIFISM

Albert Einstein Stated: *"Every man has a right over his own life and war destroys lives that were full of promise."*

Despite his legacy as being involved in the creation of the atomic bomb, Einstein was a pacifist at his core. In a letter that he wrote in 1928 to London's *"No More War"* movement, he said *"Every thoughtful, well-meaning, and conscientious human being should assume, in time of peace, the solemn and unconditional obligation not to participate in any war for any reason."*

He struggled, however, with justifying holding these views with the pressing need to stop Adolf Hitler and the Nazi regime in the 1930s. His strong feelings against Hitler prompted him to have a modified view on war as America became involved in the conflict. When the war ended, however, Einstein became a staunch supporter of the anti-nuclear movement and disarmament.

EINSTEIN DEATH AND LEGACY

Albert Einstein died of an aneurysm on April 18th, 1955 in Princeton, NJ. Over his lifetime, he had written more than 300 scientific papers on a wide variety of topics. He also published several non-scientific works, including a pacifist letter/article called *"Why War?"* (1933) which he addressed to the famous psychiatrist Sigmund Freud, who also wrote a letter in reply.

One of the most interesting things about Einstein is, since one of his famous hobbies was plumbing, that he was made an honorary member of the Steamfitters Union. His legacy to science, however, is most significant, and is still of great importance to physicists today.

Einstein became a favorite model for "mad scientists" and his white-haired look became the depiction of a "genius" in popular culture. A famous photograph of him with his tongue sticking out was taken by Arthur Sasse in 1951, and has been reproduced all over the world. His name has become synonymous with a highly intelligent person, leading to the sarcastic saying for when a person does something foolish, "Way to go, Einstein!"

AWARDS AND HONORS

Einstein received numerous awards and honors both throughout his lifetime and posthumously. One of his greatest achievements was obtaining the Nobel Prize in physics in 1921 for his discovery of the law of the photoelectric effect.

In 1999, *Time* magazine named the physicist "***the person of the century***."

MUSEUMS AND MEMORIALS

Commissioned in 1979, the Albert Einstein Memorial is located in Washington, D.C. at the National Academy of Sciences. It is a bronze statue which depicts the scientist clutching papers in his hand.

The Einstein Museum in Bern, Switzerland is a display of the scientist's life and greatest works. It documents his life both in Europe and abroad, and also has an extensive collection of personal items from the great physicist.

Located in Princeton, New Jersey, the Albert Einstein House was the home of the scientist while he worked at the Institute for Advanced Studies at the town's famed university. It is on the National Register of Historic Places, and became an official National Historic Landmark in 1976. It is, however, still a private residence. Einstein lived there until his death in 1955.

ALBERT EINSTEIN- CREATOR

In the study of a scientist's life, it is important to recognize several key elements. Scientific contributions are of utmost importance. Following mention of those, it is then possible to look at his or her life, family, and religion as well. However, for Albert Einstein, these elements must all be looked at collectively. Einstein will no doubt go down in history as a great theoretical physicist. His work is compared in importance to that of scientists such as Galileo Galilei, Nicolas Copernicus, Johannes Kepler, and Isaac Newton. Some would even say that his contributions to science were greater.

However, it is impossible to paint a complete picture of Einstein without examining his life, his religion, and his personality. His science was his life, and his religion gave him insights as to how to approach science. By observing his innate curiosity, desire for simplicity and elegance, humble outlook, and desire to seek answers, we can see what elements reached the center of his being.

Though Einstein was one of the greatest contributors to physical science of our times, he was by no means the most brilliant theorist or experimenter. Competent specialists within the field of physics could have better accomplished some of his mathematical deductions. In fact, he needed the assistance of a friend, mathematician Marcel Grossman, to wield the tools necessary to develop his general theory of relativity. Einstein shined brightest within a theoretical context, but, despite the fact that his relativistic theories were most revolutionary, the study of quantum mechanics made a larger impact on the way physics is studied today. What, then, set Einstein apart? Curiosity was the key factor. As Einstein said, "I have no special gift."

Albert Einstein's lifelong quest was to seek the answers to questions his curiosity posed. His religious inspirations and intuitive nature helped set him apart from other scientists, and aided him in finding the solutions he sought. He was just as unique a man, possessing a world view many have come to respect. In short, Einstein was a man who was much greater than the sum of his equations. It is in this light that he will be forever remembered. It is also because of this truth that Einstein is considered one of the most revolutionary men of our time.

ALBERT EINSTEIN, 1955

"It was, of course, a lie what you read about my religious convictions, a lie which is being systematically repeated. I do not believe in a personal God and I have never denied this but have expressed it clearly. If something is in me which can be called religious then it is the unbounded admiration for the structure of the world so far as our science can reveal it."

Albert Einstein, 1954

And here are some other views on religion from Einstein:

"I cannot conceive of a God who rewards and punishes his creatures, or has a will of the type of which we are conscious in ourselves. An individual who should survive his physical death is also beyond my comprehension, nor do I wish it otherwise; such notions are for the fears or absurd egoism of feeble souls. Enough for me the mystery of the eternity of life, and the inkling of the marvelous structure of reality, together with the single-hearted endeavor to comprehend a portion, be it ever so tiny, of the reason that manifests itself in nature."

Albert Einstein, 1930

"I want to know how God created this world. I'm not interested in this or that phenomenon, in the spectrum of this or that element. I want to know His thoughts; the rest are details."

Albert Einstein, 1935

"I cannot conceive of a God who rewards and punishes his creatures, or has a will of the type of which we are conscious in ourselves. An individual who should survive his physical death is also beyond my comprehension, nor do I wish it otherwise; such notions are for the fears or absurd egoism of feeble souls. Enough for me the mystery of the eternity of life, and the inkling of the marvelous structure of reality, together with the single-hearted endeavor to comprehend a portion, be it ever so tiny, of the reason that manifests itself in nature."

Albert Einstein, 1935

"It seems to me that the idea of a personal God is an anthropological concept which I cannot take seriously."

Albert Einstein, 1947

"The idea of a personal God is quite alien to me and seems even naïve."

Albert Einstein, 1952

"My position concerning God is that of an agnostic. I am convinced that a vivid consciousness of the primary importance of moral principles for the betterment and ennoblement of life does not need the idea of a law-giver, especially a law-giver who works on the basis of reward and punishment."

Albert Einstein, 1952

"My views are near those of Spinoza: admiration for the beauty of and belief in the logical simplicity of the order which we can grasp humbly and only imperfectly. I believe that we have to content ourselves with our imperfect knowledge and understanding and treat values and moral obligations as a purely human problem—the most important of all human problems."

Albert Einstein, 1954

"It was, of course, a lie what you read about my religious convictions, a lie which is being systematically repeated. I do not believe in a personal God and I have never denied this but have expressed it clearly. If something is in me which can be called religious then it is the unbounded admiration for the structure of the world so far as our science can reveal it."

Albert Einstein, 1955

"What really interests me is whether God had any choice in the creation of the world."

ERNST STRAUSS - ALBERT EINSTEIN

Einstein's idea is somewhat like a pantheist: his god was the beauty, complexity and simplicity of the universe and how the universe ticked. The universe itself was his god, but likely not in a mystical supernatural sense. Rather, it is the awe-inspiring sense of wonder when you look up in the night sky, and realize that a star just poked you in the eye, or the sense of wonder you get when Carl Sagan says "The cosmos is within us. We are made of star-stuff. We are a way for the universe to know itself."

STEPHEN HAWKING - HIS BELIEF/NON-BELIEF

The Universe

A question posed by Hawking: **Did God create the universe?**

Prof. Hawking's answer to the question: *There was no time before the big bang ... for God to exist in. What happened at the beginning of the universe is the final key for removing the need of a creator of the universe. ... There is no God who directs our fate. There is probably no heaven and no after-life either.*

Really?

Although the learned professor is a perceived authority on modern physics and cosmology, this does not make him an 'information all-rounder', i.e., an authority on theology, providence, eschatology, and immortality. Nor yet on historical or forensic science.

Indeed, most of the elements of the big bang theory as promoted by Stephen Hawking are now being challenged by many of his fellow evolutionist professors. These challenges include:

1. **Hawking's belief that everything in the universe originated from nothing**, which his peers say contradicts the principle that every effect needs a cause.
2. **Rather than there being nothing before the big bang**, as Hawking claims, many evolutionist cosmologists are now trying to fabricate ways to explain how our present universe emerged from one or more preceding universes (while at the same time avoiding saying how the first one began).
3. **Invoking infinity, as Hawking does in the idea that everything in the universe was once in an infinitely small point of infinite density (a singularity)**. This is regarded by one expert as "the same as giving up or cheating" (see Dr Param Singh, the big bounce in *What happened before the big bang?)*

However, we can't help but wonder why a man of Professor Hawking's undoubted mental acumen but limited health and strength feels the need to spend so much time and effort trying to convince the world that God does not exist. Is he perhaps trying to substantiate the wishful thinking of German atheist Friedrich Nietzsche (1844–1900), who wrote: "We deny God; in denying God we deny accountability"?

The Meaning of Life

Professor Hawking asks and then attempts to answer the question: ***"Is there a reason why we exist, a meaning to life?"*** He begins by telling us his personal axiom that everything in life is nothing more than physics.

According to Hawking, the laws of physics not only produced the universe we live in, but also our minds.

The Game of Life

In support of the latter proposition, Hawking shows a computer program called The Game of Life, invented by a John Conway in the 1970s. This is an arrangement of squares in a grid (something like a chessboard of unlimited size) that simulates a two-dimensional '*universe.*'

Some squares in the grid can reproduce themselves and then amalgamate with each other, if the starting configuration has been given the right set of instructions to cause this to happen. Hawkings explained *"it* *is possible to imagine that something like **The Game of Life,** with only a few basic laws, might produce highly complex features, perhaps even intelligence."*

We suggest that this would depend very much on who was doing the imagining! This fanciful conclusion is a repeat from Hawking and Mlodinow's book, in which it leads on to Hawking's extraordinary claim that the universe created itself:

Because there is a law like gravity, the universe can and will create itself from nothing in the manner described in Chapter 6. Spontaneous creation is the reason there is something rather than nothing, why the universe exists, why we exist. ***It is not necessary to invoke God*** to light the blue touch paper and set the universe going.

The logical errors in this statement are almost too numerous to recount. Here are some:

1. How can anything create itself before it exists?
2. What intrinsic property does nothing have that enables it to create anything?
3. Gravity is the force of attraction that arises between objects by virtue of their masses. So, before any matter existed, no gravity existed. How then could it have operated before it existed?
4. If any law of physics caused the universe to create itself, then that law must have existed before the universe began, i.e., before time began, and so that law must be outside of time. But how could that be?
5. What (or who) created the laws of physics?
6. Scientific laws do not create anything. They describe things that already exist, or processes that are observable and repeatable. They do not cause anything any more than the outline of a map causes the shape of the coastline it describes.
7. Spontaneous creation … Just how do the laws of physics achieve this?

Intellectual nonsense does not suddenly become gospel truth because it is uttered by a famous savant.

SUBJECTIVE REALITY—OR ABSOLUTE

On this subject, Hawking advances the classical evolutionist line that reality is in the mind of the beholder. To do this he shows us a girl holding a glass bowl that contains a swimming goldfish. The world that the girl sees (a market place) looks very different from the same world as seen by the goldfish through its curved glass bowl. From this, Hawking tells us that he doesn't think that one reality is more valid than another, so to him this means that reality itself is in the mind of the beholder, i.e., reality is subjective.

However, according to *The New Oxford Dictionary of English*, the primary meaning of 'reality' is: "The world or the state of things as they actually exist, as opposed to an idealistic or notional idea of them." If the girl were in a similar glass bowl, and had eyes adjusted to the refractive index of water like the fish, then she would see the market place similarly to the way the fish sees it.

And if either were blind, then they would not see the market place at all. The market place itself would not change as a result of being variously viewed, or cease to exist when not viewed at all. Nor is evidence for anything limited to sight. A market place in particular can be experienced by our other senses of smell, taste, touch, and hearing.

Hawking then introduces us to quarks as the invisible building blocks of protons, and he asks: "*Are quarks a reality*?" His own answer:

They exist only in so far as they are a model that works. This is called the concept of model-dependent reality. I believe this leads directly to the meaning of life.

This leads to the following exegesis by Hawking:

1. "The brain is responsible not only for the reality we perceive, but also for our emotions and meaning too."
2. "Love and honor, right and wrong, are part of the universe we create in our minds just as a table, a plane, and a galaxy."
3. "The meaning of life is what you choose it to be. It is not somewhere out there but right between our ears. This makes us the lords of creation."

TRADITIONAL RESPONSE

Hawking's amateurish philosophizing raises the question: what happens when different brains create different 'realities' or have different ideas of right and wrong? They can't all be right.

As Bible-believing Christians, we begin with the axiom that God does exist and has revealed Himself to man, and man has the ability to apprehend this revelation. This gives us different answers to the questions that puzzle Hawking.

1. God, as revealed in the Bible, is the eternal ***uncaused first cause***.
2. God created the universe and everything in it.
3. God chose to do this in six days about 6,000 years ago.
4. God created mankind in His own image and likeness.
5. There is another dimension to reality besides that which we can describe with our five senses, or in Hawking's case assume if it substantiates a model. This is what God says exists. It includes Heaven and Hell, and Satan. And other realities include life after death and future judgment, as well as sin, forgiveness of sin, and peace with God.
6. Because God created us, He had a purpose in doing so. Many parts of the Bible speak of this—our main purpose is simply to glorify God, and to enjoy Him forever.

Revelation 4:11 says, "Worthy are you, our Lord and God, to receive glory and honor and power, for you created all things, and by your will they existed and were created."

RICHARD DAWKINS – BELIEF/NONBELIEF

Is Richard Dawkins an atheist?

Richard Dawkins, Figure 26.6, is an atheist; he says he is and he writes vociferously against belief in God—not just any old god or gods, but he targets the Creator-God of the Bible.

However, is Dawkins *really* an atheist? What do atheists as Dawkins believe (or not believe)?

Dawkins is one of the original signatories to the *Humanist Manifesto III* (2003). This document states, consistent with Dawkins' life's work, that, '***Humans are an integral part of nature, the result of unguided evolutionary change.*** Humanists recognize nature as self-existing.' (Original emphasis)

Our existence is nothing but a cosmic ***fluke***, with no ultimate purpose. Dawkins himself says that we live in a universe which has 'no design, no purpose, no evil and no good, nothing but blind, pitiless indifference.'.

Fig.26.6: Richard Dawkins – Curtsy - Photo by Matti Á, flickr.com

In similar vein, another atheist, Professor William Provine, Cornell University, said,

'... There are no gods, no purposeful forces of any kind, no life after death. When I die, I am absolutely certain that I am going to be completely dead. That's just all—that's gonna be the end of me. There is no ultimate foundation for ethics, no ultimate meaning in life, and no free will for humans, either.'

So, life is a **farce**, basically, and when we die, we become mere worm fodder and ultimately **fertilizer** to feed plants. Such is the lot of a true atheist who believes only in nature, with no supernatural Creator.

ATHEIST PURPOSE

If we are merely a complex arrangement of atoms emanating from the cosmic fluke called the big bang, from whence come purpose and meaning for the atheist? The *Humanist Manifesto III* tries to find meaning in some nice-sounding words:

'Life's fulfillment emerges from individual participation in the service of humane ideals. ... Humans are social by nature and find meaning in relationships. ... Working to benefit society maximizes individual happiness.'

But why should anyone serve humane ideals?

If we are just atoms, mere matter, what does it matter that people are humane or not humane; whether people suffer or don't suffer? Whether we have good relationships or bad? Whether people are happy or unhappy? Ultimately, we will all become fertilizer anyway, so why should ***anything*** matter for these few short years of life?

And why should it matter whether people believe in God or don't believe in God? Ultimately, in the atheists' way of thinking, we all will end up as fertilizer, whether we believe in God or not. And the universe when it runs out of energy will die in a big whimper, so why should it matter to any *real* atheist?

METHINKS HE DOTH PROTEST TOO MUCH!

Dawkins' proselytizing ways remind me of my pimply youthful days at boarding school where those into smoking cigarettes and boozing (worse was yet to come) seemed to be really keen to get others to follow their destructive and illicit behavior. Why did they try to coerce their fellow students into doing such things? Was it 'safety in numbers', that they felt a little guilty for their behavior and would feel more secure if they could get more to join in their little rebellion?

Professor Dawkins' obsessive campaign to try to convince others to be atheists reminds me of those days. Perhaps it is not that Dawkins *really* believes that there is no Creator-God, but that he wishes there were not and if he could just convince enough others to agree with him, then he will feel more secure. Has he confused his wishful thinking for reality? Perhaps ironically, the ***Humanist Manifesto III*** says, 'We accept our life as all and enough, distinguishing things as they are from things as we might wish or imagine them to be.' Dawkins' wishful thinking comes out in his definition of biology, where he wishes away the evidence for design (a Creator):

'Biology is the study of complicated things that give the appearance of having been designed for a purpose.'

Another self-professed atheist, **Thomas Nagel**, Professor of Philosophy at New York University, put it rather candidly:

'I want atheism to be true and am made uneasy by the fact that some of the most intelligent and well-informed people I know are religious believers. It isn't just that I don't believe in God and naturally, hope there is no God! I don't want there to be a God; I don't want the universe to be like that.'

Aldous Huxley, author of *Brave New World* and grandson of T.H. Huxley, 'Darwin's Bulldog', was even more explicit:

'I had motive for not wanting the world to have a meaning; consequently, assumed that it had none, and was able without any difficulty to find satisfying reasons for this assumption. The philosopher who finds no meaning in the world is not concerned exclusively with a problem in pure metaphysics, he is also concerned to prove that there is no valid reason why he personally should not do as he wants to do, or why his friends should not seize political power and govern in the way that they find most advantageous to themselves. … For myself, the philosophy of meaninglessness was essentially an instrument of liberation, sexual and political.'

It's no wonder that **Dinesh D'Souza**, in his new book ***What's So Great About Christianity***, has a chapter called, 'Opiate of the Morally Corrupt: why unbelief is so appealing'.

To me, Dawkins seems quite insecure in his stance as he feels the need to convince others to join him. On the other hand, Christians try to convince others of the rightness of the Gospel because they believe that people, made in the image of God, have an eternal destiny and without Christ they face a truly bleak future.

Dawkins, jealously attempted to justify his atheist 'crusade,' argues that 'religion' (i.e., Christianity) is bad for society. But this flies in the face of the history of his own country, which flowered because of the Reformation and the Great Awakening. The latter spared Britain the horrors of Robespièrre's deism/ atheism-inspired bloodbath of the French Revolution, which was hardly 'humane' (a word used repeatedly in the *Humanist Manifesto III*), and engulfed Robespièrre himself.

It is also why slavery, a blot on humanity that permeated the entire world, was first abolished by British evangelical Christians like Wilberforce, in the face of pro-slavery opposition that told him to leave religion out of politics. Indeed, Dawkins has recently grudgingly implicitly admitted that Christianity has permeated his country's culture, calling himself a cultural Christian.

We could also mention the 200 million cost in human lives and untold suffering due to the atheism-inspired political movements of the last century: Communism and Nazism. Also, evolutionary ideas inspired the more recent teenage mass murderers such as Eric Harris and Dylan Klebold, and Pekka-Eric Auvinen. Is Richard Dawkins so ignorant of history that he really thinks we would all be better off if we were all atheists? No, I don't believe so; he is quite well read.

And what does it matter anyway to a real atheist if we would be better off or worse off? In the end we are only a fluky arrangement of atoms that will end up being plant food!

Is Richard Dawkins an atheist or a 'Misotheistic'?

So, is Richard Dawkins an atheist? He seems more like a God-hater than a genuine atheist.

George Orwell identified this sort of 'atheist' in the character of Bozo, in *Down and Out in Paris and London:* 'He was an embittered atheist (the sort of atheist who does not so much disbelieve in God as personally dislike Him).' Interestingly, in *The God Delusion*, Dawkins carries on about what he doesn't like about the God he doesn't believe in.

Dawkins likened himself to 'the Devil's Chaplain', the title of a compilation of his essays. However, the Devil does not *disbelieve* in God; he knows only too well that God is real (cf. *James 2:19*). The Bible identifies him as 'a liar and the father of lies' (*John 8:44*); so, what of 'the Devil's chaplain'? Perhaps Dawkins did not think very carefully about this biblical allusion!

If Dawkins really believed what atheists supposedly believe, he should just put his feet up with a bottle of scotch and just enjoy himself as he waits for his death and his consumption by worms so he can fertilize the grass. Getting his blood pressure up over people believing in God all seems rather futile, if atheism is true..

The fool says in his heart, 'There is no God'. (*Psalm 14:1*).

RICHARD DAWKINS

Clinton Richard Dawkins is probably the most famous evolutionist, anti-creationist and atheist today, and a staunch admirer of Charles Robert Darwin (1809–1882).

Dawkins was born in Nairobi, part of the then British colony of Kenya, in 1941, and moved to England aged 8. He gained his degree in Zoology at the Balliol College, Oxford, in 1962. There he was tutored by Nikolaas Tinbergen (1907–1988), who shared the 1973 Nobel Prize in Physiology or Medicine for his discoveries about instinct, learning and choice in animals. Dawkins continued to study under Tinbergen, at the University of Oxford, receiving his M.A. and D.Phil. (the Oxford equivalent of Ph.D.) degrees in 1966.

After this, Dawkins took a position of assistant professor of zoology at the University of California, Berkeley, before returning to Oxford as a lecturer in 1970. In this time, he researched animal decision-making. Since the 1970s, he has concentrated on writing for popular audiences, for which he is far more famous than for his scientific research on animal behavior.

EVOLUTION ADVOCATE

Dawkins 'first book, *The Selfish Gene* (1976), advocated a gene-centered view of evolution. That is, life first began from a 'replicator' that could make approximate copies of itself, therefore would predominate in a primordial soup. Those copies that could make machines to help them copy better would reproduce more. Dawkins claims that these replicators are basically our genes, and our bodies are just 'gigantic lumbering robots' which are their 'survival machines.' This book also independently introduced the idea of the 'meme,' a set of ideas that is replicated in other minds.

Dawkins regards his second book, *The Extended Phenotype* (1982), as his most important contribution to evolutionary biology. This is a kind of sequel and defense of *The Selfish Gene*; whereas in his first book, Dawkins argues that the organism is the gene's survival machine, in his second he extends the genes' influence to the environment modified by the organism's behavior. If this behavior helps the organism's survival, then the genes 'for' that behavior will reproduce best. His examples include beaver dams and termite mounds, as well as animal behavior that benefits a parasite afflicting it, hence the genes of that parasite.

ANTI-CREATIONIST

In 1986, Dawkins wrote *The Blind Watchmaker*, an attack on the argument that design in the living world demonstrates an intelligent Designer; rather, the apparent design is the result of evolution by natural selection. He regards that as a vital argument for his own atheistic faith:

"An atheist before Darwin could have said, following Hume: 'I have no explanation for complex biological design. All I know is that God isn't a good explanation, so we must wait and hope that somebody comes up with a better one.' I can't help feeling that such a position, though logically sound, would have left one feeling pretty unsatisfied, and that although atheism might have been *logically* tenable before Darwin, Darwin made it possible to be an intellectually fulfilled atheist."

Earlier that year, on Valentine's Day, Dawkins participated in the Huxley Memorial Debate at the Oxford Union, opposing the proposition, "That the doctrine of creation is more valid than the theory of evolution." With him was the leading English evolutionist John Maynard Smith (1920–2004), and opposed were two biblical creationist scientists: triple doctorate organic chemist and pharmacologist A. E. Wilder-Smith (1915–1995) and Edgar Andrews (1932–), then Professor of Materials at the University of London. The audience vote of Oxford students was a modest win for the evolution side, 198–115. Yet Dawkins was not happy—in his closing comments, he had "implored" the audience (his word) not to give a single vote to the creationist side, since every such vote "would be a blot on the escutcheon of the ancient University of Oxford." (Alternatively, it would be a return to Oxford's roots, since it was founded by creationists.) After that, he is on record refusing to debate any biblical creationist.

Simonyi "Chair of Atheism" for "Darwin's Rottweiler"

In 1995, he was appointed the Simonyi Professor for the Public Understanding of Science at Oxford. This was an endowment by leading Microsoft software designer and billionaire Charles Simonyi (b. Simonyi Károly, 1948) explicitly for Dawkins. One report said:

"Evolution's first great advocate, 1860s biologist Thomas Henry Huxley, earned the nickname 'Darwin's bulldog' from his fellow Victorians. In our own less decorous day, Dawkins deserves an even stronger epithet: 'Darwin's Rottweiler, perhaps,' Simonyi suggests."

Dawkins retired from this post in September 2008. I am unable to find any example of Dawkins aiding the public understanding of such *real* science as physics or chemistry, or even of the history or philosophy of science. But during this professorship, Dawkins wrote seven books on evolution/atheism. It's not surprising that British author Paul Johnson called it "Oxford's first Chair of Atheism."

PROLIFIC EVOLUTIONARY AUTHOR

For example, *Climbing Mount Improbable* (1996), one of Dawkins' own favorites among his books, aimed to defend slow and gradual evolution. The title is a parable: many structures in living organisms are so complex that there is a vanishingly small probability of producing them in a single step—this corresponds to leaping the high Mt Improbable in a single step. But, says Dawkins, this mountain has a gently upward-sloping terrain on the other side, where a climber can ascend gradually, constantly progressing to the top. This corresponds to the neo-Darwinian mechanism of evolution—mutations natural selection. Mutations produce gradual improvements, and natural selection means that organisms which have them are slightly more likely to leave offspring. So a later generation of organisms is slightly more complex, or higher up the slope of Mt Improbable. This process is repeated until the dizzy peaks are scaled by this ever-so-gradual process.

In his largest book, *The Ancestor's Tale: A Pilgrimage to the Dawn of Evolution* (2004, 688 pages hardcover), Dawkins aimed to illustrate the history of life on Earth. This was a series of 40 tales, from the point of view of man's alleged evolutionary precursors, and the name is a play on the Middle English classic *The Canterbury Tales* by Geoffrey Chaucer (c. 1343–1400).

This has made him probably the best-known exponent of evolution in the world. Yet Ernst Mayr (1904–2005), one of the most influential evolutionists as far as biologists are concerned, says:

"The funny thing is if in England, you ask a man in the street who the greatest living Darwinian is, he will say Richard Dawkins. And indeed, Dawkins has done a marvelous job of popularizing Darwinism. But Dawkins' basic theory of the gene being the object of evolution is totally non-Darwinian. I would not call him the greatest Darwinian."

APOSTLE OF ANTITHEISM

Richard Dawkins not only regards Darwinism as compatible with atheism, but that atheism is a logical outcome of evolutionary belief. He has long promoted atheism both individually and as part of atheistic organizations. Dawkins is an Honorary Associate of the National Secular Society, a vice-president of the British Humanist Association (since 1996), a Distinguished Supporter of the Humanist Society of Scotland, a Humanist Laureate of the International Academy of Humanism, and a fellow of the Committee for Skeptical Inquiry. In 2003, he signed Humanism and Its Aspirations, published by the American Humanist Association.

In his 1991 essay "Viruses of the Mind", Dawkins singled out theistic religion as one of the most pernicious of these viruses. I.e., he regards theism as a kind of disease or pathology, and parents who teach it to their children are, in Dawkins' view, supposedly practicing mental child abuse. But the sorts of criteria Dawkins applies have led critics to wonder whether Dawkins' own strident atheism itself could be a mental pathology—or 'atheopathy.'

After the 11 September 2001 terrorist attack, Dawkins argued:

"Many of us saw religion as harmless nonsense. Beliefs might lack all supporting evidence but, we thought, if people needed a crutch for consolation, where's the harm? September 11th changed all that. Revealed faith is not harmless nonsense; it can be lethally dangerous nonsense. Dangerous, because it gives people unshakeable confidence in their own righteousness. Dangerous, because it gives them false courage to kill themselves, which automatically removes normal barriers to killing others. Dangerous, because it teaches enmity to others labelled only by a difference of inherited tradition. And, dangerous, because we have all bought into a weird respect, which uniquely protects religion from normal criticism. Let's now stop being so damned respectful!"

Dawkins somehow overlooked the record-breaking tens of millions killed by atheistic regimes last century. This was thoroughly documented by Rudolph Rummel (b. 1932), Professor Emeritus of Political Science at the University of Hawaii, who coined the term *democide* for murder by government.

This antitheism continued with presenting a Channel 4 program in the UK, called *The Root of All Evil?* (2006). This title was Channel 4's choice, not Dawkins', but he argued that humanity would be better off without belief in God. The victims of the democides catalogued by Prof. Rummel might not agree. In this program, Dawkins interviewed a number of Christian leaders, and visited several holy sites and communities of major religions. However, some critics attacked the program for not having informed Christian responses. For example, Dawkins' fellow Oxford Don, Alister McGrath (1953–), Professor of Historical Theology (with a D.Phil. in molecular biophysics), claimed that after his responses Dawkins seemed uncomfortable, so was not surprised that his own contribution remained on the cutting room floor.

Dawkins' defense of atheism produced his best-seller to date, *The God Delusion* (2006), with 1.5 million copies sold. Many high-profile atheists praised it, and naturally Christians criticized it. For example, Philip Bell, M.Sc. and former cancer researcher, published a detailed review, and there are other books responding to it. However, leading logician and Christian philosopher Alvin Plantinga (1932–), currently "John A. O'Brien Professor of Philosophy at the University of Notre Dame", was not impressed with Dawkins' excursions outside biology into philosophy, claiming that they could be called sophomoric were it not a grave insult to most sophomores.

Prof. McGrath himself responded to the book (co-authored with his wife). This also revealed that Dawkins' support among atheists was not universal—famous evolutionary philosopher Michael Ruse writes in the blurb, "*The God Delusion* makes me embarrassed to be an atheist, and the McGraths show why." Ruse also said that the "new atheists" led by Dawkins are "a bloody disaster", and said the following about the book:

"**Question:** What do you think of *The God Delusion* by Richard Dawkins? Your approach is a lot milder? (The book lays open on his bed in the hotel room in Amsterdam where Ruse is interviewed.)

"**Answer:** I am just as critical of this book as of the work of Intelligent Design authors like Michael Behe, despite the fact that I, as an agnostic, am closer to Dawkins, and am 99% in agreement with his conclusions. But this book is stupid, politically disastrous and bad academics. If someone spoke about biology and evolution as he does on theology, Dawkins would react without mercy.

"A good academic will inform himself in depth in a subject he is writing about. Dawkins did not. He is neither a philosopher nor a theologian. I am not a biologist myself, but at least I study the subject in depth before I write about it. And that arrogance and that pedantic attitude of his. ...

"Dawkins' book confirms my analysis of evolution as pseudo-religion. His secular humanism has quasi-religious characteristics."

Another atheist, Terry Eagleton, Professor of Cultural Theory at the National University of Ireland, Galway, began his review of *The God Delusion* with these words:

"Imagine someone holding forth on biology whose only knowledge of the subject is the *Book of British Birds*, and you have a rough idea of what it feels like to read Richard Dawkins on theology."

Eagleton continues:

" … does he imagine like a bumptious young barrister that you can defeat the opposition while being complacently ignorant of its toughest case? Dawkins, it appears, has sometimes been told by theologians that he sets up straw men only to bowl them over, a charge he rebuts in this book; but if *The God Delusion* is anything to go by, they are absolutely right."

Dawkins publicly debated his book with John Carson Lennox, Professor of Mathematics at Oxford. Lennox is also a Christian apologist and Intelligent Design supporter, and teacher of Science and Religion at Oxford, and the author of several books on the relations of science with religion and ethics. This debate did not cover evolution, but the wider Christianity vs atheism topics covered in *The God Delusion.* Dawkins seemed quite red-faced and uncomfortable during the debate.

However, Dawkins refuses to debate best-selling author Dinesh D'Souza, author of *What's So Great about Christianity* among others, even though D'Souza is a theistic evolutionist not a creationist. Yet many of Dawkins' fellow 'new atheists' such as Christopher Hitchens and Daniel Dennett have been willing. In an open letter, D'Souza contrasted Dawkins' eagerness to entrap non-scientist Christians on his TV shows with a refusal to debate a strong opponent on level terms:

"To be honest, I find your behavior extremely bizarre. You go halfway around the world to chase down televangelists to outsmart them in an interview format that you control, but given several opportunities to engage the issues you profess to care about in a true spirit of open debate and inquiry, you duck and dodge and run away. …

" If you are so confident that your position is right, and that belief in God is an obvious delusion, surely you should be willing to vindicate that position not only against Bible-toting pastors but also against a fellow scholar and informed critic like me!

"If not, you are nothing but a showman who takes on unprepared and unsuspecting opponents when you yourself control the editing, but when a strong opponent shows up you manufacture reasons to avoid him."

MAJOR DEFENSE OF EVOLUTION

Now we come to Dawkins' current book, *The Greatest Show on Earth* (2009), deliberately published on the bicentennial of Darwin's birth. There is some irony right at the start of the preface (p. vii): many evolutionists have long touted Dawkins' books about evolution as its proof, yet Dawkins says:

"*The Selfish Gene* and *Extended Phenotype* … didn't discuss the evidence for evolution itself. … My next three books … *The Blind Watchmaker*, *River Out of Eden* 22 and … *Climbing Mt Improbable* … although they cleared away stumbling blocks, did not present the actual evidence that evolution was a fact. … *The Ancestor's Tale* … again assumed that evolution is true. …

"Looking back on these books, I realized that the evidence for evolution is nowhere explicitly set out, and that it seemed like a good gap to close."

In a brief review, Dr Henry Gee, senior editor of the leading science journal *Nature*, wrote:

"Even some of Dawkins' admirers felt that *The God Delusion* was an embarrassment. *The Greatest Show on Earth* is 'not intended as an anti-religious book. I've done that, it's another T-shirt, this is not the place to wear it again,' [p. 6]. And so he moves on, with disarming lucidity."

In this book, Dawkins is irate at what he calls "history deniers," who deny evolution, and intends to show how wrong they are, as well as use the 'guilt by association' with 'Holocaust deniers.' He is especially distressed that polls show that at least 40% of Americans accept biblical creation, including the <10,000-year time scale, so his other name for 'history denier' is 'forty percenter.'

DAWKINS' - GOD FORGIVES SIN – HOW?

In *The God Delusion*, author Richard Dawkins asks: "If God wanted to forgive our sins, why not just forgive them, without having himself tortured and executed in payment … ?"

The answer depends on three things:

- What is sin?
- Why does God oppose it?
- How can God justly forgive it?

Dawkins begins with the axiom that God does not exist. We shall begin with the axiom that God does exist and the Bible is His written Word.

I. What is sin?

When God created Adam and Eve, He made human beings who were not only dependent on Him for existence and life, but who He intended to enjoy a relationship with Him of sharing in His life and love. Sin, in essence, is the desire of mankind to be free from this dependence on God, and indeed from any relationship with God at all.

When Satan tempted Eve to disobey God, the 'bait' he used was the assertion "you will be like God." Thus, when Adam and Eve ate the fruit that God had forbidden them, they were defying God, repudiating His authority over them, and elevating their own wills above God's will.

Sin does not primarily refer to isolated acts (sins), for they are only the outworking of human self-will. It refers primarily to the rebellion of men and women *against God*, which may range all the way from careless indifference to the hell-bent hostility of which Dawkins' posturing is an extreme example. Since sin is defined by this opposition to God and his standards, if God doesn't exist, then the concept of sin becomes meaningless.

II. Why does God Oppose Sin?

The Creator God of the Bible (*YHWH* in Genesis1) is the great "I AM WHO I AM" (*Yahweh* in *Exodus 3:14*), who claims to be the one and only holy, true, and loving God, whose word and authority are binding on us all absolutely. Cf. "I am the Lord your God … you shall have no other gods before me" (*Exodus 20:2, 3*).

SIN - ATTACK ON THE 'GOODNESS' OF GOD

Sin opposes God's holiness, repudiates His authority, and rejects His self-giving in love. God remains God whatever happens, and so by His very 'God-ness,' i.e., His eternal will as God to be who He is, God must and does oppose sin. If He did not, He would not be God, and there would be no ultimate difference between God's will and the sinner's will, or between good and evil. Hence sin merits God's 'curse' (*Genesis 3:14–19*) and God's wrath.

GOD'S WRATH

God's wrath is not petulance, but is His holy anger against man's rejection of the truth about Him (*Romans 1:18*). It is a measure of the gravity of sin. Inasmuch as sin opposes God's infinite holiness, God's perfect justice requires the exercise of His holy wrath. Otherwise, He would cease to be God.

NOTHING TRIVIAL ABOUT SIN

When we see sin as deliberate rebellion against an infinitely holy and loving God, it is obvious that God cannot "just forgive" it, as Dawkins naïvely suggests. Sin is not just a matter of things we have done, but of what we are in our attitude to God in the light of His perfect holiness, i.e., our polarization against Him. Dawkins' 'wave-of-the-hand forgiveness' implies that all this is so trivial that it doesn't really matter. Such indifference on God's part would only encourage us to continue in our rebellion against God, confident that we could do so with impunity.

One result of Adam and Eve's sin was that they produced offspring with a tendency to sin. We have all been born with a sinful nature—so sin belongs irrevocably to the nature of mankind (*Romans 5:12*). And as long as we are in this condition, God can no longer admit any of us to His presence, apart from judgment.

I. How can God Justly Forgive Sin?

Christ the Son of God

Since the whole of the human race is under God's condemnation, we can't initiate reconciliation; it must come from God. And, in His love for us and His grace, He has done just that. His plan to redeem us involved God Himself, in the person of His Son, the Lord Jesus Christ, entering our humanity in order to deliver us from our captivity to sin.

The deity of Jesus Christ is essential for our salvation, for unless salvation is an act of God it would be worthless. Furthermore, our Redeemer must be fully divine to endure God's infinite wrath. A mere creature could not withstand it, and the death of anyone else would have no redeeming value. If Jesus is not God, then there is no Gospel.

CHRIST OUR SUBSTITUTE

Christ also needed to be 'one of us'—fully human, yet without sin, a descendant of the first Adam (*Genesis 3:20*) via Mary; hence He is called "the last Adam" (*1 Corinthians 15:45*). Thus, in addition to His divinity, He is a blood relative of every human who has ever lived (and so is qualified to be the "kinsman-redeemer" prophesied in *Isaiah 59:20*). Because of His life of perfect holiness and obedience to God whereby sin had no power over Him, He was uniquely qualified to be the substitute for all humanity. He is not like the priests of Israel who first had to offer sacrifices for their own sins (*Hebrews 7:27*).

Fig.26.7: The Word of God – Curtsy ©istockphoto.com

This necessary dual nature of the Redeemer fits perfectly with *1 Timothy 2:5*: "For there is one God, and there is one mediator between God and men, the man Christ Jesus." An effective mediator between two groups should ideally be a member of *both*. Jesus is such a mediator, because He is the only member of both groups: 'God' and 'man' Figure 26.7.

God's provision for our salvation had to pay the price for our transgression of His law. So, Christ, the sinless Son of God, had to die on the Cross as our substitute (His life for ours), to pay the just penalty for our sin. The Son of God took our curse upon Himself. There was no mitigation or lessening of judgment, because any such would have meant that God had not really opposed sin, and that sin had not been totally dealt with. So, because our judgment has been paid in full, God can *justly* forgive all those who exercise faith and repentance, and we can know that we are truly forgiven.

ATONEMENT

Atonement is the divine work of covering and putting away sin, thus creating reconciliation between God and us. The Bible says: "God … loved us and sent His Son to be the propitiation for our sins" (*1 John 4:10*). The Greek word translated "propitiation" in the KJV, NASB and ESV, and "expiation" in the RSV, is rendered "atoning sacrifice" in the NIV. It means that when the Lord Jesus died by shedding His blood on the Cross, *He* was the means whereby the cost of our reconciliation with God was fully met (expiated), and God's wrath upon us by reason of our sin was turned away (propitiated).

The Bible does not say that Jesus propitiated God, or that Jesus somehow persuaded a reluctant God to forgive us. It was God in His love who supplied the atoning sacrifice, i.e., Jesus Himself *is* the propitiation for our sins (*1 John 2:2*). So, it is God who performs the act of propitiation, expiation, atonement, forgiveness, reconciliation, and removal of sin (cf. *Isaiah 43:11*).

THE IMPORTANCE OF THE RESURRECTION OF JESUS

The Cross was no mere act of amnesty, but an action in which our sins are utterly dealt with, overcome, and undone, Figure 26.8. Unless salvation vindicates God's holiness, righteousness and justice, as well as His love, mercy and grace, it would not be valid. *Romans 3:25* tells us: "This was to show God's righteousness" cf. " … to demonstrate His justice" (NIV). Furthermore, this was ratified by God through the resurrection of Jesus, "who was … raised for our justification" (*Romans 4:25*).

Fig.26.8: The Tomb is Empty – Curtsy Flickr/James Emery

Dawkins wonders why Christ had to die. In addition to all the above, one further reason was so that He could overcome death by rising from the dead (before ascending into Heaven). Thus, the overcoming/undoing of sin in atonement was confirmed by the overcoming/undoing of death (sin's consequence) in the resurrection of Jesus. As theologian Thomas Torrance explains: "If death is not actually overcome, then the act of forgiveness has not ultimately touched sin at its very root and undone it."

Dawkins would also do well to ponder the fact that the Bible says the resurrection of Jesus is proof that there is a coming Day of Judgment of the living and the dead, at which Christ will be the Judge (*Acts 17:31, 10:42*). And, "Whoever believes in the Son has eternal life; whoever does not obey the Son shall not see life, but the wrath of God remains on him" (*John 3:36*).

THE AUTHOR'S CHRISTIAN DOCTRINE

The author of this book "The Origin of The Universe," believes everything in the Bible, which is inseparably bound up with its first book of the Bible, Genesis. This is because Genesis gives us the origin and initial explanation of all major biblical doctrines.

Obviously not *everything* that God took 66 books of the Bible to tell us over some 15 centuries is contained in just the first book. There is a progress of doctrine throughout the Bible. From the first verse of Genesis to the last verse of Revelation, we learn more about God, ourselves, sin, redemption, etc. with each successive book. All the major doctrines of the Bible are like rivers that become deeper and broader as they flow from the initial watershed of Genesis.

We will examine the major Christian doctrines and their connection with Genesis.

I. About God (Theology)

Genesis tells us about God, not just as the *Creator,* as seen in chapter 1, but also as the One who has a plan and purpose for mankind, that is, for us. This plan and purpose involve our living in a relationship of obedience to God (as well as of trust and love for Him). Thus, God is seen as *Lawgiver* in His command to Adam not to eat from the Tree of the Knowledge of Good and Evil in the Garden of Eden (*Genesis 2:17*).

Then God is seen as *Judge* following Adam's disobedience (*Genesis 3*), as well as in His judgment at the Flood, at Babel, and on Sodom and Gomorrah (Genesis chapters *6–9, 11, 19*). God is also seen as *Savior,* prophesied in *Genesis 3:15*, and then in action in His saving Noah and his family from the judgment of the Flood, and Lot and his daughters from the judgment on Sodom (*Genesis 18, 19*).

As the Creator of all things, God has the absolute right to rule over all things, and He exercises this authority in the world—demonstrating His sovereignty. This is seen in Genesis in four outstanding events: the Creation, the Fall, the Flood, and Babel. It is also seen in God's choice, call and direction of four outstanding people: Abraham, Isaac, Jacob and Joseph.

The Trinity can be seen in Genesis. The Hebrew word for God, *Elohim,* in Genesis chapter 1 is plural. In *Genesis 1:26*, God says, 'Let us make man in our image . . . .' **The Spirit of God** is mentioned 'hovering over the waters' in *Genesis 1:2*. **Christ** is mentioned prophetically as the 'seed of the woman' in *Genesis 3:15*. This passage also prophesies the *virginal conception of Christ*—that is why He is the seed of the *woman,* in contrast to the usual biblical pattern of listing *only fathers* in genealogies. Adam, the Ark, Melchizedek, Isaac, and Joseph, are all commonly regarded as 'types of Christ.'

In Genesis chapters 1 and 2 we also see two very important things about God—attributes that atheists have tried to demolish with spurious arguments. The first is God's omniscience/omnipotence in that everything that God did He got right the very first time. Contrary to Carl Sagan's claim that God is a 'sloppy manufacturer,' in everything that God created there was no experimentation, no trial and error, no 'Oops'! The second is that everything that God created was 'very good' (*Genesis 1:31*). Contrary to the criticism of David Attenborough, concerning a parasitic worm that lives in the eyeballs of children in Africa, everything that God created demonstrated the goodness of God. In the world before sin had entered there was no death, no suffering, no disease, no carnivory, no detriment, and no lack of any good thing.

II. About Us—Mankind (Anthropology)

The first man, Adam, and the first woman, Eve, appear in Genesis as special creations of God—Adam made from the dust, Eve from Adam's rib—both made by God in the image of God (*Genesis 1:26–27*). Therefore, we are not evolved animals, or mere agglomerations of chemicals, but beings with a spiritual or God-conscious nature.

Eve was created to be a 'companion' for Adam (*Genesis 2:20–22*). From this follows the **doctrine of marriage** (*Genesis 2:24–25*—confirmed by Jesus in *Matthew 19:4–6*), as the union of one man and one woman for life (***not of the union of two men or of two women, or something else***). Clearly, also, the whole human race is descended from a single pair (*Genesis 3:20*).

III. About Sin (Hamartiology)

With the first man came the first sin—seen in Genesis as violation of the law of God (*Genesis 3:6–11*), and as depravity both imputed and imparted to the whole human race (cf. *Genesis 4:8; 6:5*). When God created Adam and Eve, they had the ability not to sin, as well as being able to sin. When they chose to reject God's rule over them, they and mankind lost the ability not to sin; instead, we have an innate sinful nature. The first sin brought the first *guilt* (*Genesis 3:8*).

The first sin also brought the first judgment (*Genesis 3:14–19*). There would be enmity between Satan's seed (unbelievers and possibly demons) and the woman's seed (believers but specifically Christ). Women and men would suffer in their respective roles. All humanity would now be subject to death.

IV. About Salvation (Soteriology)

The Bible teaches that God in His mercy and grace forgives our sin, but only when the penalty is paid by a substitutionary sacrifice. Thus, God has provided salvation from the guilt, the power, the eternal penalty, and ultimately the presence of sin, by means of the person and work of His Son, the Lord Jesus Christ. The enactment and fulfilment of this salvation through the death, burial, and resurrection of Jesus is not seen until the Gospels; however, the prediction and promise of what was to come is first seen in the promise that the seed of the woman would crush the head of the serpent (*Genesis 3:15*).

Further, this Seed is a descendant of the first man Adam (*Luke 3:38*), and is called 'the Last Adam' (*1 Corinthians 15:45*). This is essential, because Isaiah spoke of this coming Savior as literally the 'Kinsman-Redeemer', i.e., one who is *related by blood* to those he redeems (*Isa. 59:20*, which uses the same Hebrew word לאוג (*gôēl*) as is used to describe Boaz in relation to Naomi in *Ruth 2:20, 3:1–4:17*). The Book of Hebrews also explains how Jesus took upon Himself the nature of a man to save mankind, but not angels (*Heb. 2:11–18*). This vital kinsman-redeemer concept is sourced in Genesis.

The beginning of the Jewish nation within which the Messiah would be born, die and rise from the dead is seen in the call of Abraham (*Genesis 12:1–3; 17:19; 49:10*).

The substitutionary nature of sacrifice is first seen in *Genesis 22:1–13*, where Abraham is directed to offer a ram as a burnt offering instead of his son Isaac.

V. About Angels (Angelology)

Just when God created the angels is not mentioned in the Bible, but it was probably before He created the earth (*Genesis 1:1*), or at least before the dry land appeared (*Gen. 1:9*), because according to *Job 38:4–7*, when God laid the foundations of the earth 'the sons of God shouted for joy.'

As God is not the author of evil, and because He pronounced His whole creation to be 'very good' at the end of Day 6 of Creation Week (*Genesis 1:31*), we take it that the being we now call Satan had not fallen into sin at that time.

In *Genesis 3:1–14* we read the first reference to this being who slanders God and who tempted Eve to rebel against God, and whose ultimate destiny is foretold by God (*Genesis 3:15*). Elsewhere in the Bible we learn that the name of this creature is Satan, which means 'slanderer' (cf. *Revelation 12:9; 20:2*).

The first reference to good angels is in *Genesis 3:24* where cherubim are placed in the Garden of Eden by God to guard the way to the tree of life.

VI. About the Church (Ecclesiology)

The doctrine of the Church is revealed in the New Testament. It is one of the things that the Apostle Paul calls a mystery, meaning a previously unrevealed truth, now divulged. However, the very fact that Paul calls the Church the Bride of Christ (*Ephesians 5:23–32*) brings us back to the first divinely-ordained husband-wife relationship, in *Genesis 2:24*.

Also, the church is surely foreshadowed in Genesis, with Abraham being called out to form (through his descendants) the nation of Israel, which God blessed and was also to be a blessing to all people on earth (*Genesis 12:1–3*). This blessing culminated in a unique Seed of Abraham, Jesus Himself (*Galatians 3:16*), who was to be the source of blessing to all the nations (*Galatians 3:14*). Paul tells us, 'If you belong to Christ, then you are Abraham's seed, and heirs according to the promise' (*Galatians 3:29*). Those who belong to Christ are His true Church.

VII. About the Last Things (Eschatology)

The principal aspects of what are called 'the last things' are the second coming of the Lord Jesus Christ, the future resurrection(s) of the dead, the judgment of all mankind, and the final state of the redeemed and of the wicked.

By their very nature (being the last things) we would not expect these matters to be detailed in Genesis. However, they are the outworking of God's ultimate plan and purpose for mankind, the earth, and the universe. He purposed to provide an eternal 'bride' for His Son, the Lord Jesus Christ, from redeemed humanity, and He set this plan into action when He created the heavens and the earth, and mankind, as recorded in Genesis chapter 1.

What we see in Genesis is God beginning the process which will ultimately bring about this purpose—a plan which was in the mind of God from before the foundation of the world (*Ephesians 1:4; 1 Peter 1:20*).

Also, while the 'last things' are not detailed in Genesis, the places where they *are* detailed make no sense without it. In the Eternal State, there will once again be no death or suffering of any sort, as *Revelation 21:4* says—and the reason is that 'there shall be no more curse' (*Revelation 22:3*). There will also be a return to an Eden-like state with a return of the Tree of Life (v. 2) and to a state like Days 1–3 of Creation Week (v. 5, cf. *Genesis 1:16–19*), where God provided His own Glory as light without the sun and moon, because God is Light; "1 John 1:5; "This is the message we have heard from Him and announce to you, that God is Light, and in Him there is no darkness at all."

All major Christian doctrines have their source, directly or indirectly, in the book of Genesis. Preachers, missionaries and theologians who fail to see this have lost the foundation for what they teach. Conversely, those who do see this have the God-given proper basis for all their Christian witnessing, preaching, counselling, and teaching.

THE AGE OF THE EARTH

The author strongly rejects the notions of deep time and rejects "the uniformitarian naturalism of secular scientists." The author is in favor of a relatively recent creation date. However, some professed christians mistakenly follow a common view among a number of biblical creationists two or more decades ago, that the Bible allows for the earth and universe to be as old as 10,000 years or more: "The earth is relatively young—*perhaps* less than ten thousand years old."

However, upon the author careful examination of the biblical data (e.g., chrono-genealogies, as stated in chapter 23 of this book, it is clear that Scripture leaves no room for an age of creation of 10,000 years or more. In fact, no matter which textual assumptions are granted for dating certain events and deciding which manuscripts are more reliable, the date of creation only falls in the range of somewhere between **5,665** and **3,822 BC**, thus yielding a maximum possible creation date of ~**7,700 years ago**. Therefore, it would have been more appropriate to state that the earth is *definitely* less than 10,000 years old.

GOD CREATED EVERYTHING FULLY FUNCTIONALLY MATURED

Another point that may require a little background knowledge. The two distinctly Young Earth Creationism models for interpreting the Genesis creation account—and the relevant scientific data—are the fiat creation model and the apparent age theory. Both models affirm the basic precepts of young-earth creationism. However, the apparent age theory postulates that God created everything with the appearance of age so much so that the earth and universe *look* or *appear to be* old while in reality only being thousands of years old, while the fiat creation model says that God created everything not with the appearance of age but with *functional maturity* so that trees were created already bearing fruit, and Adam and Eve were created with the physiological maturity to reproduce.

6-DAY CREATION

The author believes in 6-Day Creation. If you think about it, an infinite Creator God could have created everything in no time. Why, then, did He take as long as six days? The answer is given in *Exodus 20:11*. Here we find that God tells us that He deliberately took six days and rested for one as a pattern for man—this is where the seven-day week comes from. The seven-day week has no basis for existing except from Scripture. If one believes that the days of creation are long periods of time, then the week becomes meaningless.

The Bible tells us that Adam was created on the sixth day. If he lived through day six and day seven, and then died when he was 930 years old, and if each of these days was a thousand or a million years, you have major problems!

On the fourth day of creation (*Genesis 1:14-19*), we are given the comparison of day to night, and days to years. If the word 'day' doesn't mean an ordinary day, then the comparison of day to night and day to years becomes meaningless.

APPENDIX

i. Some have accused him of Arianism, but this is rejected by Pfizenmaier, T.C., Was Isaac Newton an Arian? biography.pdfdirectory.org, *Journal of the History of Ideas* **68**(1):57–80, 1997. A very detailed defense of Newton's Trinitarianism is Van Alan Herd, The theology of Sir Isaac Newton, gradworks.umi.com, Doctoral Dissertation, University of Oklahoma, 2008. This documents much evidence, including Newton's words refuting tritheism and affirming Triniarian monotheism:
    a. That to say there is but one God, ye father of all things, excludes not the son & Holy ghost from the Godhead becaus they are virtually conteined & implied in the father. … To apply ye name of God to ye Son or holy ghost as distinct persons from the father makes them not divers Gods from ye Father. … Soe there is divinity in ye father, divinity in ye Son, & divinity in ye holy ghost, & yet they are not thre forces but one force.
ii. The well-known story that Newton hit upon the idea of universal gravitation after observing an apple fall to the ground in his garden is not known for certain to be true. The anti-religious French philosopher and skeptic Voltaire first circulated the story after reputedly hearing it from Newton's grandniece. Return to text.

iii. Conducting experiments to observe effects is called operational science, whereas hypotheses about the unobserved and unobservable past involve historical or forensic science.

iv. An axiom is an assumption made without proof for the purpose of argument.

v. Chapter 6 of *The Grand Design* book deals principally with Hubble's concept of an expanding universe that was smaller in the past, Eddington's concept of this as a bubble, Friedmann's concept of inflation proceeding at greater than the speed of light, and the claimed uniform cosmic background radiation.

vi. From Greek μισέω miseō = hate; θεός theos = God. Others invented the term independently.

vii. An axiom is a self-evident principle or one that is accepted as true without proof as the basis for argument.

viii. Animals do not sin because they were not created to have a relationship with their Creator, which is a fundamental disconnect between humans and animals.

ix. The word "atonement" was introduced by William Tyndale (from "at-one-ment") in his 1526 new Testament translation of the Greek words *katallassō* (καταλλάσσω) and *katalagē* (καταλλαγή), rendered "reconcile" and "reconciliation" in 2 Corinthians 5:18–19 (KJV). See Gypsyscholarship.blogspot.com/2008/12/william-tyndales-atonement. This is reflected in the Anglican Homily for Good Friday which reads: "Without payment God the Father would never be at one with us." Return to text.

x. "I, even I, am the Lord: and beside me there is no Savior." So calling Jesus 'Savior' is logically calling Him YHWH (*Yahweh*) since YHWH is the only Savior. No wonder that the great Trinitarian Church Father Athanasius (c. 293–373) noted: "Those who maintain 'There was a time when the Son was not' [i.e. was only a created being] rob God of his Word, like plunderers."

xi. "God … commands all people everywhere to repent, because he has fixed a day on which he will judge the world in righteousness by a man whom he has appointed; and of this he has given assurance to all by raising him from the dead" (Acts 17:31). "He [Jesus Christ] is the one appointed by God to be judge of the living and the dead" (Acts 10:42b).

xii. "Jesus Christ is the full reality of God and the full reality of man in one person, in such a way that his divinity and humanity cannot be divided or separated from one another, but each remain fully what they are without any change or confusion with the other." Torrance, T., *Atonement: The person and Work of Christ*, IVP Academic, Illinois, 2009, p. lxxiii.

xiii. Note that when Joseph was propositioned by Potiphar's wife (*Genesis 39:7–9*), his reply was, 'How then could I do such a wicked thing and sin against God?', even though the commandment against adultery was not explicitly given in writing until the time of Moses (*Exodus 20:14*), several centuries later. We conclude that God put the knowledge of right and wrong, i.e., God's law, into man's conscience, when He made him 'in *the image of God*', as recorded in *Genesis 1:26–27*. This was likely derived from knowledge that marriage was something God had made between one man and one woman (*Genesis 1:27, 2:24*), thus, man must not break (cf. *Matthew 19:3–6*). See also *Romans 2:15*.

REFERENCES

1. *Principia*, Book III; cited in; *Newton's Philosophy of Nature: Selections from his writings*, p. 42, ed. H.S. Thayer, Hafner Library of Classics, NY, 1953.
2. *A Short Scheme of the True Religion*, manuscript quoted in *Memoirs of the Life, Writings and Discoveries of Sir Isaac Newton*, p. 347, by Sir David Brewster, Edinburgh, 1855.
3. Storr, A., Isaac Newton, *British Medical J. (Clinical Research Edition)* 291(6511):1779–1784, 1985; p. 1779.
4. Brewster, D., *The life of Sir Isaac Newton*, Harper & Brothers, New York, p. 18, 1840.

5. Keynes, J.M., Newton, the man; in: Johnson, E. and D. Moggridge (Eds.), *The collected writings of John Maynard Keynes*, Royal Economic Society, Cambridge, pp. 363–364, 1978.
6. Bedford, A., Animadversions upon Sir Isaac Newton's book, intitled the chronology of ancient kingdoms amended, printed by Charles Ackers, London, p. 1, 1728.
7. Russell, B., *The history of western philosophy*, Simon & Schuster, Inc., New York, p. 535, 2008. Russell claims that Newton's "triumph" in the sciences was "so complete", that he "was in danger of becoming another Aristotle."
8. Rickey, V.F., Isaac Newton: man, myth, and mathematics, *The College Mathematics J.* 18(5):362–389, 1987; p. 362.
9. Manuel, F.E., *Isaac Newton: Historian*, Belknap Press, Cambridge, MA, p. 2, 1963.
10. In the original: "Collocavit igitur Deus planetas in diversis distaniis a sole, ut quilibet pro gradu densitatis calore solis majore vel minore fruatur" (Newton, I., *Philosophiæ naturalis principia mathematica*, 1st edn, Streater, Joseph, London, p. 415, 1687).
11. Newton, I., *The Principia: Mathematical principles of natural philosophy*, University of California Press, Berkeley, CA, p. 814, 1999.
12. Cohen, I.B., *Introduction to Newton's 'Principia'*, The University Press, Cambridge, pp. 155–156, 1971.
13. Newton, I., *Newton's Principia: The mathematical principles of natural philosophy*, ed. Motte, A., Daniel Adee, New York, p. 504, 1846.
14. Newton, I., *The third book of Opticks (1718)*, University of Sussex, East Sussex, UK, p. 345, 2006; newtonproject.sussex.ac.uk/view/texts/normalized/NATP00051.
15. Newton, I., Opticks: Or a treatise of the reflections, refractions, inflections and colours of light, 4th edn, Printed for William Innys, London, p. 344, 1730.
16. Newton, I., *A short schem of the true religion*, Publisher, Cambridge, p. 1r, 2002; newtonproject.ox.ac.uk/view/texts/normalized/THEM00007.
17. Manuel, F.E., *The Religion of Isaac Newton*, Oxford University Press, Oxford, p. 35, 1974.
18. Newton, I., *Original letter from Isaac Newton to Richard Bentley, dated 10 December 1692*, Trinity College Library, vol. 189.R.4.47, folio no. 4A-5, Cambridge, p. 4r, 2007; newtonproject.ox.ac.uk/view/texts/normalized/THEM00254.
19. Newton, I., *Seven statements on religion*, The Newton Project, Cambridge, p. 1, 2002; newtonproject.sussex.ac.uk/view/texts/normalized/THEM00006.
20. Snobelen, S.D., The theology of Isaac Newton's Principia Mathematica: a preliminary survey, p. 377, 2010. Similarly, Westfall: "Having studied the entire corpus of his theological papers, I remain unconvinced that it is valid to speak of a theological influence on Newton's science." (Westfall, R.S., Newton's theological manuscripts; in: Bechler, Z. (Ed.), *Contemporary Newtonian Research*, Studies in the history of modern science, D. Reidel Publishing Company, London, p. 139, 2012)
21. Snobelen, S.D., Isaac Newton and Apocalypse Now: A response to Tom Harpur's "Newton's strange bedfellows", University of King's College, Canada, p. 2, 2004.
22. Hearne, T., *Remarks and Collections of Thomas Hearne*, Salter, H.E. (Ed.), vol. 11, Clarendon Press, Oxford, pp. 100–101, 1921; See also: Hearne, T., *Remarks and Collections of Thomas Hearne*, Salter, H.E. (Ed.), vol. 10, Clarendon Press, Oxford, p. 412, 1915; Pfizenmair, T.C., Was Isaac Newton an Arian? *J. History of Ideas* 58(1):57–80, 1997; p. 80.
23. Snobelen, S.D., Isaac Newton, heretic: the strategies of a Nicodemite, *The British J. History of Science* 32:381–419, 1999; pp. 383, 388.

24. Austin, W.H., Isaac Newton on Science and Religion, *J. History of Ideas* 31(4):521–542, 1970; p. 53. Keynes, ref. 3, pp. 368–369. Mandelbrote, S., A duty of the greatest moment: Isaac Newton and the writing of biblical criticism, *The British J. History of Science* 26(3):281–302, 1993; pp. 283–284. Pfizenmair, ref. 27, p. 79. Rickey, ref. 9, p. 384. Storr, ref. 1, p. 1780. Westfall, ref. 25, p. 130. Hughes, M., Newton, Hermes and Berkeley, *The British J. Philosophy of Science* 43(1):1–19, 1992; pp. 1–4.
25. Manuel, ref. 21, p. 58. In a fairly recent dissertation on the theology of Isaac Newton, Van Alan Herd has argued that Newton was not an Arian, but instead a "mainstream Puritan" (Herd, V.A., *The theology of Isaac Newton*: a dissertation submitted to the graduate faculty in partial fulfillment of the requirements for the degree of Doctor of Philosophy, The department of the history of science, University of Oklahoma, OK, p. 2, 2008; https://pqdtopen.proquest.com/pubnum/3304232.html?FMT=AI). But as will be evident from the ensuing discussion, this thesis is hardly compelling.
26. Newton, I., *Miscellaneous notes and extracts on the Temple, the Fathers, prophecy, Church history, doctrinal issues, etc.*, National Library of Israel, Publisher, Jerusalem, pp. 25r, 173r–173v, 2012; newtonproject.sussex.ac.uk/view/texts/normalized/THEM00057. Newton has been quoted as saying to a Mr Hopion Haynes that, "the time will come, when the doctrine of the incarnation as commonly received, shall be exploded as an absurdity equal to transubstantiation!" (Baron, R., The preface, *A cordial for low spirits*, vol. 1, London, pp. xviii–xix, 1763.)
27. Newton, I., *Irenicum, or Ecclesiastical Polyty tending to Peace*, Publisher, Cambridge, p. 33, 2002; newtonproject.ox.ac.uk/view/texts/normalized/THEM00003.
28. Newton, I., *Treatise on Revelation (Section 1)*, National Library of Israel, Jerusalem, pp. 19v–21v, 2002; newtonproject.sussex.ac.uk/view/texts/normalized/THEM00216. Manuel, ref. 10, p. 149. Manuel, ref. 21, p. 84. Snobelen, ref. 28, p. 388. Snobelen, ref. 26, p. 160. It is possible that Newton's demonology was derived from Thomas Hobbes (1588–1679). (Hobbes, T., *Hobbe's Leviathan*, Clarendon Press, Oxford, p. 502, 1929.)
29. King, L., *The life and letters of John Locke*, Henry G. Bohn, London, p. 230, 1858. Popkin, R.H., Newton, Spinoza and the biblical scholarship of the day; in: Osler, M.J. (Ed.), *Rethinking the scientific revolution*, Cambridge University Press, Cambridge, pp. 299–300, 2000.
30. Hoffecker, W.A., Enlightenments and awakenings: the beginnings of modern culture wars; in: *Revolutions in Worldview: Understanding the flow of western thought*, Hoffecker W.A. (Ed.), P&R Publishing Company, New Jersey, p. 244, 2007. Korab-Karpowicz, W.J., *A history of political philosophy: from Thucydides to Locke*, Global Scholarly Publications, New York, p. 291, 2010.
31. Newton, I., Letters from Sir Isaac Newton; in: King, L. (Ed.), *The life and letters of John Locke*, Henry G. Bohn, London, p. 217, 1858. King, ref. 35, p. 231. Westfall, R.S., *Never at rest: a biography of Isaac Newton*, Cambridge University Press, Cambridge, pp. 490–491, 2010.
32. Like Newton, Le Clerc disliked the trinitarian formulation in Nicene Creed and probably shared Newton's preference for Arius over Athanasius (Klauber, M.I., Between protestant orthodoxy and rationalism: fundamental articles in the early career of Jean LeClerc, *J. History of Ideas* 54(4):611–636, 1993; pp. 630–631. Manuel, ref. 21, p. 58).
33. Israel, J.I., Radical enlightenment: philosophy and the making of modernity 1650–1750, Oxford University Press, New York, p. 464, 2001.
34. Burnet, T., The sacred theory of the earth: containing an account of the original of the earth, and of all the general changes which it hath already undergone, or is to undergo, till the consummation of all things, 6th edn, vol. 1, Printed for J. Hooke, London, pp. xxiii, pp. 206–207, 1726.
35. Burnet, T., *Correspondence with Thomas Burnet*, The Newton Project, Cambridge, p. 2, 2008; newtonproject.sussex.ac.uk/view/texts/normalized/THEM00014.

36. Newton, I., *Correspondence with Thomas Burnet*, The Newton Project, Cambridge, pp. 10–12, 2008; newtonproject.sussex.ac.uk/view/texts/normalized/THEM00014.
37. Westfall claims that Newton considered Revelation to be the "fundamental book" of the Bible by which the other 65 books should be interpreted (Westfall, ref. 25, p. 131). Although I have not been able to corroborate this claim from Newton's writings, Newton did consider the study of Revelation to be "a duty of the greatest moment" for the church living in these last days (Newton, I., *Untitled treatise on Revelation (section 1.1)*, National Library of Israel, Publisher, Jerusalem, p. 1r, 2004; newtonproject.ox.ac.uk/view/texts/normalized/THEM00135).
38. *Contra* Drake, E.T. and P.D. Komar, Speculations about the earth: the role of Robert Hooke and others in the 17th century, *Earth Sciences History* 2(1):11–16, 1983, p. 12; Jones, F.N., *The Chronology of the Old Testament: A return to the basics*, 21st edn, Master Books, Arizona, p. 22, 2019; Morris, H.M., *Men of Science, Men of God: Great scientists of the past who believed the Bible*, Master Books, Arizona, p. 32, 2012.
39. *Contra* Janiak, A., The book of nature, the book of Scripture, *The New Atlantis* 44:95–103, 2015; pp. 95–96, 99.
40. Burnet, T., Archeologiae Philosophicae: Or, the ancient doctrine concerning the originals of things, written in Latin: to which is added Burnet's theory of the visible world by way of commentary on his own theory of the earth, being the second part of his archiologiae philosophicae, J. Fisher, London, pp. 11–12, 17–18, 21–23, 35, 50, 1736.
41. Although Newton affirms the general historicity of Babel, he postulates that the division and dissemination of nations was primarily caused by "the rebellion of Nimrod". Newton, I., *The Chronology of Ancient Kingdoms Amended*, Printed for J. Tonson in the Strand, and J. Osborn and T. Longman in Pater-noster Row, London, pp. 186–187, 1728.
42. Newton, I., *Of the Chronology of the First Ages of the Greeks & Latines*, Jerusalem, p. 38r, 2013; newtonproject.ox.ac.uk/view/texts/normalized/THEM00406. Return to text.
43. Gaukroger, S., *The collapse of mechanism and the rise of sensibility: science and the shaping of modernity, 1680–1760*, Oxford University Press, Oxford, p. 376, 2010; Jones, ref. 53, p. 22; Lincoln, B., Isaac Newton and Oriental Jones on myth, ancient history, and the relative prestige of peoples, *History of Religions* 42(1):1–18, 2002; pp. 4–5.
44. Westfall, R.S., Newton's scientific personality, *J. History of Ideas*, 48(4):551–570, 1987; pp. 568–569.
45. Although he concedes that the "Greeks and Latines have made their first kings a little older than the truth."
46. Israel, ref. 41, p. 605. Redwood, J.A., Charles Blount (1654–93), Deism, and English free thought, *J. History of Ideas* 35(3):490–498, 1974; p. 490.
47. Whiston, W., A discourse concerning the nature, style, and extent of the Mosaic history of the creation; in: *A New Theory of the earth*, Printed for John Whiston at Mr Boyle's Head, London, pp. 52, 55, 87–109, 325, 1737.
48. Snobelen, S.D. and L. Stewart, Making Newton easy: William Whiston in Cambridge and London; in: Knox, K.C. and R. Noakes (Eds.), *From Newton to Hawking: A history of Cambridge University's Lucasian professors of mathematics*, Cambridge University Press, Cambridge, p. 140, 2003.
49. Halley, E., A short account of the cause of the saltness of the ocean, and of the several lakes that emit no rivers; with a proposal, by help thereof, to discover the age of the world. Produced before the Royal-Society by Edmund Halley, R. S. Secr., *Philosophical Transactions (1683–1775)* 29:296–300; p. 296.
50. Cohen, I.B., Newton in light of recent scholarship, *Isis* 51(4):489–514, 1960; p. 491.
51. Sibley, A., Deep time in 18th-century France—part 1: a developing belief, *J. Creation* 32(3):85–92, 2018; pp. 86–87. Rappaport, R., Fontenelle interprets the Earth's history, *Revue d'histoire des sciences* 44(3/4):281–300, 1991; p. 282.

52. Voltaire, The important examination of the holy Scriptures: attributed to Lord Bolingbroke, but written by M. Voltaire, R. Carlile, London, pp. 8, 12, 1819.
53. Buffon, G.-L.L.C.d., Buffon's Natural History: Containing a theory of the Earth, a general history of man, of the brute creation, and of vegetables, minerals, etc., vol. 1, Printed for the proprietor, London, p. 112, 1797.
54. Mayr, E., *The Growth of Biological Thought: Diversity, evolution, and inheritance*, Belknap Press of Harvard University Press, Cambridge, p. 336, 1982.
55. Frame, J.M., *A History of Western Philosophy and Theology*, 1st edn, P&R Publishing, New Jersey, p. 252, 2015.
56. Kant, I., Universal natural history and theory of the heavens or essay on the constitution and the mechanical origin of the whole universe according to Newtonian principles; in: Watkins, E. (Ed.), *Natural Science*, Cambridge University Press, Cambridge, pp. 200–201, 2012.
57. Israel, J., The early Dutch and German reaction to the *Tractatus Theologico-Politicus*: foreshadowing the Enlightenment's more general Spinoza reception?; in: Melamed, Y.Y. and M.A. Rosenthal (Eds.), *Spinoza's 'Theological-Political Treatise': A critical guide*, Cambridge University Press, Cambridge, p. 20, 2010.
58. Ford's assessment corroborates this: "Young Earth literalism was a minority viewpoint from 1660 onwards. A historical view of science and religion which does not embrace this view is unrealistic, and the founding of geology in the 1790s cannot be seen in its proper context." (Ford, B.J., Shining through the centuries: John Ray's life and legacy. A report of the meeting 'John Ray and his successors', *Notes and Records of the Royal Society of London* 54(1):5–22, 2000; p. 8.)
59. Halley, E., Ode on this spendid ornament of our time and our nation, the mathematico-physical treatise by the eminent Isaac Newton; in: Cohen, I.B. and A. Whitman (Eds.), *The Principia: Mathematical principles of natural philosophy*, University of California Press, Berkeley, p. 380, 1999.
60. Manuel, F.E., *The Enlightenment*, Prentice-Hall, Inc., New Jersey, p. 4, 1965.
61. Popkin, ref. 35, pp. 309–310. This point is made *contra* Manuel who maintains that it was Newton's science rather than his religion which had a lasting impact on western society (Manuel, ref. 21, p. 4).
62. Hawking, S., & Mlodinow, L, *The Grand Design*, Bantam Press, London, 2010.
63. Nietzsche, F., *Twilight of the Idols*, Chapter 6, The Four Great Errors, section 8, trans. by R.J. Hollingdale.
64. See Bell, P., Atheist with a mission; a review of *The God Delusion*, by Richard Dawkins; creation.com/delusion.
65. *The Humanist Manifesto III* http://www.americanhumanist.org/3/HMsigners.htm.
66. Provine, W.B., *Darwinism: Science or Naturalistic Philosophy? The Debate at Stanford University*, William B. Provine (Cornell University) and Phillip E. Johnson (University of California, Berkeley), videorecording © 1994 Regents of the University of California. (See also: *Origins Research* 16(1):9, 1994; arn.org/docs/orpages/or161/161main.htm.)
67. Dawkins, R., *The Blind Watchmaker*, W.W. Norton & Company, New York, p. 1., 1986.
68. Nagel, T., *The Last Word*, Oxford University Press, New York, 1997, p. 130.
69. Sarfati, J., Anti-slavery activist William Wilberforce: Christian hero, *Creation* 29(4):12–15, *J. of Creation* 21(2):121–127, 2007; creation.com/wilberforce.
70. Catchpoole, D., Inside the mind of a killer: The Finnish high school tragedy once again shows that ideas have consequences; creation.com/killer, 9 November 2007.
71. See also Bell, P., The humanist apostles creed—that which a Devil's Chaplain might write, 4 October 2004.
72. Dawkins, R., *The God Delusion*, W.F. Howes edition, Leicester, UK, 2006, p. 373.

Printed at Repro India Ltd.